AF443068

UNIVERSITY OF SUNDERLAND
WITHDRAWN

Glucuronidation of Drugs and Other Compounds

Author:

Geoffrey J. Dutton
Department of Biochemistry
Medical Sciences Institute
University of Dundee
Dundee, Scotland

CRC Press, Inc.
Boca Raton, Florida

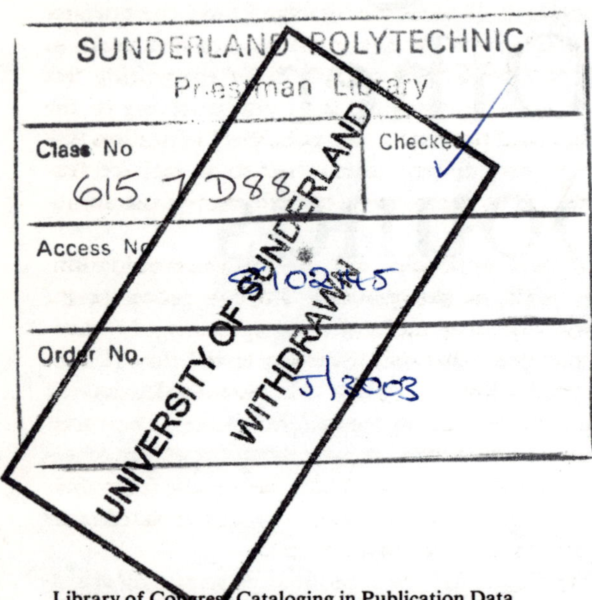

Library of Congress Cataloging in Publication Data

Dutton, Geoffrey J
    Glucuronidation of drugs and other compounds.

    Bibliography: p.
    Includes index.
    1.   Drug metabolism.   2.   Glucuronic acid synthesis.
I.   Title. [DNLM: 1.   Glucuronates--Metabolism.
2.   Drugs--Metabolism.   QU84 D981g]
RM301.55.D87   615'.7   80-36895
ISBN 0-8493-5295-9

This book represents information obtained from authentic and highly regarded sources. Reprinted material is quoted with permission, and sources are indicated. A wide variety of references are listed. Every reasonable effort has been made to give reliable data and information, but the author and the publisher cannot assume responsibility for the validity of all materials or for the consequences of their use.

Direct all inquiries to CRC Press, Inc., 2000 N.W. 24th Street, Boca Raton, Florida, 33431.

International Standard Book Number 0-8493-5295-9

Library of Congress Card Number 80-36895
Printed in the United States

PREFACE

In a previous publication[1] on glucuronic acid both free and conjugated, the author expressed the hope that glucuronic acid studies over the following few years might expand vigorously. They have expanded, and none more vigorously than the study of biosynthesis of simple glucuronides.

Until recently this expansion appeared an extension of facts, without any underlying principle. Lately, however, some framework of ideas has emerged, and the purpose of the present book is twofold — to extract from the enormous literature as much as considered useful, and to outline this possible framework. As expected, writing this book has been onerous, and the result, as also expected, far from satisfying to the author; but because its compilation appeared necessary, the publisher' invitation was accepted. For at present a shorter review must either confine itself to an isolated feature of glucuronidation, or so condense the literature as to be little more than an author-subject index.

We treat only the glucuronidation of small molecules, and not the glucuronidation, actual or possible, of polysaccharides, lipids, or polypeptides. The two processes appear conveniently distinct metabolically. The small molecules are primarily of "xenobiotic" origin, and we define, for our purposes, that unsatisfactory term below (Chapter 1, Section V). However, it is impossible to ignore the glucuronidation of endogenous molecules such as bilirubin, the steroids, or the catecholamines. Their routine glucuronidation in the tissues must be taken into account when the effect of xenobiotic aglycons are considered, and important recent advances have come from their experimental use. The limited number of endogenous compounds glucuronidated must therefore be included in the titular phrase "and other compounds".

Also relevant to the study of glucuronidation are the analogous processes of glucosylation and sulfation, and the hydroxylation systems preceding conjugation. Their influence on glucuronidation is treated in a separate chapter.

Finally, we have assembled some common assay techniques for studying glucuronidation in vitro. At present these are widely scattered in the literature and often out of date. As mere recipes are uninformative, the comments of cook, or of consumer, have been added.

It will be clear to the reader that the major glucuronidating enzyme(s), UDPglucuronyltransferase, is now not only seen as a key "scavenger" of reactive products from xenobiotic metabolism, but is increasingly studied in its own right as an example of that current biochemical problem, the membrane-bound microsomal enzyme. With its associated enzymes, anabolic and catabolic, it also provides an excellent model for studying integration of metabolism in the intact cell or organ. Interest in glucuronidation is therefore understandably wide and productive of literature.

Library search was concluded in autumn 1977. However, many important findings have since appeared and the temptation to include, or mention, them has sometimes proved too strong. Particularly, some reports from the symposium on Conjugation Reactions at Turku, Finland, in July 1978 have been added at the last moment with the authors' permission, because that meeting was especially concerned with the subject of this book. Coverage over late 1977 to 1978 is therefore somewhat erratic, but a line, even a wavering one, must be drawn if a book is to get to press. Additional Notes have been appended to some chapters at the proof stage.

Many colleagues throughout the world have helped to reduce inaccuracies in the following pages, but especial thanks are due to Drs. Brian Burchell, Mary Campbell, Gerard Mulder, and Graham Wishart, who have commented on certain sections. The inevitable errors remain the author's own. It is hoped that this unavoidably imperfect book will prove helpful to the increasing number of workers on this important subject.

AUTHOR

G. J. Dutton is Professor of Pharmacological Biochemistry in the University of Dundee, Scotland. He graduated B.Sc., Ph.D. from Edinburgh University in 1953, working there initially in G. F. Marrian's department with Dr. I. D. E. Storey, where his papers with Dr. Storey, reporting the first transglycosylation from a sugar nucleotide in mammalian tissues, described the isolation and characterization of UDPglucuronic acid and some properties of the associated microsomal enzyme UDPglucuronyltransferase. Since then he has worked in Edinburgh and Dundee principally with this enzyme system, particularly its distribution, induction and development, and its relation to other "detoxication" processes, this research being largely supported by the Medical Research Council of Great Britain. He has a D.Sc. from the University of Dundee and honorary doctorates from the Universities of Kuopio and of Nancy. He is a Fellow of the Royal Society of Edinburgh. Professor Dutton is responsible for many reviews on the subject, including the multi-author standard work *Glucuronic Acid — Free and Conjugated,* published in 1966 by Academic Press.

TABLE OF CONTENTS

Chapter 6
Purification of UDPGlucuronyltransferase and its Heterogeneity

PART II — FACTORS AFFECTING GLUCURONIDATION IN VIVO

Chapter 10
The Effect of Age on Glucuronidation 99

Chapter 13
The Influence of Hormones and Xenobiotics on Glucuronidation

Chapter 16
Relation of other Drug-Metabolizing Pathways to Glucuronidation.............169

Part I
Glucuronidation, Glucuronides, and Studies on UDPGlucuronyltransferase In Vitro

Chapter 1

INTRODUCTION — THE BIOLOGICAL FUNCTION OF GLUCURONIDATION

I. WHAT GLUCURONIDATION IS

Glucuronidation is the most widespread form of "conjugation" in mammalian metabolism. "Conjugation" is a synthetic reaction involving the coupling in vitro of two molecules, usually with elimination of water. There are ten major conjugation reactions in mammals,[2] and many others exist, in both animals and plants.[3] In glucuronidation, the sugar acid, D-glucuronic acid (Figure 1), is coupled with a wide variety of compounds (see Chapter 2, Sections I and II) to form glycosides, the β-D-glucopyranosiduronic acids or, as termed throughout this book except where confusion could arise, the glucuronides.[4]

II. "DETOXICATION" REACTIONS

Conjugation is best known as a "detoxication" reaction. The validity of the term "detoxication" is discussed below, but the process can be considered as a progressive increase in polarity of a molecule. This increased polarity leads to its solubility in bile or urine and its consequent excretion from the body. The authority on detoxication, R. T. Williams, conveniently divided detoxication into two phases.[5] Phase 1 consists of reactions such as oxidations, reductions, and hydrolyses whereby the molecule achieves a "handle", a polar group such as ⁻OH, ⁻NH₂, or ⁻COOH. It is then able to enter Phase 2, where the "handle" takes part in conjugation. In Phase 2, conjugation with a highly polar molecule or ion such as glucuronic, sulfuric, or acetic acids usually ensures rapid excretion.[5]

A classical example, the detoxication of the drug phenacetin, is shown in Figure 2. It is quoted by Williams[5] from the work of Smith and Williams.[7]

In Phase 1, the neutral lipid-soluble phenacetin is oxidatively deethylated to 4-acetamidophenol. This compound is sparingly soluble in water, with a pK_a of about 10, and is 0.25% dissociated at the body pH of 7.4.[7] In Phase 2, the 4-acetamidophenol is conjugated with glucuronic acid to give the glucuronide, a strong acid of pK_a 3.5, highly water soluble, and 99.9% dissociated at pH 7.4. 4-Acetamidophenyl glucuronide is readily excreted by the kidney. The body is, therefore, cleared of the drug by metabolism through Phases 1 and 2. A given compound (e.g., a phenol) may be ingested in a form already suitable for metabolism by Phase 2 enzymes. In that case, detoxication need not involve any Phase 1.

III. DISTINCTIVE ASPECTS OF THE PHASE 2 REACTIONS OF DETOXICATION

Phase 2 reactions are distinct from Phase 1 reactions. First, different sets of enzymes are involved in the two phases. Second, the products of Phase 1 reactions are still partially lipid soluble; many possess biological activity, either toxic or therapeutic. Thus 4-acetamidophenol, the Phase 1 metabolite of phenacetin noted above, is an active drug, the activity of administered phenacetin being largely due to this metabolite.[7] Products of Phase 2 are usually, but not always, much less active, or quite inactive, biologically. This is due to two factors: (1) increased water solubility and decreased lipid solubility, allowing rapid removal from the body; (2) masking of biolog-

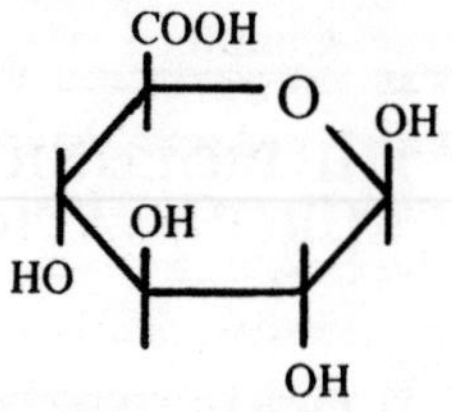

FIGURE 1. Structure
of D-glucuronic acid.

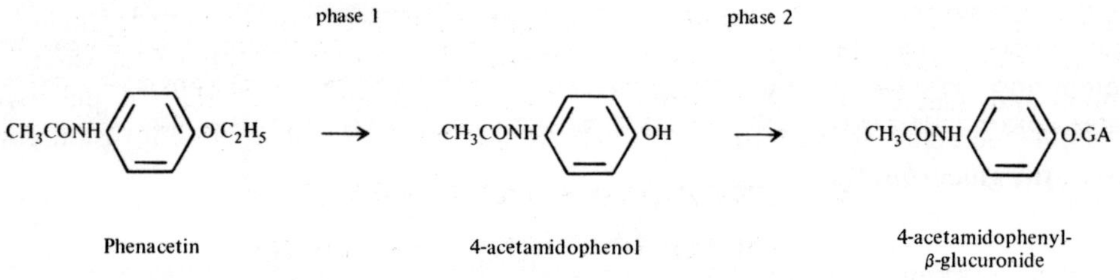

phase 1   phase 2

CH$_3$CONH — O C$_2$H$_5$ →   CH$_3$CONH — OH   →   CH$_3$CONH — O.GA

Phenacetin   4-acetamidophenol   4-acetamidophenyl-
                                                             β-glucuronide

FIGURE 2.   The detoxication of phenacetin.

ically-active groups by superimposition of, or stereochemical hindrance by, the conjugating molecule. 4-Acetamidophenyl glucuronide in Figure 1 exhibits none of the antipyretic or analgesic properties of the administered phenacetin or its Phase 1 product.

Third, Phase 2 reactions are all synthetic reactions involving expenditure of energy. This energy can be applied by two methods to ensure conjugation. The first, and principal, method forms a "high-energy" endogenous molecule which donates the conjugating group, e.g., in formation of acetyl-coenzyme A, which later donates an acetyl group to the molecule, conjugating it as an acetate. Another example is the formation of UDPglucuronic acid, the donor of glucuronic acid.

The second method involves activation of the xenobiotic molecule to be conjugated. This is most common in conjugations with amino acids, as that of administered benzoic acid with glycine to form benzoylglycine (hippuric acid).

IV. THE CONCEPT OF DETOXICATION

We must examine more closely the term "detoxication". The use of the word "detoxication" presumably implies that toxicity of a molecule or ion to a particular species has been lessened or abolished. It would apply equally to molecules of endogenous and of exogenous origin, to conversion of ammonia to urea as much as to conversion of phenacetin to 4-acetamidophenyl glucuronide. Generally, the term is reserved for originally exogenous, usually lipid soluble, molecules. Such compounds, if not utilizable as fuel or building material by the species in question (i.e., if they are anutrients), will be metabolized by Phase 1, Phase 2, or both. The resulting metabolites will usually be more water soluble and less lipid soluble. They will, therefore, when once in blood or bile, remain there and not readily pass through lipid membranes into cells. From blood, they will be filtered by the kidney and pass down the tubules without reabsorption by the tubule cells, to appear almost quantitatively in the urine. The increased water-solubility and pK$_a$ values of some administered compounds and their glucuronides have been listed.[6,8]

We can now reconsider the concept of detoxication. Any detoxication of compounds

X and Y that we see is merely an accompaniment of the metabolism of X and Y. X and Y enter the body and are oxidized, reduced, hydrolyzed, conjugated, or otherwise treated according to their chemical structure. Their structure alone determines what Phase they enter and by which of the host's enzymes they are accepted. Their structure determines the structure of their metabolite, and so its solubility in water, its ease of excretion, and its toxicity. There is no guarantee that these metabolites will be less toxic than X or Y, and hence, no guarantee that X and Y will be truly detoxicated by the biphasic metabolism loosely termed "detoxication".

This might be thought obvious. Yet much confusion, and several unfortunate incidents, have resulted from the assumption that the molecule is invariably detoxicated by the normal body. This assumption, at one time formulated teleologically as "Chemical Defense" (see Section V), sprang from the known detoxication of endogenous compounds and the demonstrably low toxicity exhibited by the glucuronides and sulfates isolated after early experiments in administering toxic phenols to animals.[8]

V. XENOBIOTICS AND EVOLUTION

Let us examine how far this "Chemical Defense" — or true detoxication — can operate. First, we must consider the definition of "xenobiotic". This useful term, due to Mason et al.,[9] is loosely equated with "foreign compound", a term of similarly elusive meaning. "Foreign" implies an origin exogenous to the organism. An exogenous compound is obviously not necessarily a xenobiotic. Foodstuffs are exogenous. A xenobiotic is an exogenous compound of no functional value to the organism.[9] It cannot yield energy, contribute to structure, or confer evolutionary advantage. It is encountered by the species in the diet or elsewhere in the external environment. It may not be directly toxic,[10] but it will occupy space, and by definition it cannot be beneficial to a healthy species. Therefore, its continued presence is undesirable and, so, toxic. All "xenobiotics" should be considered essentially toxic to a normal healthy species.

As an existing species is successful in its environmental niche, it follows that the xenobiotics it encounters there are successfully detoxicated for at least the breeding life of the species. The species is on efficient metabolic terms with its environment, both external and internal.[11] Bilirubin, a molecule toxic to an infant mammal is, because of its structure, conjugated with the polar molecule glucuronic acid and so rendered water soluble and available for biliary excretion. We can agree that mammalian evolution has selected out strains of a species which can so metabolize bilirubin and eliminate it. Mutants unable to conjugate bilirubin with glucuronic acid, such as the Gunn rat and the Crigler-Najjar child (Chapter 12, Section III), still crop up, and do not survive long. In the same way, natural selection has removed those strains or species which did not possess, or develop, enzyme systems capable of detoxicating the toxic compounds present in their environment. Simple phenols, for example, must always have occurred in the diet of land animals, and one would expect detoxication of phenols to be efficient in existing species. The Felidae, deficient in glucuronidation of these compounds, exist with a compensatory sulfation pathway just adequate for their environmental niche. It is important to name the species when using the term "xenobiotic". One species' food may be another's xenobiotic. Again, one species may have manufactured a compound which is toxic to other species though not, through physical or metabolic sequestration, toxic to itself: e.g., insect sting poisons or plant toxins. These compounds are elaborated within their owners and possess evolutionary usefulness. They are "xenobiotics" only to other species encountering them.

Over a long period, environmental changes would be balanced by selection of enzymic mutants. Indeed, Brodie and Maickel[12] suggested that drug-metabolizing enzymes evolved during transition of life from oceans to land. The concept of chemical defense would therefore appear itself defensible. However, during the last century there appeared the greatest toxinogen yet evolved — the organic chemist.

As a result, the recent environment has been deluged with toxic molecules, many of a kind never before encountered by living organisms. Metabolized according to their structure, these molecules may be rendered either less or more toxic. There has been no time to select out biologically valuable changes in the molecule. Consequently, the term "detoxication" can be highly misleading.

This book cites many examples of xenobiotics whose metabolism has greatly increased their toxicity. The polycyclic hydrocarbons, for example, are metabolized to epoxides and dihydrodiols or other intermediates which can bind to protein or nucleic acid, causing necrosis or mutagenesis. Again, the compound may be of such a structure that the species possesses no enzyme to metabolize it readily; dieldrin and hexachlorobenzene, for example, persist for many months in adipose tissue because of their slow metabolism. Highly polar compounds are usually not detoxicated either, for they will not easily enter cells, are unlikely to be made more polar, and are rapidly excreted unchanged. An example is ethylenediaminetetraacetic acid. Exceptions are the quaternary ammonium compounds, some of which form glucuronides.

Even when a conjugate is formed with these new xenobiotics, it may not be nontoxic. Stilbestrol monoglucuronide is not readily soluble in water[13,14] and, like glucuronides of the other synthetic estrogens, hexestrol and dinestrol,[14,15] is believed to possess estrogenic action. The glucuronides of certain *N*-hydroxy compounds are highly toxic (Section IX.B.2 below). The acetylated metabolites of several sulfanilamides[5] are notoriously insoluble, and the degree of renal obstruction they cause is directly related to their insolubility.[5]

However, Phase 2 processes are more likely than Phase 1 processes to yield nontoxic products. The added polar groups normally ensure rapid excretion, and the conjugating moiety itself may render toxic groups less toxic by conjugating with them directly or by attaching so near them that their biological effect is diminished through steric hindrance. A third possibility is that which has been lately suggested to occur in bilirubin conjugation:[16] attachment of the conjugating moiety incurs a steric change in the toxic molecule. The ridge-tile structure of bilirubin IXα, with several intramolecular hydrogen bonds,[17] stretches on conjugation, and the H-bonds are broken. The molecule then becomes water soluble and more readily excretable, and also less biologically active.

The effect of glucuronidation in lessening toxicity of metabolites from Phase 1 or other pathways is well illustrated by the work of Durston and Ames[18] and similar independent observations of Commoner et al.[19] Durston and Ames injected rats with small amounts of the carcinogen, 2-acetylaminofluorene, collected the urine, and added this to a series of Petri dishes containing a bacterial strain sensitive to mutagens. Certain urine samples had been previously treated with a fortified rat-liver homogenate capable of carrying out Phase 1 metabolism, or with β-glucuronidase, or with both homogenate and β-glucuronidase. The results indicated that only a low concentration of mutagens was present in the urine, but their concentration increased three-fold after β-glucuronidase treatment. When the urine had been exposed to the homogenate, the number of revertants per plate increased almost ten-fold and when β-glucuronidase was also present, it increased further by some 20-fold. The increase with β-glucuronidase is presumably due to release from their glucuronides of mutagenic metabolites formed (a) by the rat in vivo from administered carcinogen and (b) by homogenate from released aglycons.

Funneling of the products of Phase 1 metabolism into glucuronidation is discussed later (Chapter 16, Section IV). This funneling should be efficient in vivo, otherwise,

Phase 1 products may cause toxic effects directly or through side reactions (see also Reference [20]).

The balance is delicate between toxification and detoxication of Phase 1 products, and may be upset by genetic changes, age, hormones, diet, and drug treatment. In this balance, glucuronidation, the relatively safe outlet, plays a key role, and the effects of these conditions upon it are understandably important.

VI. COMPETITION BY XENOBIOTICS IN ROUTINE GLUCURONIDATION

A. General

We have seen that when results of conjugation have not been tested through natural selection, glucuronidation itself may in rare cases incur toxicity. Another factor through which glucuronidation may cause stress to the organism is competition.

Competition operates in two ways: for materials and for active site of an enzyme.

B. Competition for Materials

Phase 2 reactions require energy and an endogenous molecule. In the case of glucuronidation, the organism must supply ATP, a source of uridine to be phosphorylated, and a source of carbohydrate:

$$ATP + UDP \rightarrow UTP + ADP \tag{1}$$

$$ATP + Glucose \rightarrow Glucose\text{-}6\text{-}P + ADP \tag{2}$$

or

$$ATP + Glucose \rightarrow Glycogen + Glucose\ 1\text{-}P \tag{3}$$

Before the subsequent steps:

$$UTP + Glucose\text{-}1\text{-}P \rightarrow UPD\text{-}glucose + PP \tag{4}$$

$$UDP\text{-}glucose + NAD^+ \rightarrow UDP\text{-}glucuronic\ acid + NAD \cdot H \tag{5}$$

$$UDP\text{-}glucuronic\ acid + ROH \rightarrow R \cdot O \cdot glucuronide + UDP \tag{6}$$

Glucuronidation involves (Section VIII below), less of a strain on bodily resources than, for example, conjugation with sulfate or amino acids, but sometimes ATP and carbohydrate cannot readily be spared for conjugation processes additional to those already carried out routinely with endogenous compounds as aglycons. Consequently, an additional load of xenobiotic for glucuronidation may, in starving, exhausted, or diseased animals, diminish the efficacy of the routine glucuronidation of, for example, bilirubin and steroids. There could be sharing of limited resources between glucuronidogenic compounds under these conditions, and one might expect this sharing to depend on metabolic priorities selected out during evolution. Intrusion of the modern biologically unfamiliar types of xenobiotic could upset the sharing and more readily provoke damage.

Diversion of UDPglucuronic acid (UDPGlcUA) towards conjugatory glucuronidation and away from its role in polysaccharide synthesis has been proposed as a reason for pathological changes in the alimentary tract.[21] As conjugatory glucuronidation has not been detected at the site of this polysaccharide synthesis, the hypothesis has not been confirmed.[21]

C. Competition for Enzyme Sites

This aspect, dealt with more fully later, can be introduced here. Competitive inhibition by xenobiotics for active sites on the glucuronidating enzyme, UDPglucuronyltransferase, depends on the degree of heterogeneity of the enzyme (Chapter 6, Section II). If only one transferase exists, any glucuronidogenic xenobiotic will compete with all endogenous substrates. It is unlikely that only one UDPglucuronyltransferase exists, and therefore, important how many UDPglucuronyltransferases there are and what their specificities may be. As the specificities have been evolved to suit endogenous or frequently met environmental substrates, they are not likely to distinguish between the newer types of xenobiotic substrates. We can, therefore, expect overlapping among specificities towards these xenobiotics, any of which may competitively inhibit more than one UDPglucuronyltransferase activity.

The transferase(s) probably require(s) allosteric modulators in vivo (Chapter 5), and xenobiotics could bind reversibly at the allosteric sites, again exhibiting competitive behavior (if not competitive kinetics) towards endogenous compounds.

VII. WHY GLUCURONIDATION IS A MAJOR PATHWAY

Glucuronidation is the principal conjugatory pathway in all vertebrate species examined (Chapter 12), over a wide range of tissues (Chapter 14), and accounts for most of the conjugated detoxicatory material in bile and urine.[8,22] There are two main reasons for this widespread utilization.

Glucuronic acid is readily obtained. It derives fairly directly from the universal vertebrate fuel, glucose, and its store, glycogen. Supply of these molecules is less likely to run low than that of amino acids or sulfate. Both UDPglucose and its dehydrogenase occur widely in the body (Chapter 9, Section II.B), usually together with UDPglucuronyltransferase. Sulfation (Chapter 16, Section III) or conjugation with glycine[23] may predominate when small doses of a phenolic drug are administered, but yield to an increasing proportion of glucuronidation as dosage is increased beyond the capacity of the organism to dispose of the drug as sulfate or glycine conjugate. The frequently quoted example of "saturation" of glucuronidation seen when salicylamide and salicylate are administered together to men in sufficient quantity to limit their excretion as glucuronides,[24-26] though unequivocally "saturation" (Chapter 9, Section III.I), is capable of explanations other than shortage of transferase or of UDPGlcUA.

The second reason for glucuronidation being the major pathway is the capacity of glucuronic acid to be conjugated with a remarkably wide range of molecular groupings. These are outlined in Chapter 2, Section I, but summarized in Table 1 which shows the versatility of glucuronidation in vivo compared with the other Phase 2 reactions.

VIII. ANALOGOUS GLYCOSIDATION

Glucuronidation is a major detoxicatory pathway in all vertebrates examined, but not in invertebrate animals or plants, glucuronidation there being limited and only of endogenous aglycons (Chapter 12, Sections II.B and II.C). Glucuronidation as a detoxicatory mechanism may therefore be a very ancient evolutionary step.[27]

Many organisms not utilizing glucuronidation to conjugate xenobiotics use glucosidation (or "glucosylation") instead (Chapter 16, Section II.B). Glucosidation in many properties resembles glucuronidation, and possesses some of its advantages; a variety

TABLE 1

Range of Xenobiotic Structures Glucuronidated

Group	Structure
Linkage through *O:*	
Aryl-OH	Ar.O.GA[a]
Aryl or alkyl enolic	$-$CH$=$CO.GA.
Alkyl-OH (primary, secondary, tertiary)	$-\dot{\text{C}}$O.GA
Acyl-OH (aryl or alkyl)	$-\dot{\text{C}}$.COO.GA
Hydroxylaminic	$-$N.O.GA
Linkage through *S:*	
Thiolic	$-$S.GA
Carbodithioic	$-$C.S.S.GA
Linkage through *N:*	
Amino (aryl)	Ar.NH.GA
Ureido (carbamate)	$-$NH.CO.NH.GA
1-Thioureido	$-$NH.CS.NH.GA
Sulfonimido	$-$SO$_2$.N.GA
Heterocyclic	$=$N.GA
Linkage through *C:*	
(See Figure 3)	$-\dot{\text{C}}$.GA

[a] GA indicates the glucuronyl moiety.

of aglycons, and readily available raw material, the glucose being derived from UDPglucose, the source of glucuronic acid in glucuronides.

IX. METABOLIC ROLE OF GLUCURONIDATION

A. General

Conjugation with glucuronic acid confers greater polarity and water solubility on the aglycon, leading to readier excretion and accounting for glucuronidation being a "true" detoxication process.

However, conjugation does not always end the metabolic activity of a molecule, but sometimes merely changes it. Aglycons conjugated by Phase 2 reactions may enter metabolic reactions previously denied to the unconjugated aglycon, as might be predicted from their presumably specific transport through membranes in the conjugated form during excretion. This first became clear with sulfation; the picture with glucuronidation is not yet so clear, but may in many respects be similar. In both cases, the first reports described the metabolic activity of conjugated steroids and the reactive nature of conjugated *N*-hydroxylated aromatic compounds.

B. Metabolic Role of Steroid Glucuronides

Early work on the reactivity of steroid sulfates was reviewed in 1965.[28] Later findings were reviewed in 1973,[29] e.g., for ten different C_{18}, C_{19}, and C_{21} steroids, of qualitatively and quantitatively different metabolic pathways existed in human liver microsomes for the free and for the sulfated compounds.[30]

A few examples can be given of the role of steroid glucuronides. It is as fully documented as that of the sulfates. First, simple "detoxication". Glucuronides of androsterone,[31] aldosterone,[32] and estriol,[33] appear to be excreted in the urine without further metabolism. Again, 5β-androsterone and pregnanediol, once glucuronidated, no longer induce porphyrin biosynthesis in cultured liver cells.[34]

Second, "direct" metabolism. Dehydroepiandrosterone glucuronide is metabolized to 5-androstenediol glucuronide,[35] estrone 3-glucuronide to estradiol 3-glucuronide,[36] and testosterone glucuronide to 5β-androstane-3α,17β-diol 17-glucuronide.[37] Direct conversion of 17β-estradiol 3-glucuronide, as well as of the 3-sulfate, to its 17-keto form occurs in human kidney homogenates,[38] being cited[29] as an example of morphological compartmentalization, a term under scrutiny in xenobiotic glucuronidation studies (Chapter 5, Section I). Free testosterone can be reduced to either 5α- or 5β-androstanediol.[38] Testosterone glucuronide, however, never gives rise to 5α-reduced metabolites in the urine; it is not split in vivo.[38] As the 5α-reductase is microsomal and the 5β-reductase cytoplasmic, the glucuronidated testosterone (being polar) presumably never reaches the 5α-reductase and is directed into the 5β product by reacting with the 5β-reductase which it can reach.[38] This supposition is likely because[39] testosterone 17-glucuronide is a poor substrate for the microsomal Δ⁴-5α-reductase, but a better substrate than testosterone for the cytoplasmic 5β-reductase. The 17-N-acetylglucosaminide of testosterone similarly yields in man selectively 5β-metabolites.[40] Free testosterone can suffer many other fates not open to the glucuronide,[29] for testosterone glucuronide, unlike testosterone, selectively gives rise to 5β-androsterone, an inducer of δ-aminolevulinic acid synthetase.[29] Therefore, porphyrin biosynthesis could be regulated by the glucuronide of testosterone without production of the actively androgenic free aglycon.[29]

Other examples of preferential enzymic reaction with glucuronides are known. A phenolic steroid 17α-dehydrogenase more readily accepts the 3α-glucuronide of 17α-estradiol than the free steroid.[41] It possesses[42] six forms of which three displayed greater activity towards the glucuronide than towards the free steroid. One form was "essentially specific" for this conjugate. It was 30 times more active towards it than towards the free steroid,[42] and glucose or galacturonic acid could not replace glucuronic acid at the C-3 position.[42a] The glucuronide is oxidized at C-17, conjugated there with N-acetylglucosamine, and excreted as a double conjugate.[42a]

There seems, however, no interconversion of estrone 3-glucuronide and estradiol 3-glucuronide with an estrogen-glucuronidating enzyme from pig liver, kidney, and intestine.[43] In rat liver, the glucuronyl group remains on the C-2 phenolic hydroxyl of 2-hydroxyestrone during its enzymic methylation in vitro and in vivo, and directs methylation to the C-3 position.[44, 45]

Third, steroid glucuronides can exhibit regulatory properties, acting as competitive and, possibly, feedback inhibitors of glucuronidating enzymes and so enforcing regulatory control (there will be examples later in the volume).[29]

Secretion of steroid glucuronides into tissue fluids is not as well documented as that of sulfates, but does occur (e.g., testosterone glucuronide secretion by the human testis[46]), and further suggests participation in vivo of glucuronides in metabolism.

C. Metabolic Role of other Glucuronides
1. Absorption

Generally, glucuronides of nonsteroids have been reported inactive biologically on administration because of poor absorption into the cell. Examples are common.[8] Where activity can be demonstrated, hydrolysis has usually occurred first, as shown with triiodothyronine glucuronide.[47] Little absorption of labeled aglycon occurred when the conjugate was offered to isolated loops of rat jejunum, but absorption from colonic loops was similar to that of the free hormone, presumably from prior hydrolysis of the glucuronide by colonic β-glucuronidase.[47]

Narcotic glucuronides are also usually inactive. The 3-glucuronides of levorphanol and morphine are inactive towards guinea pig ileum when compared with the free drugs.[48] The analgesic action of levorphanol 3-glucuronide, injected intracerebrally, may be due to prior hydrolysis. Labeled levorphanol was found in the brain after such injection.[48] However, some[49] claim analgesic activity of morphine 6-glucuronide as a true property of the conjugate itself. Despite its high polarity, they found it penetrated the blood-brain barrier of rats after intraperitoneal (i.p.) injection, though more slowly than free morphine, and consider that it reacts with the receptor without prior hydrolysis. If injected intracerebrally, it was indeed a stronger analgesic than the free drug, and at the period of greatest analgesia, no free morphine, but only the 6-glucuronide, was found in brain (a tissue with low glucuronidating ability, Chapter 14, Section V).[49] Morphine 3-glucuronide was not analgesic. As it also penetrated the blood-brain barrier, it presumably did not react with the receptors.[49]

2. Binding of Toxic Glucuronides

The above results suggest binding of conjugates either to enzymes or to receptors. This is now substantiated. The Millers[50] early proposed that the ultimate carcinogenic metabolites of aromatic amines were the O-conjugates of their N-hydroxy derivatives, and subsequent work is typified in a report[51] that the electrophilic sulfuric acid ester of N-hydroxy-N-methyl-4-aminoazobenzene yields products with nucleophiles such as methionine and guanine which are identical with products derived from protein and nucleic acid in livers of rats fed the carcinogenic N-methyl-4-aminoazobenzene. Irving[52, 53] has reviewed such metabolic activation of N-hydroxy compounds and published much on the subject. He confirmed[53] that the glucuronide of N-hydroxy-2-acetylaminofluorene (also known as N-hydroxy-N-2-fluorenylacetamide) was carcinogenic in the rat. This glucuronide[54] caused mutations and loss of transforming activity of DNA in bacteria exposed to it. Commoner et al.[19] demonstrated the mutagenicity of the biologically synthesized N-O-glucuronide of N-hydroxy-2-acetylaminofluorene,[19] this latter glucuronide being readily deacetylated to the O-glucuronide of N-2-fluorenylhydroxylamine, which reacts with the guanine residue of DNA or RNA even more rapidly. This reaction occurs in guinea pig, rabbit, or rat liver microsomes,[55] with formation of tRNA adducts. In guinea pig, deacetylation and adduct formation were more efficient with conjugate than with free compound, indicating that the glucuronide was the more metabolically active.[55]

The relative order of binding in vitro between polynucleotides and various glucuronides of N-arylacethydroxamic acids is (aglycons being denoted)[56] (1) N-hydroxy-2-acetylaminofluorene, (2) N-hydroxy-4-acetylaminostilbene (3) N-hydroxy-4-acetylaminobiphenyl, and (4) with very slight binding, N-hydroxy-2-acetylaminophenanthrene. Only the first of these glucuronides was tested for carcinogenicity.

Recently,[57,58] Gillette's group investigated toxic products of phenacetin, which appear to be the N-O-sulfate and N-O-glucuronide of the N-hydroxy derivative. Both are bound covalently to liver protein at pH 7.4, the glucuronide binds slower than the sulfate but both more strongly than the free compound. A similar pattern was observed with other N-O-conjugates, including those of N-hydroxy-2-acetylaminofluorene, when the conjugates were generated in vitro. Binding of sulfate was immediate, that of glucuronide slower, so that free glucuronide was detectable during the reaction. Binding occurred to added albumin as well as to microsomal protein. Degree of binding varied with instability of the glucuronide. The N-O-glucuronide of N-hydroxy-p-chloroacetanilide, relatively stable, was not bound. Damage in liver and kidney following phenacetin abuse may be caused by intermediates formed during breakdown of the bound conjugate.[58] With N-hydroxy-phenacetin conjugates, the toxic compound is probably the N-acetylimidoquinone, which subsequently binds to thiol groups. Ascorbate inhibited binding of this intermediate by reducing it to acetoaminophen.[58] Hydrolysis of the intermediate yielded acetamide.[58]

Interestingly, phenacetin itself can be formed by breakdown of the glucuronide of its N-hydroxy derivative,[58] from internal reduction of the derivative to phenacetin by the glucuronic acid moiety before it leaves. The glucuronic acid, belatedly an effective detoxicant, is possibly oxidized to glucaric acid. The sulfate conjugate cannot be reduced, and so, cannot yield phenacetin. Moreover, rearrangement of the *N*-hydroxyphenacetin glucuronide in vitro yielded 2-hydroxyphenacetin glucuronide.[58]

N-O-Glucuronidation of *N*-hydroxyphenacetin may therefore result in production of acetaminophen, acetamide, phenacetin, and 2-hydroxyphenacetin glucuronide. If this occurs in vivo, it will modify interpretation of the observed fate of *N*-hydroxy compounds.

From the above, glucuronides can no longer be looked on as biologically inert excretion products, and the term "detoxication" is at times most unsuitable as a description of the process of glucuronidation.

3. Role of the Hydrolysis of Glucuronides in Their Metabolism

This aspect will be covered later when the enterohepatic circulation is considered (Chapter 14, Section F) and the role of β-glucuronidase evaluated (Chapter 9, Section IV). It should be clear that, for example, many steroid hormones may travel towards their target organs in the conjugated form and there be released as active hormone by β-glucuronidase. This principle has been utilized in the design of an anticancer drug as a glucuronide. Therapeutically inactive until hydrolyzed, it might be expected to exert its effect on exposure to the high β-glucuronidase activity of neoplastic tissue.[59] A similar enzymic hydrolysis probably accounts for the greater carcinogenicity of the phenolic glucuronide of 2-amino 1-naphthol over the free aglycon when both are implanted as pellets in the bladder,[60] and for delayed pharmacological activity of orally administered glucuronides e.g., of trichloroethanol.[5]

ADDITIONAL NOTES

Section V: A useful brief review summarizes current information on the detoxification-toxification of xenobiotics through conjugation.[60a] Further examples of prior glucuronidation modifying carcinogen-DNA interactions are given in a recent overview article.[60b]

Chapter 2

STRUCTURE AND PROPERTIES OF GLUCURONIDES

I. CHEMICAL STRUCTURES GLUCURONIDATED

As outlined in Table 1, glucuronides contain the D-glucopyranuronosyl radical linked to −O.R, −SR, −N.R′R″, or −C.R groups. The linkage in animals and, for simple conjugates, in plants always appears as β. The previous report of an α-glycoside in urine[61] has not been confirmed.[62] β-D-Furanosiduronic acids[62] (of which at least one has been synthesized chemically[63]) may be too unstable to be isolable from biological sources, except possibly in lactone form.[62]

II. CHEMICAL PROPERTIES OF GLUCURONIDES

This subject, reviewed[63,63a] in detail, need only be summarized. An exhaustive chemically classified list of glucuronides known up to 1965 with their degree of characterization and full references, has been collected by Marsh.[63] A less comprehensive collection,[64] excluding much of Marsh's list, includes many examples up to the early 1970s.

A. *O*-Glucuronides

Linkage is through an oxygen atom. These are the commonest glucuronides, whether with xenobiotic or endogenous aglycons. "R" can be aryl, alkyl, or acyl. The hydroxylaminic link, also through an oxygen atom, is treated in Section II.C below.

1. Aryl-O- (Phenolic) Glucuronides

$$Ar \cdot O \cdot G \cdot A$$

This bond, the "ether" bond, is remarkably stable chemically, a factor of practical importance. Stability helps quantitative isolation of the glucuronide, but its subsequent chemical cleavage is difficult without alteration of aglycon or of glucuronic acid itself. Complete hydrolysis of phenyl glucuronide required 1 hr with 10 M sulfuric acid or 3.5 hr with M hydrochloric acid at 100°C.[65] Estriol glucuronide[66] never yielded more than 50% free estriol, however long refluxed with 15% hydrochloric acid and despite the estriol moiety being stable under these conditions. Strength of the linkage is influenced by inductive effects of neighboring constituent groups of the aglycon.[63] Recent advances in techniques now permit assay and characterization of ether glucuronides without prior hydrolysis (Chapter 7, Section IV).

This bond is also relatively stable to alkali. Lists of aglycons forming this link[8,63] include, among well-known drugs, morphine, paracetamol, and salicylamide.

2. Aryl- or Alkyl-O-(Enolic) Glucuronides

$$-CH = CO.GA$$

Here, the aglycon does not possess a free hydroxyl group and is conjugated with glucuronic acid through an enolized keto group. This link is much less stable than the phenolic link, e.g., $\Delta^{3,5}$-androstadiene-3,17-dione, 3-enol β-glucuronide was rapidly hydrolyzed at pH 4 and 37°C.[67] Further information on this glucuronide is given by Matsui et al.[68] who found it fairly stable at neutral or alkaline pH. Prolonged hydrolysis of the ^{14}C-labeled compound with β-glucuronidase still left 20 to 30% of the ^{14}C unextractable by ether, probably from formation of nonextractable polar steroids, which are readily produced from the enol glucuronide nonenzymically on hydrolysis. The glucuronide of 4-hydroxycoumarin[69] is another example of an enolic glucuronide.

3. Alkyl-O-Glucuronides

$$-CH_2O.GA, \quad \rangle CHO.Ga, \quad -\rangle CO.GA$$

These conjugates are usually[63] more readily hydrolyzed by acid than are aryl-*O*-glucuronides. Examples are glucuronides of aliphatic alcohols and of alcohols of heterocyclic and reduced benzenoid ring systems. They may be of primary, secondary, or tertiary alcohols, e.g., of chloramphenicol, trichloroethanol, and *tert*-butanol.[63]

4. Acyl-O-Glucuronides ("Ester Glucuronides")

$$Ar.COO.GA, \quad -CH_2COO.GA, \quad \rangle CH.COO.GA, \quad -\rangle C.COO.GA$$

These conjugates are glucuronic acid l-esters of carboxylic esters, and the "ester" link is readily hydrolyzed by dilute alkali. Storage of aqueous solutions at pH 11 to 12 and room temperature for 5 to 30 min. effected complete hydrolysis in some cases.[63,70] "Ester" glucuronides are therefore positive for reducing sugar in clinical urinary tests based on reduction of cupric ion in alkaline media. The aglycon can be aryl, primary, secondary, or tertiary aliphatic, or heterocyclic. Examples are salicylic acid, indomethacin, iodopanoic acid, trimethylacetic acid, and nicotinic acid, respectively.[63]

B. S-Glucuronides

These can be thiolic

$$-S \cdot G A$$

and carbodithioic

$$-C.S.S.G A$$

They appear to resemble the corresponding aryl (or alkyl) and acyl *O*-glucuronides in stability.[63] Examples of common thiolic glucuronides are those of thiophenol, 2-mercaptobenzothiazole, and diethyldithiocarbamic acid.[63]

C. N-Glucuronides (see also Reference 63a)

These conjugates, anomalous as glucuronides in several ways, are discussed by Marsh,[63] who prefers to term them glucuronosylamines because they usually have little resemblance in chemical properties to normal glycosides. The amine glucuronides are very unstable below neutral pH values. The ureido (carbamate), $-NHCO \cdot N-GA$; l-thioureido, $-NH.CS.NH-GA$; and the sulfonimido, $-SO_2.N-GA$ links are more stable. The instability of the oxygen-linked hydroxylaminic link $-N.O-GA$, has profound biological consequences (Chapter 1, Section IX.C.2). *N*-(β-1-Glucosiduronyl)oxy-2-naphthylamine is very unstable and at neutral pH forms adducts with nucleic acid. The direct *N* link, as in *N*-(β-1-glucosiduronyl)-*N*-hydroxy-2-naphthylamine is a little more stable and will not form these adducts.[71] The half-life of phenacetin *N-O*-glucu-

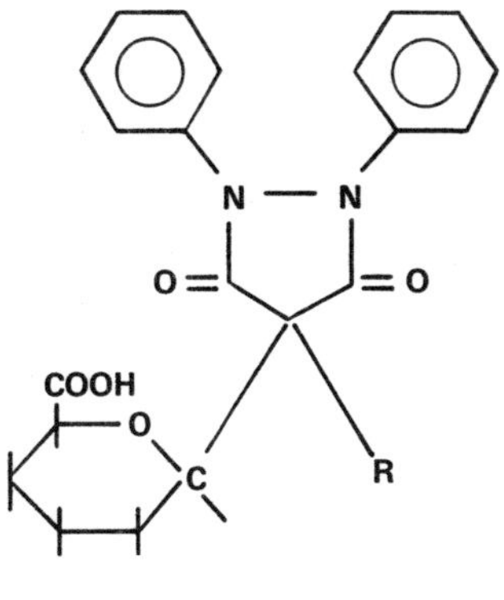

FIGURE 3. General structure of
C—C glucuronides with substrates
containing a pyrazolidine ring.

ronide at pH 7.4 and 37°C is some 9 hr.[58] Examples of drugs forming various N links,[8] are dapsone (aromatic amino), sulfisoxazole (heterocyclic imino), meprobamate (ureido), sulfathiazole (sulfonimido). N-Glucuronidation in primates is discussed by Walker and Williams.[72] A new type of glucuronide possibly only formed by primates has the glucuronic acid moiety linked to the N of a pyrazole nucleus. This link is unexpectedly stable to hydrolysis.[73] Hydrolysis by β-glucuronidase of N-glucuronides, once thought unlikely,[74,75] appears possible in some cases.[71] Certain N-glucuronides can be formed spontaneously from the aglycon and glucuronic acid in solution at physiological temperatures and pH (Section III.C.2 below).

D. C-Glucuronides

Although C-glycosyl compounds have been known for some time,[76] the first C-glucuronides have only recently been reported.[77-79] As much as 40 and 12% of total urinary activity following a single oral dose of 400 mg of ^{14}C-phenylbutazone in a human male was due to the C(4)-β-glucuronides of phenylbutazone and γ-hydroxyphenylbutazone, respectively. Structures were established by IR, UV, mass spectroscopy, and by NMR. The pyrazolidine rings are directly attached to glucuronic acid. The proton at C-4 is part of a 1,3-dicarbonyl system and so fairly acidic. A similar C-glucuronide was obtained from sulfinpyrazone. The general structure is shown in Figure 3. pK$_a$ values of 2.8 and 4.5 were found for phenylbutazone and sulfinpyrazone glucuronides, respectively. The glucuronides are stable chemically and are not hydrolyzed by β-glucuronidase. For several other compounds (e.g., oxyphenylbutazone), this direct C-glucuronidation may also be an important metabolic pathway.[78]

III. PHYSICOCHEMICAL PROPERTIES OF GLUCURONIDES

This subject has been mentioned from its functional aspects in Chapter 1, and is discussed by Smith and Williams.[8] When the apparent dissociation constants of some 15 O-glucuronides were determined[6] the pK$_a$ values lay between 3.0 and 4.0. These glucuronides are, therefore, relatively strong acids. Their sodium salts are in most cases highly water soluble.

Use of the physicochemical properties of glucuronides in their extraction from biological material and their assay in vitro is touched upon by Marsh,[63] described in detail for steroid glucuronides by Jayle and Pasqualini,[80] and is further dealt with in Chapter 17.

Chapter 3

MECHANISM OF GLUCURONIDATION

I. HISTORICAL BACKGROUND

The earliest work on glucuronidation is detailed in the first edition of Williams' classical *Detoxication Mechanisms.*[5] Other useful reviews of work up to 1950 are available.[4, 81-83] These early studies are summarized below, from References 4 and 5.

A. Work up to 1950

The first glucuronide was isolated in 1855 by Schmidt from urine of cows fed mango leaves. After acid hydrolysis, this compound, the pigment Indian Yellow, yielded a reducing agent, later shown by Baeyer in 1870 to be acidic and to possess the formula $C_6H_{10}O_7$. Xenobiotics were at this period being administered to animals and the resulting urines examined. In many cases the urine gave rise to this acidic levorotatory-reducing compound. Jaffe, in 1874 and 1875, suggested that the compound was a carbohydrate and, because not fermentable by yeast, probably glucose with an alcoholic group oxidized to a carboxyl group with the aldehyde group still intact. Schmiedeberg and Meyer isolated *Glykuronsaure* in 1879 from the urine of dogs fed camphor and suggested the structure

$$(CHOH)_4 \begin{cases} CHO \\ COOH \end{cases}$$

a good deduction, despite the fact that the solution was dextrorotatory and they had isolated glucuronolactone. They proposed that camphor had trapped the acid, itself an intermediate of the normal oxidative breakdown of glucose. Others more convincingly suggested that camphor glucoside had been formed first and had been subsequently oxidized to the glucuronide, the reactive aldehyde group being preserved during oxidation.

Prior formation, and subsequent oxidation, of a glucoside remained popular but although injection of glucosides did yield urinary glucuronides, Williams marshaled evidence against direct oxidation of the aglycon glucoside. Injection of the glucoside phloridzin resulted in its excretion with glucuronic acid added on to, not derived from, the glucoside molecule. Feeding of either phenol or phenyl β-glucoside increased ethereal sulfate excretion identically, indicating complete hydrolysis of glucoside in the gut. Also, whereas addition of borneol to perfused or sliced liver preparations gave rise to bornyl glucuronide, addition of bornyl β glucoside did not.

Subsequent work with whole animals suggested that carbohydrate depletion lowered glucuronidation and that glucuronidation required not glucose itself, but a common precursor.

With tissue preparations, liver appeared a major site. In perfused liver,[84] cyanide abolished glucuronidation, suggesting an aerobic process. In liver slices,[85-87] glucuronidation was inhibited by cyanide, fluoride, dinitrophenol, or iodoacetate and lessened by low glycogen content, and therefore required carbohydrate, oxidative processes, and phosphorylation. Free glucose, gluconate, glucarate, glucuronolactone, or glucuronic acid did not stimulate conjugation in these systems[85-87] and isotopic evidence (summarized in Reference 4) suggested that whereas glucose could be converted to conjugated glucuronic acid fairly directly, free glucuronate or its lactone was split, probably to trioses, before incorporation.

If β-glucuronidase catalyzed the biosynthesis of glucuronides as well as their hydrolysis, free glucuronic acid should be a direct precursor of the conjugated form. A major biosynthetic role for β-glucuronidase was popular for some time, but appeared to require very high glucuronide concentrations.[88] Moreover, biosynthesis was unaffected by glucarolactone,[89] the specific inhibitor of β-glucuronidase and in tissues proceeded at a rate inversely proportional to their content of β-glucuronidase.[89] Glucuronyl *transfer* can be mediated by β-glucuronidase (Section III.C.1 below), but this enzyme appears to play no part in the primary synthesis of glucuronides.

B. Isolation of UDPGlucuronic Acid and Identification of UDPGlucuronyltransferase

A full account exists of work leading to the discovery of the "active" form of glucuronic acid and of the enzyme, UDPglucuronyltransferase (EC 2.4.1.17), responsible for catalyzing the last step in glucuronidation.[4]

This work became possible with the advent of the "homogenate" technique. Unfortified liver homogenates[89,90] or liver homogenates fortified with a wide range of possible sources of phosphorylation, energy, or conjugatable glucuronic acid (such as a postulated glucuronic acid 1-phosphate[87]) failed to glucuronidate 2-aminophenol at above 15% of the rate in liver slices.[90] To allow for unknown factors, the homogenate was therefore fortified with a concentrated, boiled extract of liver,[91] when glucuronide synthesis increased to that in the intact cell.[90,91] It was the last step in glucuronidation, being, unlike the overall process, anaerobic and unaffected by inhibitors of phosphorylation.[91]

Further work isolated the responsible "active factor" and established it as UDP-glucuronic acid (UDPGlcUA).[92,93] UDPGlucose had been characterized shortly before,[94] but its role as glycosyl donor was not yet known. UDPGlcUA appeared to be a cosubstrate in the following reaction with a phenol, R.OH:

$$\text{UDP-glucuronic acid} \ + \ \text{ROH} \ \rightarrow \ \text{UDP} \ + \ \text{R}\cdot\text{O}\cdot\text{glucuronic acid}$$

Although UDP was not isolated, its degradation products were identified from the above reaction which, catalyzed by "uridine diphosphate glucuronate β-glucuronosyltransferase" (UDPglucuronosyltransferase, UDPglucuronyltransferase, EC 2.4.1.17), was the first to record glycosyl transference brought about with sugar nucleotides, a mechanism which, brilliantly extended by Leloir's group, revolutionized carbohydrate biochemistry.

As UDPGlcUA is biosynthesized by the NAD-dependent oxidation of UDPglucose, in its turn derived from UTP and glucose-1-phosphate (Chapter 1, Section VI.B), dependence of glucuronidation upon oxygen, carbohydrate, and an energy source is understandable. The early hypotheses were to a great extent borne out; a glucoside is indeed formed and oxidized (Section I.A above), but not the aglycon glucoside; glucose itself is not an immediate precursor, but readily passes unbroken to conjugated glucuronic acid. Free glucuronic acid or its lactone can only indirectly pass to the "glucuronic acid 1-phosphate" moiety combined in UDPglucuronic acid. Glucuronidation has been demonstrated in vitro from glycogen and UTP or from glucose, ATP, and UTP, utilizing the necessary enzymes.[95]

More detailed consideration of the biosynthesis, structure, and metabolic fate of UDPGlcUA follows in Chapter 9. The role of the sugar nucleotide in glucuronidation may now be discussed.

II. GLUCURONIDATION INVOLVING SUGAR NUCLEOTIDES

A. Specificity of Sugar Nucleotides as Glucuronyl Donors

Nothing further on the specificity of sugar nucleotides as glucuronyl donors has appeared since the 1966 review.[4] There[4] it was noted that some glucuronyl transference occurs to 4-nitrophenol from the 5-fluoro and 6-azo analog of UDPGlcUA[96] (N.D. Goldberg, personal communications, 1964—1965). No transference occurs with the β-linked analog of UDPGlcUA,[97] nor with either the 5α- or 5β-ribofuranosyluracil diphosphate glucuronic acid.[98] Possible activity of the α-linked thymidine diphosphate glucuronic acid[99] has not been further explored.

Substrate specificity of UDPglucuronyltransferase towards the sugar moiety of sugar nucleotides not containing glucuronic acid is discussed in Chapter 16.

B. Lipid Acceptors of Glucuronic Acid

UDPGlcUA appears too polar to enter the membrane passively.[100] If, as according to one viewpoint (Chapter 5), UDPglucuronyltransferase is buried in the microsomal membrane, some means of transferring the glucuronyl group through the membrane lipid must be available. Certain workers[101,102] have suggested a porter system for UDPGlcUA itself. Others have looked for a lipid-soluble intermediate analogous to dolichol monophosphate glucose through which UDPGlcUA can transfer glucuronic acid to the enzyme. Dolichol monophosphate, present in the endoplasmic reticulum,[103] does not bind glucuronic acid, and none of its analogs investigated appear to do so.[104] Labeled glucuronic acid could be transferred by a microsomal enzyme from UDPGlcUA to a water-insoluble, lipid-soluble component of rat-liver microsomes,[105,106] mitochondria,[106] or erythrocyte ghosts.[106] The bound glucuronic acid could not be transferred, however, to phenolic acceptors;[105, 106] it did not appear part of a phospholipid, and was released by β-glucuronidase. The search continues, for UDPGlcUA does transfer glucuronic acid to form a glycolipid in Pseudomonad preparations.[107,108] It presumably helps to form the lipid N-acetylglucosamine-glucuronic-acid derivative found in transformed human lung fibroblasts[109] and rat fibrosarcoma.[110] Reports continue of "endogenous" glucuronide formation by washed rat-liver microsomes not exposed to added UDPGlcUA[4,111,112] (J. Marniemi and M. Laitinen, personal communication, 1975); this "endogenous conjugation" could reach, on short incubations, some 30% of that occurring with saturating amounts of UDPGlcUA. A similar observation in steroid glucosidation may be due to a lipid-soluble glucosyl donor in the membrane.[113] Progress is likely soon.

C. Transglucuronylation with UDPGlucuronyltransferase

Early,[4,114] it was noted that if the reaction

$$Y + UDPGA \rightarrow UDP + Y.GA$$

was reversible, then the possibility existed of a "transglucuronylation" of glucuronic acid (GA) from aglycon Y to aglycon X:

An analogous cycling of PAP and PAPS occurred with nitrophenyl sulfates.[115] Early workers found no reversibility of UDPglucuronyltransferase with 2-aminophenol but when the reaction with 4-nitrophenol was shown to be reversible,[116,117] demonstration of transglucuronylation became possible. The addition of 4-nitrophenyl glucuronide, UDP, and 2-aminophenol to guinea-pig-liver microsomes allowed[118] the UDP-dependent production of 2-aminophenyl glucuronide,[118] identified chromatographically (C. Berry and T. Hallinan, personal communication, 1975). Implications of this transglucuronylation are discussed in Chapter 4, Section II.

III. GLUCURONIDATION NOT INVOLVING SUGAR NUCLEOTIDES

A. Transglucuronylation with other Enzymes

Transglucuronylation from one aglycon to another can occur without participation of UDPGlcUA, as first demonstrated by Fishman and Green.[119,120] In presence of a glucuronide and a suitable glucuronyl acceptor, they found less glucuronic acid appearing on incubation than liberated aglycon, and a new glucuronide being formed. The amount of this new glucuronide agreed closely with the unaccounted-for glucuronic acid. If the donor glucuronide were Y−GA, the acceptor, X.OH, and glucuronic acid, GA, then hydrolysis:

$$Y.GA + HOH \rightarrow Y.OH + GA$$

would be accompanied by transfer:

$$Y.GA + Y.OH \rightarrow Y.OH + X.GA$$

so that in sum we would observe:

$$Y.GA + HOH + X.OH \rightarrow Y.OH + GA + X.GA$$

Donors (Y.GA) used were glucuronides of stilbestrol, 2,4-dichloro-1-naphthol, 4-chlorophenol, phenol, biphenyl, phenolphthalein, *l*-menthol, and 8-hydroxyquinoline at about 1 mM. Acceptors, at very high concentration (up to 2.8 M depending on solubility), consisted of methanol and other aliphatic alcohols, ethylene and other glycols, cyclopentanol, cyclohexanol and benzyl alcohol, but not phenols. The enzyme was a purified β-glucuronidase preparation from mammalian livers, from snail, and from *Escherichia coli*.

Transfer could reach 80% of the liberated glucuronic acid in the best instance, with propylene glycol as acceptor, and as it paralleled hydrolytic activity during purification, β-glucuronidase is probably responsible. Susceptibility of transfer to heat, pH values, concentration of substrate and enzyme, and inhibition by glucarolactone preparations all supported this conclusion.[119-121] Fishman,[121] however, suggested that high efficiency of transfer and its detection at quite low substrate concentrations (0.01 M) could indicate a group-transferring enzyme (a "β-glucuronylase") rather than a hydrolytic enzyme.

This work was followed[122,123] with more labile, and therefore possibly more "active", glucuronides as donors. With partly purified β-glucuronidase as enzyme, 1 mM concentrations of the ester glucuronides of 4-aminobenzoic or 4-aminosalicylic acids transferred glucuronic acid to methanol, ethanol, propanol, and butanol. The resulting glucuronides were chromatographically identified. Although the *E. coli* enzyme was more active, β-glucuronidase from rabbit liver achieved 23.5% transference. A phenol (3-aminophenol) also acted as acceptor in this system, but not — confirming Fishman — in a parallel system when ether glucuronides were the potential donors. This 3-aminophenyl conjugate contained equimolar aglycon and glucuronic acid, behaved like authentic 3-aminophenyl β-glucuronide on electrophoresis, but not on chromatography, and was not hydrolyzed by β-glucuronidase; its identity remains unknown. The transferase was clearly distinguished[123] from UDPGlcUA-dependent transferase.

A β-glucuronidase activity, inhibited by glucarolactone, was claimed to transfer the glucuronyl group from phenyl glucuronide to β-naphthylamine at pH 8.4 (Okubo and Takarashi, 1964, quoted by Wakabayishi[74]), but no further report followed. With purified rat preputial-gland β-glucuronidase, transference occurred from aryl glucuronides to alkyl acceptors. No transference could be found to phenol, benzoic acid, ethylthiol, or bilirubin.[124]

The responsible enzyme is probably β-glucuronidase, but its "β-glucuronylase" activity in vivo remains unknown. Transfer requires higher concentrations of acceptor than are usually attained, or at least measured, in vivo. Also, the tissue distribution of β-glucuronidase is anomalous. Whenever glucuronidation is low in vivo, a similar deficiency of UDPglucuronyltransferase is evident and β-glucuronidase levels bear no consistent relation to glucuronidation and often appear inversely proportional (Chapter 9, Section IV). "β-Glucuronylase" is inhibited by glucarolactone,[121] but administration of glucarolactone tends to raise rather than diminish glucuronide excretion (Chapter 9, Section IV). We await confirmation of this attractive role for β-glucuronidase in vivo. Perhaps the use of more reactive donors with β-glucuronidase, such as UDP-β-glucuronic acid or β-glucuronic acid 1-phosphate, will broaden the range of acceptors, at least in vitro.[74] Their occurrence in vivo is doubtful. Investigation of the microsomal β-glucuronidase (Chapter 9, Section IV) might prove more fruitful. In that microenvironment pH and substrate concentration may be suitable. "Endogenous glucuronidation" (Section II.B above) of microsomes could be relevant here.

Jansen et al.[125] consider the diglucuronide of bilirubin is formed from the monoglucuronide by a "transglucuronylation" not involving UDPglucuronic acid, and occurring in plasma membranes:

$$2 \text{ bilirubin monoglucuronide} \longrightarrow \text{bilirubin diglucuronide} + \text{bilirubin}$$

The evidence is discussed in Chapter 7, Section III.B.

From the above, use of the term "glucuronyltransferase" without the qualifying prefix "UDP" is not only imprecise, but misleading if applied only to enzyme EC 2.4.1.17.

B. Nonenzymic Glucuronidation

Nonenzymic glucuronidation in vivo seems only likely for certain N-glucuronides. When various aromatic amines, aliphatic amines, or amino acids are mixed with the free acid, its lactone, or its amide under physiological conditions in vitro, N-glucuronides can be identified[126] (Ohgiya, 1959, quoted by Wakabayishi[74]). This phenomenon has been extensively studied in Japan.[4] The observed[127] pharmacological action of glucuronic acid and its (unconjugated) derivatives could arise from their combination with endogenous toxic amines liberated in fatigue or disease.[127] When glucuronide was injected into animals simultaneously with cyclohexamine, however, toxicity of the amine was not lessened, and the physiological value of nonenzymic glucuronidation remains unproved.[8] It should be stressed that the role of UDPglucuronyltransferase, or of any enzyme, in forming certain of the more labile N-glucuronides found, or assumed found, in biological fluids, is not certain. Instability of these conjugates weakens any hypothesis of their formation or evidence of their isolation.

IV. ADDITIONAL NOTES

Section III.A: Rat-liver microsomal β-glucuronidase transfers the glucuronyl group from phenyl-β-D-glucuronide to certain pyranoses.[127a] With $1 M$ acceptor, transfer is twice the degree of hydrolysis. The products are β-glucuronyl($1 \rightarrow 3$) glycosides), and the reaction is of unknown physiological relevance.[127a]

Chapter 4

LOCATION OF UDPGLUCURONYLTRANSFERASE IN THE CELL

I. INTRODUCTION

Glucuronidation catalyzed by UDPglucuronyltransferase appears to be the only known major mechanism for the formation of glucuronides in living tissues, and it satisfactorily accounts for the glucuronidation of almost all the aglycons known. We will, therefore, consider in detail the properties of this enzyme or family of enzymes, beginning with its location in the cell.

II. LOCATION OF UDPGLUCURONYLTRANSFERASE AMONG CELLULAR FRACTIONS

A. General

UDPGlucuronyltransferase was first found in the cytoplasmic granules of a liver homogenate.[90] Subsequent evidence, virtually all from centrifugal fractionation of homogenates, places it predominantly in the microsomal fraction, whether prepared classically or rapidly by calcium-ion aggregation.[128] Amar-Costesec et al.[129] considered it a "true" microsomal enzyme, its activity towards 4-methylumbelliferone and bilirubin being distributed among other centrifuged cell membranes exactly as glucose-6-phosphatase. However, its appearance in other membranes is not always due to their contamination with microsomes (Section I.E below).

B. Distribution Among Rough and Smooth Endoplasmic Reticulum of the Liver Cell

We must assume that "rough" and "smooth" microsomes are derived without significant ribosomal loss respectively from "rough" and "smooth" endoplasmic reticulum.[129]

One of the first studies noting the nonuniform distribution of transferase activities in the microsomal subfractions[130] placed activities towards 2-aminophenol and 4-nitrophenol predominantly in the rough, and that towards phenolphthalein equally between rough and smooth fractions, a distribution generally supported,[131-135,145a] with most of the activity towards simple phenols being reported in the rough, and none or little in the smooth subfraction. Activities to phenolphthalein[135] and bilirubin[135,137] have also been reported predominantly in the rough fraction, but most evidence[135a,145a] places them, with activities to steroids,[138,145a] in the smooth. According to Halac et al.,[136] only bilirubin monoglucuronide is formed by the smooth microsomes, the diglucuronide being formed by both subfractions (but see Chapter 7, Section III.B).

Too much stress need not be laid on apparent contradictions. Mulder[139] drew attention to the changed submicrosomal distribution of liver transferase activities after pretreatment of the animal with phenobarbital. Using 4-nitrophenol as substrate, he showed a greater total increase of activity in the smooth than in the rough subfraction; as protein also increased in the smooth subfraction, the specific activity increased similarly in both subfractions. This work emphasizes that what is isolated centrifugally is density dependent, and that density can increase by extra protein or by less phospholipid. As we believe that age, dietary change, hormones, or drugs, can alter the endoplasmic membrane composition, and so its density, we could expect a variable distribution of transferase activities between subfractions isolated centrifugally as "rough" and "smooth" procedures. Moreover (Chapter 5, Section II.D), centrifugation itself results in an often unsuspected activation of the transferase.

A different approach used morphometric stereology of the cell together with transferase assay in vitro.[143] In chick-embryo liver cultured at 5 and at 11 days and exposed to phenobarbital, transferase activity towards 2-aminophenol and endoplasmic reticulum always increased together in the ratio of 2.2×10^{-9} enzyme units/μm^2 of membrane. Cultured 5-day liver possesses little or no smooth membrane, so its transferase is virtually all in the rough membrane. In 11-day liver, "smooth" falls and "rough" rises over the first 3 days of culture; later, the "rough" is replaced by "smooth".[143] The authors suggest[143] that synthesis and degradation of the enzyme are coupled to synthesis and degradation of the membrane, and that new membrane, with its transferase, is made in the rough endoplasmic reticulum and then converted, still with its transferase, into smooth endoplasmic reticulum. The membrane might maintain a protein framework of constant composition in the constant environment of culture.[143] In vivo, however, the composition will continually change,[143] exhibiting the "lateral asymmetry" described by DePierre and Dallner.[144]

Indeed, "rough" and "smooth" may be an over-simplification. Transferase activities develop in separate clusters at birth (Chapter 10, Section III.E.1.), and may be associated, as clusters, with different membrane environments; they therefore might separate differentially as clusters under some conditions of centrifugation. We should search for groups of activities which stay together, or apart, under a variety of sedimenting procedures.

For example, activities towards bilirubin and 4-methylumbelliferone with 92 to 97% recovery were found in the microsomal subfraction *c*. This subfraction was one of four, *a* to *d*, of increasing median density, and was "less rough" than *b*.[145] Again, Wishart et al.[145a] found evidence of predominantly "Group 1" transferase activities (e.g., to small molecular weight phenols, Chapter 10, Section III.E.1) in "rough" fractions, and predominantly "Group 2" activities (e.g., to the bulkier phenols, bilirubin, and steroids) in "smooth" fractions. As Group 1 activities usually form urinary, and Group 2 biliary, glucuronides, these observations, besides extending earlier work, suggest topologically different excretory sites on the membrane.

C. Occurrence in "Microsomal Fractions" of Extrahepatic Tissues

Extrahepatic UDPglucuronyltransferase is discussed in Chapter 14. It appears predominantly in that fraction of homogenates which centrifuges similarly to the hepatic microsomal fraction. Morphology has not usually been investigated; one should not assume equal centrifugal characteristics for fragmented cell organelles from different tissues.

D. Occurrence in Nuclear Envelope

Presumably, the reported 14 to 18% of total recoverable transferase activity attributed to the "nuclear" fraction from rat liver[145] was believed due to contamination from the microsomal pellet, which contained 63 to 75% of the recoverable activity. The actual amount of transferase in nuclear envelopes expressed as percentage total homogenate activity is less than this,[146,147] but until recently it has not been possible to distinguish between the intrinsic and the contaminating transferase activities of nuclear membranes.

Absence of suitable microsomal marker enzymes in preparations from nuclei, and the tendency of nuclear envelopes to form microsomelike vesicles, can overestimate contamination if the nuclear-envelope fraction alone is characterized.[148] Fry and Wishart[146] used electron-microscopic morphometry to characterize the nuclear and nuclear-envelope preparations from chick-embryo liver and to indicate the degree of microsomal contamination. Microsomal contamination of the nuclear fraction was always below 20%, well over 80% of the membrane being recognizably nuclear. Mitochondria were less than 3% of the total membrane area. In the nuclear-envelope fraction, less easily characterized, over 50% of the membrane present was associated with nuclear-pore complexes. Transferase induced in the embryo liver by phenobarbital yielded comparable specific activities in both nuclear envelope and microsomal preparations, suggesting that the enzyme existed intrinsically in the envelope. Confirming this, transferase was five times more concentrated in envelopes than in nuclei. Of the total recovered homogenate activity of 62%, microsomal activity was 28%, and nuclear activity some 0.8%, a quantitatively minor contribution.

In female-rat liver,[147] purified nuclear fractions contained less than 10% nonnuclear membrane, and true nuclear transferase comprised 2.5% of the homogenate activity, much less than suggested by crude centrifugal fractionation. Nuclear envelope retained over 80% of total nuclear activity and so appears the primary location of the hepatic nuclear enzyme in rat, as in chick embryo. Maximal nuclear, nuclear envelope, and microsomal transferase activities were comparable (approximately 230 nmol glucuronide formed per milligram phospholipid per hour) with 2-aminophenol, serotonin, or bilirubin as substrate. Gorski and Kasper[149,150] have also examined the transferase (towards 4-nitrophenol) in rat-liver nuclear envelopes, and likewise concluded it to be a true component of this membrane with a specific activity similar to that of the microsomal enzyme. Their evidence of characterization was morphological, not morphometric. UDPGlucuronyltransferase, therefore, like glucose-6-phosphatase and other "microsomal" enzymes[151] contributes evidence for the morphological continuity of endoplasmic reticulum and nuclear envelope.

Although only a little of the liver-cell transferase exists in the nuclear envelope, it may be important there as part of a lipid-sequestering membrane protecting genetic material from mutagenic compounds. Its inducibility by phenobarbital and 3-methylcholanthrene[146,147,149] is therefore significant.

E. Occurrence in Other Cell Fractions

1. Mitochondria

Some 5 to 10% of the total recovered transferase activity towards 4-methylumbelliferone and bilirubin was in the mitochondrial fraction.[145] This and similar findings could have been from microsomal contamination, but Breuer's group[152-154] conclude that transferase activity exists in the outer mitochondrial membrane from gastrointestinal (GI) tract cells. Their preparations were free of the microsomal marker, glucose-6-phosphatase, but contained the mitochondrial marker, monoamine oxidase. The activity was specific for the 3-hydroxyl group of estrone and 17β-estradiol and appeared identical with that from the microsomal fraction.

2. Cytoplasm

Transferase activity in the high speed supernatant of centrifuged homogenates is minimal.[4,145] Activity towards steroids previously reported by Breuer's group[4] in this fraction from the GI tract and liver probably arose artifactually from the endoplasmic reticulum.[155]

3. Golgi Apparatus and Plasma Membranes

Transferase activity in the Golgi apparatus appears low[134,135] and in plasma membranes "negligible".[134] Possibly the enzyme occurs to some extent in all cell membranes, but proof requires careful morphometric treatment.

An enzyme not UDPglucuronyltransferase and which forms the diglucuronide of bilirubin exhibits its highest specific activity in the plasma-membrane fraction of rat-liver homogenates.[125]

F. Occurrence in Nonparenchymal Liver Cells

After perfusion of rat liver with Ca-free buffer and treatment with collagenase and hyaluronidase, the nonparenchymal cells were isolated and examined for transferase activity to 4-methylumbelliferone.[155a] They possessed 10 to 15% of the activity of parenchymal cells per mg wet wt and 50% per mg protein. Their transferase exhibited no latency,[155a] possibly because of the isolation procedure. This work agrees with earlier reports that hepatocytes are the principal sites of liver UDPglucuronyltransferase.[194,740] Nothing is known of the specificity of any transferase activity in nonparenchymal (e.g., Kupffer) cells, or of distribution of transferase activities among various zones of the liver.

III. LOCATION OF UDPGLUCURONYLTRANSFERASE WITHIN THE MEMBRANE

The preceding section makes clear that the transferase is membrane bound; the following chapter considers the constraint of its activity by the membrane. We discuss here the position of the enzyme in, or on, the membrane, the structure of the endoplasmic reticulum being established sufficiently for us to consider its transverse[156] as well as its lateral asymmetry.

There are two schools of thought regarding topology of the transferase. One school considers it embedded deeply within the membrane, the other that its catalytic sites are exposed freely to the cytoplasm. We discuss the former first.

Hänninen's group first postulated location of the enzyme behind a lipophilic membrane.[157] Inhibition of its activity towards 4-nitrophenol in guinea-pig-liver microsomes by aliphatic alcohols increased exponentially with the length of the alkyl chain and linearly with lipid solubility of the alcohol.[157] This suggested improved accessibility to the active center of the more lipid-soluble competitors.[157] Similar increase of inhibition with increase of chain length of alcohols was noted by Vainio,[158] who found also that more lipid-soluble compounds (e.g. chloroform or acetone) could activate the enzyme, presumably by removing the lipid barrier. Moreover, increased permeability of the barrier by phospholipase A action is prevented by albumin, which also prevents transferase activation by phospholipase A (Chapter 5, Section II.L). The complexity of the situation is indicated by work[159] with Triton® X-100 and homogenates which suggested that the active center lay behind a lipid-impenetrable barrier. However, homogenates are crude preparations and better evidence that this barrier is not simply lipid is the observation of Hänninen's group[160] that transferase in untreated microsomes was activated by trypsin, although not solubilized. The enzyme in digitonin-treated microsomes was inactivated and partly solubilized by trypsin. Digitonin is believed to perforate the vesicles,[160] and may have allowed the trypsin to enter and inactivate the transferase by a fuller digestion. They proposed that the trypsin-sensitive bonds were covered by a lipid membrane and that part of the transferase extends to the inner surface of the microsomal vesicle.[160] Guinea-pig-liver transferase[161] is, if aged[162] or treated with perturbants,[160] readily inactivated by trypsin and so may lie near the outer surface of the membrane. The rat-liver monooxygenase system, believed to be very superficial,[144] is for example readily destroyed by trypsinization.[144]

Berry and Hallinan[102] favor the deep imbedding of the transferase with its active center exposed to the luminal surface of the vesicle, and in this way account for the high (40%) efficiency of their transglucuronidation[118] and its overall inhibition by detergents, which activate each of the two component steps separately. They find 2-aminophenol is glucuronidated twice as fast with UDP and 4-nitrophenyl glucuronide than with added UDPGlcUA itself. They suggest that added UDPGlcUA has difficulty in entering the lumen from outside, but is rapidly generated inside by transfer of glucuronic acid to UDP from 4-nitrophenyl glucuronide.[102] However, Zakim and Vessey[163] would explain this by the 4-nitrophenyl glucuronide activating the transferase which glucuronidates 2-aminophenol. Evidence suggesting that UDPGlcUA is not produced within the lumen and that the transferase is not wholly embedded in the membrane comes from work by Wishart,[163a] who performed the reverse reaction, centrifuged, and found the UDGlcUA not in the microsomes or in subsequent extracts from them, but in the supernatant. Yet the hour from beginning the experiment could have allowed transport of the UDGlcUA out of the lumen. He also found that[163a] transglucuronidation did not proceed, as Berry and Hallinan suggested,[102] faster than the two steps taken separately. If the reverse reaction (with 4-nitrophenyl glucuronide) was stopped by boiling, and fresh microsomes and 2-aminophenol then added, the resulting glucuronidation of 2-aminophenol was the same or greater than when transglucuronidation had been allowed to continue undisturbed in the microsomes. Factors such as product activation or inhibition, cannot be controlled in such experiments, and to support intracisternal location of the transferase, Hallinan's group[164] cites its comparable behavior after perturbation to that of nucleoside diphosphatase, believed to be exposed to the microsomal lumen. Intracisternal location incurs, of course, not only entry of the polar UDPGlcUA but also egress of the almost equally polar glucuronides. Egress via the lumen, an elegant solution, seems incompatible with much evidence[100,165,166] that glucuronides of 1-naphthol, desmethylimipramine, and 4-nitrophenol leave the hepatocyte through the cytoplasm, and not, like albumin, through the lumen of the endoplasmic reticulum. A porter system for glucuronides would also therefore need to be invoked. Support for a deep imbedding of the enzyme without, however, exposure to either side comes from Nilsson and Dallner,[156] who found transferase activity to 4-nitrophenol unaffected by degrees of trypsinization from either side of the vesicle sufficient to inactivate most other endoplasmic reticulum enzymes studied. AMPase was similarly unaffected. However, they admit that their method of allowing access of trypsin to the lumen (a concentration of deoxycholate disorganizing the membrane sufficiently to allow ingress of large molecules without disrupting it) is open to criticism, and they did not report any activation or inhibition of the transferase under these conditions.[156]

After treatment with phenobarbital, rat liver transferase to certain substrates increases its latency (Chapter 5, Section I.A), which may be due to its even deeper embedding in the membrane, for following such pretreatment the protein liberated by trypsin during activation increased.

Evidence for compartmentation is further argued by Hallinan.[167a] Vessey and Zakim[167b] adduce more reasons for rejecting compartmentation, from their recent studies on sonicated rat liver microsomes (Chapter 5, Section II.A).

Topographical models (e.g., Reference 168) are discussed in Chapter 16, Section IV. Evidence for compartmentation involving activation energy and use of fluorescent probes is discussed in Chapter 8, Section III.

The other school of thought is typified by the elegant arguments of Zakim and Vessey and their colleagues (summarized in Reference 163). They believe that no embedding of the active centers for aglycon or nucleotide exists, and therefore, that topological compartmentation, with its apparatus of permeases and/or lipid-acceptors, cannot occur. They cite, for example, competitive inhibition of the transferase by (unspecified) Sepharose®-bound compounds which obviously could not enter the membrane.[169] As many of the hypotheses of this group depend on interpreting kinetics of microsomal-bound transferase as if its catalytic centers were indeed exposed to the cytoplasm the belief is understandable. Although they admit the membrane sequesters highly lipid-soluble substrates,[170] Zakim and colleagues believe the major role of the membrane is to determine the conformation of the transferase. They hold that the transferase protein specific for any one substrate exists in several different conformational isomers, each demonstrably possessing its own kinetic properties. Altering lipid-protein interactions mechanically, or by detergents or enzymes, would redistribute the conformational isomers, accounting for the multiple differential effects they have noted. How far the multiplicity of effects reported reflects the complexity of the membrane or the complexity of the kinetics endeavoring to interpret it, remains to be seen. A "complete characterization"[214] after each perturbing procedure may engender more publications than is useful. Zakim and Vessey[163] reject the "compartmentation" theory for several reasons, among which is its inability to account for this range of kinetic "isomers" except by an equal range of "compartments"; as different perturbing procedures probably provoke different changes in the transferase's microenvironment, the range of kinetic effects noted is however not surprising and equally consistent with compartmentation.[248,254] Either change in conformation or removal of compartmentational barriers, could be responsible for the phenomenon known as "activation", which will now be discussed.

IV. ADDITIONAL NOTES

Section III: Recent work with [31]P n.m.r. on the nucleotide products of the microsomal UDPglucuronyltransferase reaction finds UDP broken down very rapidly to UMP and phosphate.[170a] This is consistent with the coupling of the transferase and nucleoside diphosphatase in the transglucuronidation theory. However, most of the UDP was broken down outside the microsome and not, as the theory requires, within it; possible very rapid specific transport of the nucleotide across the membrane takes place, as suggested by the greater inhibition of the transferase by UDP with sealed microsomes than with leaky microsomes.[170b]

Chapter 5

FACTORS AFFECTING UDPGLUCURONYLTRANSFERASE ACTIVITY IN VITRO

I. LATENCY AND ACTIVATION: INTRODUCTION

Assay of UDPglucuronyltransferase in broken-cell preparations depends on the activation of the enzyme. Early work[4] did not appreciate this important factor, and reports still appear where activation during preparation or assay is either ignored, ill-defined, or otherwise not related to "maximal" activation under the conditions employed. The problem is difficult because of the wide variety of activation procedures, and the progressive inactivation of the enzyme as they are increased in time or intensity.

"Activation" or "inhibition" is defined as previously:[171] an increase in enzyme activity under conditions precluding protein synthesis which directly reflects change in the activity of preformed protein molecules whose catalytic potency as UDPglucuronyltransferase is partially or wholly unexpressed.

A. Latency
1. Latency — an Artifact or Not?

UDPGlucuronyltransferase, like many enzymes in the endoplasmic reticulum, is "latent" in freshly prepared homogenates or microsomes. i.e., its measurable activity is increased after membrane-perturbing processes such as mechanical disruption, aging, freezing, thawing, or exposure to detergents, chaotropes, organic solvents, alkali, certain ions, proteases, or phospholipases. An enormous and potentially confusing amount of literature has accumulated since Lueders and Kuff[172] originally observed activation of the enzyme towards 4-nitrophenol in rat, mouse, and guinea pig liver during storage of microsomes at 0° to 4°C for several days, or after exposure to deoxycholate or Triton® X-100. These authors noted that high concentrations of the detergents progressively inhibited the enzyme and that all concentrations of detergent inhibited the "spontaneously" activated enzyme: activations were not additive. This nonadditive property is, generally true for the membrane-perturbation procedures, which must be clearly distinguished from the relatively few specific activations of the transferase, e.g., by UDP-*N*-acetylglucosamine (UDPGlcNAc). Lueders and Kuff[172] also noted that "solubilization" of the enzyme need not occur, therefore, its activation reflects its relation to other membrane constituents.

If transferase activity is determined by other microsomal membrane constituents, what relationship has this activity observed in microsome or homogenate to that in the undisturbed endoplasmic reticulum of the cell? Is latency an artifact of homogenization?

Before proceeding to examine in detail the findings with each type of activator, we shall outline the current (1977) views on the causes of latency. These views stem directly from opinions held on the topology of the transferase in endoplasmic reticulum.

If one holds that the catalytic site for one or both substrates is embedded in the hydrophobic membrane, then latency is due to restricted entry of reactant or egress of product. Activation removes these barriers. If one holds that the catalytic sites are exposed to the cytoplasm, then latency and activation reflect conformational change in the transferase protein as the membrane is perturbed or degraded. This change in conformation may follow simple relaxation of pressure or readjustment of hydrophobic groups within the enzyme protein following exposure of more of the protein to the cytoplasm.

To test these hypotheses, and to investigate the degree of transferase latency existing in the undisturbed endoplasmic reticulum, two approaches have been made. One (a) compares the rates of glucuronidation in intact organ, slice, or cell with that in broken-cell preparations fortified with "physiological" concentrations of UDPGlcUA and of various endogenous modifiers of transferase activity. The other (b) embarks on kinetic analysis of latent and activated transferase in presence of the endogenous modifiers. Both approaches suffer drawbacks: (a) cannot fully allow for factors limiting uptake or secretion of reactants and product in the experimental tissue preparations; (b) will rely too heavily on the basically unsatisfactory kinetics associated with membrane-bound enzymes. Both (a) and (b) must assume that the concentrations of UDPGlcUA and of regulators reported to occur in extracts of whole liver also exist in the immediate environment of the transferase in vivo, i.e., they must assume virtually no "compartmentation" of the enzyme. However, both approaches reach the conclusion that the transferase in the cell is neither "fully" activated nor "fully" latent; the latent enzyme is partially activated by endogenous factors. We now quote representative evidence from each approach.

2. Evidence from Comparison of Transferase Activity with Overall Glucuronidation

With rat liver, fresh homogenate activities of transferase measured at the physiological UPGlcUA concentration of 0.25 mM could not, even at V_{max} and with the "endogenous" activator UDPGlcNAc present, adequately account for the glucuronidation observed in sliced liver from the same animals.[173] This was probably not due to the rapid destruction of nucleotide by rat liver homogenates, for under similar conditions the maximally activated enzyme yielded greater glucuronidation than did slices. With Gunn rats, whose transferase activity to the substrate tested (2-aminophenol) is negligible in homogenates or microsomes unless its specific activator diethylnitrosamine is also present, the fresh homogenate activities yielded 1/10 the glucuronidation seen in slices, whereas "fully activated" (diethylnitrosamine + UDPGLcNAc) homogenates glucuronidated four times faster than did slices.[173]

Use of sliced tissue is open to objections. Hamada and Gessner[174] compared glucuronidation of 4-nitrophenol in perfused rat liver with that in microsomes. Results suggested a higher transferase activity in the microsomes than in the intact cell, but the degree of activation of the microsomal preparation was not stated. That it was to some extent activated in the microsomes was suggested by their later work,[175] which equated the rate of glucuronidation in perfused liver with that in unactivated homogenates. Bock's group[176-177,178] found that the glucuronidation of 1-naphthol by perfused rat liver corresponded approximately to the glucuronidation in fresh microsomes, but whereas pretreatment of the rats with inducers increased glucuronidation in the perfused liver, it only increased it in the microsomes if they had been activated. Hamada and Gessner[174] had also noted similar enhancement of 4-nitrophenol glucuronidation in both perfused liver and their (activated(?)) microsomal preparations from pretreated rats. The results of Bock's group might therefore confirm that the transferase exists activated in the pretreated liver, but the situation is complicated by the rise of UDPGlcUA levels in the pretreated perfused liver. Bock[178] estimated the rate of 1-naphthyl glucuronide formation in the intact hepatocyte to be 1 μmol/min/10 g liver, corresponding rather to that of the nonactivated microsomal enzymes. His group[179] (see also Reference 180) have conducted an elegant series of investigations on the effect of galactosamine and carbon tetrachloride on glucuronidation by perfused liver and by microsomes with and without UDPGlcNAc. They found that galactosamine decreased both UDPGlcUA level and glucuronidation of 1-naphthol in the perfused organ. Glucuronidation of bilirubin was unaffected. Lowering UDPGlcUA levels in vitro decreased glucuronidation of both aglycons in "native" microsomes, but only that of 1-naphthol in the UDPGlcNAc-fortified microsomes. The latter thus corresponded more closely with conditions in the intact organ. This conclusion was also reached from the similar decrease of 1-naphthol glucuronidation in both perfused liver and in UDPGlcNAc-fortified microsomes from rats pretreated with carbon tetrachloride, as compared with the quite opposite increase seen in fresh microsomes from these animals, which appear to have been nonspecifically activated by the pretreatment. This is excellent evidence for the specific activation in vivo of the transferase by UDPGlcNAc, a possibility discussed later (Section II.O).

Moldéus et al.,[166] comparing Phase 1 and Phase 2 metabolism in suspensions of isolated rat hepatocytes and in microsomes, found glucuronidation of 4-nitrophenol to be somewhat higher in the suspensions than in the microsomes. Although isolation of hepatocytes may at times involve intracellular activation of the transferase,[182] Orrenius et al.[181a] quote similar rates of glucuronidation of 4-methylumbelliferone and harmol in rat hepatocytes and nonactivated liver microsomes.

Membrane perturbation by detergents prevents the rise of transferase activity seen in stored broken-cell preparations.[183] As the same rise due to spontaneous activation was seen in stored whole liver (assayed subsequently for transferase activity to several substrates) and was likewise prevented by detergents,[183] a considerable degree of latency of the enzyme should exist in the intact cell.[183]

In conclusion, although certain results suggest that the maximum excretory rates of bilirubin glucuronide in vivo correspond better in vitro with the highly activated than the unactivated form of transferase activity towards bilirbuin,[184,185] most investigators at present assume a degree of specific activation by endogenous effectors such as UDPGlcNAc acting on a transferase subject to constraint in vivo. They differ only in assigning this constraint to "compartmentational" or "conformational" factors.

3. Evidence from Kinetic Studies

As seen above, UDPGlcNAc is believed to modulate the degree of latency of the transferase in vivo. Winsnes[101,186] was the first to suggest such a role, but the theory has been stimulatingly developed by Zakim and Vessey and their colleagues.[163,187] According to it[101,163,187] the unspecifically activated enzyme is an artifact. In vivo the transferase is latent, i.e., as in fresh microsomes, and is specifically activated by endogenous modulators such as UDPGlcNAc and divalent metal ions. These ideas, which have received wide outline acceptance and which, as we have seen, are borne out by work from the other approach, have been based on transferase kinetics in fresh and activated microsomes. Some of the detailed conclusions are likely to be invalid, but this does not detract from the usefulness of the approach.

We shall outline kinetic evidence on nonspecific activation. Sections dealing with specific surfactants (Section II.E), UDPGlcNAc (Section II.O), and phospholipases (Section II.L) cover other implications more fully.

The basic work of Zakim's group was performed with one animal, one tissue, and one substrate: guinea pig, liver, and 4-nitrophenol. Nevertheless, the concept of an essentially latent enzyme specifically activated in vivo has been more widely supported by evidence from the sources noted above.

Studies of UDPglucuronyltransferase in vitro in presence of nonspecific activators which perturb the membrane cannot, they suggest, yield useful information on its regulation in vivo, and therefore, on its degree of latency there, because the membrane environment is an integral part of the regulatory mechanism. Removal from this environment will produce a spectrum of modified physical forms of the transferase.[163]

In the membrane environment, the transferase would be under constraint, and the constraint conformational, not compartmentational. The constraint allows the exposed catalytic site for UDPGlcUA to accept this sugar nucleotide only from among a large number of these compounds examined.[163] Such high specificity is valuable because the overall concentration of UDPGlcUA in liver is low.[188,189] However, this active site is inhibited by UDP, which is, produced by the glucuronidation reaction and by UTP.[117,186] As both UDP and UTP are present in liver, the latter at a high level, the former approaching that of UDPGlcUA,[188,189] end-product inhibition of the transferase could inhibit glucuronidation in the intact cell. However, inhibition by UDP virtually disappears if the untreated transferase is assayed together with UDPGlcNAc and divalent metal ions.[190] UDPGlcNAc acts, in the presence of Mn^{++} or Mg^{++}, as a positive K type of allosteric effector for the binding of UDPGlcUA[190,191] and as a negative K type of allosteric effector for the binding of UDP. The concentration of UDPGlcUA required for half-maximal rates of glucuronidation then falls to values approaching those found overall in liver.[190,191] The presence of UDPGlcNAc and the metal ions, therefore, converts the transferase of fresh microsomes into a form kinetically efficient in vivo.[163]

"Activation" of the microsomal transferase by membrane perturbation destroys this capacity for efficient regulation in vivo. After perturbation by phospholipase A treatment, the specificity of the UDPGlcUA-binding site is lost; for all the UDP-sugars tested by Zakim's group, including UDPGlcNAc, inhibited transferase action by binding at that site. Moreover, inhibition by UDP was increased.[163] Activation by Triton® X-100 does not potentiate end-product inhibition by UDP, nor abolish specificity for UDPGlcUA; It does, however, abolish the capacity for regulation of transferase activity by UDPlcNAc.[163] Activation by either agent should, therefore, diminish the efficiency of transferase action in vivo.[163]

In vitro, transferase activated by either phospholipase A or Triton® X-100 can exhibit high activities because concentrations of UDPGlcUA can be made very high and activation may decrease $K_{UDPGlcUA}$. Although the increased affinity for UDPGlcUA shown on activation may result from several mechanisms[163] (Section II.E and II.I below), we can now perceive theoretical grounds for the remarkable "activation" of transferase activity in vitro.

The role of partial activation of the enzyme in vivo by agents other than UDPGlcNAc, Mg^{++}, or Mn^{++} is discussed later (Chapter 15, Section VIII). Changes in the characteristics of its membrane environment by accumulation of bile salts, by diet, and by induction with drugs and hormones provide a (perhaps too) fertile ground for speculation on the changes in its in vivo activity under such conditions.

4. Evidence from Nuclei

In preparations of rat-liver nuclei and nuclear envelopes which were microsomes, nonvesicular by electron microscopy, [163a] the transferase is not "latent" not being activated by detergents or UDPGlcNAc.[163a,163b] Latency may, therefore, be an accompaniment of vesiculation.[163a]

II. ACTIVATION AND INHIBITION PROCEDURES

These may be divided into three groups: (1) those specifically, (2) those nonspecifically, affecting the transferase enzyme itself; (2) operates by abolishing latency through membrane perturbation.

Activation can reach 40-fold in vitro[192] and does not result from destruction of any inhibitor present in untreated microsomes.[172,192]

Perturbation procedures differ in efficacy with age, dietary or hormonal state, and with tissue or species; in some instances (e.g., digitonin on chicken-liver transferase[193]), it may yield no discernible activation. In general, however, perturbation procedures are additive in their effects, and continued perturbation results eventually in inactiva-

TABLE 2

Some other references (not quoted in context in Chapter
5, Section II) to procedures or compounds changing
UDPglucuronyltransferase activity in vitro

Storage, ageing or preincubation	101, 130, 172, 186, 196, 275
Surfactants	116, 173, 182, 185, 193, 206, 208, 230, 235, 249, 251, 287—289, 311—319
EDTA or dialysis with EDTA	136, 184, 227, 273, 320, 321
Diethylnitrosamine	322, 323
Phospholipases and trypsin	191, 235, 244, 324-326
Thiol reagents	327
UDP-N-acetylglucosamine	173, 185, 193, 206, 227, 273, 275

tion (inhibition) of the enzyme. Pretreatment with drugs (Chapter 13, Section II.E) may profoundly influence latency of the enzyme, and hence, its activation characteristics.

Table 2 lists a number of references not quoted in the text for those wishing further access to details.

A. Mechanical Disruption

Activation of UDPglucuronyltransferase by sonication, first clearly shown by Henderson[194,195] in rat liver, has been noted in bovine liver,[116] human and mouse liver,[196,197] and chicken liver,[193] and is dose- and age-dependent.[182,196] Berry et al.[164] activated guinea pig-liver transferase by subjecting microsomes to several mechanical procedures — sonication, Ultraturrax "blending", and grinding with glass. The activation might have resulted from liberation of (1) lipid peroxides (Section II.N. below), or (2) free fatty acids or partially hydrolyzed glycerides (Section II.M below). As activation occurred in presence of α-tocopherol and under nitrogen, (1) is unlikely. As few fatty acids were released, and as activation still occurred when bovine serum albumin was present to bind them, (2) was also unlikely.[164] Berry et al.[164] proposed that these mechanical procedures activated the enzyme by simply rupturing vesicles and improving substrate accessibility, quoting the simultaneous activation of latent nucleoside diphosphatase and of both the forward and reverse reactions of the transferase. Activation by conformational change due to increased membrane fluidity (itself due to decreased vesicle size) they thought less likely.[164]

Vessey and Zakim[167b] activated microsomes by sonicating them for 5 min at 0°C with intermittent cooling. If activation had been due to destruction of compartmentation, they argued, it should have arisen either from (1) leakiness or (2) a turning inside-out of the vesicles. (1) Was discounted by the continued impermeability of the sonicated microsomes to inulin, although the vesicles were reduced in size. (2) Was discounted from evidence with trypsin. Trypsin quantitatively releases the known externally bound NADPH-cytochrome-c reductase from untreated microsomes into solution without affecting internal proteins. After sonication, the reductase was still released by trypsin. As inulin could not enter, the larger trypsin molecule should certainly not have entered. Therefore, the sonicated microsomes were still "outside out". Moreover, as the trypsin removed the transferase activity, the transferase is, they contend, on the exterior of the microsome.

B. Temperature

Temperature change itself does not activate the transferase apart from its role in storage (II.C below), and the gross physical change in the membrane following freezing and thawing. Rise in temperature may appear to activate the enzyme in the presence of UDPGlcNAc because the latter activates the transferase towards 4-nitrophenol above 16°, but not below 16°C.[198] The break is sharp; UDPglucose, UDPmannose, and UDPxylose are reported to inhibit below, but not above, this temperature.[198] It may be related to an abrupt phase transition in the membrane. Eletr et al.[199] noted sharp breaks at 19° (not, this time, 16°) and 32°C in glucuronide conjugation by intact guinea pig microsomes, and Pechey et al.[200] found this transition point at 20 to 25°C. The Phase 1 microsomal system exhibits a similar property.[201] Evidence that these changes concern the phospholipid bilayer is discussed in Chapter 8, Section III, as are results on temperature-dependent kinetic changes.[202]

C. Storage

As remarked elsewhere,[203] it is curious that early investigators failed to notice activation of the fresh transferase on storage. Activity to 4-nitrophenol in rat-liver microsomes, for example, increases eight times when stored for 8 days at 0°C. By 10 days, optimal activation has been passed.[160] However, if already-activated transferase is stored, only a progressive fall is noted.[133,160] The activation is not additive, and cumulative perturbation inactivates the enzyme. Stimulation of transferase activity in stored homogenates and in stored mouse liver itself at 0 and 37°C, is activation not induction, being unaffected by cycloheximide.[183] After an initial fall within the first hour or so, activities to 2-aminophenol, 4-nitrophenol, and bilirubin peaked to three to four times the fresh levels at around 12 hr, falling slowly thereafter towards zero.[183] Detergents abolished the activation.[183] Rat liver clamped in vivo, ischemic for up to 2 hr, lost much of the activity of glucose-6-phosphatase and other microsomal enzymes,[204] but not that of transferase activity towards 4-nitrophenol (beyond a slight initial fall).[204] Additional perturbation during preparation of microsomes from the ischemic liver could obscure any activation resulting from the ischemia. Storage at 4 and 37°C, or ischemia, did not significantly solubilize microsomal membranes, except possibly from the lumen, and electron microscopy showed aggregation and membrane rupture.[205]

Activation on storage could result from endogenous bile salts, liberation of peroxides, lysolecithin, and free fatty acids, and/or disruption of the membrane by proteases and phospholipases. It appears markedly temperature-dependent.[160,183] Its relevance to pathological changes is discussed in Chapter 15, Section VIII.

As Winsnes reported,[160] activation characteristics of transferase activity may vary with substrate. In rat liver preincubated at 37°C, he found no initial fall with 4-methylumbelliferone and phenolphthalein.[160] Activity towards 4-nitrothiophenol, unlike that towards 4-nitrophenol, was not significantly stimulated by storage at 0 to 4°C for several days, but gradually declined.[206]

D. Centrifugation

Centrifugation is an activation procedure.[207] It involves storage, up to 2 to 3 hr if washing of microsomes is excessive, at 0 to 4°C. It may incur exposure to activatory ions. Conflict of results from different laboratories has been attributed to centrifugal procedures.[208,209] Mulder[139] considers accidental activation of transferase by repeated homogenization and centrifugation procedures of other workers to explain the comparatively low values of his fresh preparations. Centrifugation in presence of Ca⁺⁺ to reduce centrifuging time results in loss of activatability of rat-liver transferase by digitonin, though not by trypsin.[128]

E. Surfactants

Certain endogenous surfactants such as fatty acids and bile salts are treated further below (Section II.M and Chapter 15, Section VIII). The effect of surfactants on UDPglucuronyltransferase has been generally studied since Lueders and Kuff in 1967[172] showed activation up to tenfold by deoxycholate or Triton® X-100. As the transferase can be activated before being solubilized,[160,192] and as UDPGlcUA pyrophosphatase and β-glucuronidase remain unaffected by the concentrations of activator used,[172,192,210] the activation is truly of the membrane-bound transferase. The anionic surfactant deoxycholate, the cationic cetylpyridinium chloride, and the nonionic digitonin and Triton® X-100 were compared.[160,211] All activated the transferase towards 4-nitrophenol, though the nonionic ones possessed a broader optimum concentration (on a basis of grams of surfactant per grams of microsomal protein). Under the conditions employed, deoxycholate released most microsomal protein into solution, and cetylpyridinium chloride (possessing very similar activation characteristics for the transferase as deoxycholate) released virtually none. Phase 1 enzymes were inhibited by all concentrations. The authors interpreted these findings and confirmed their belief that the transferase was buried deeper than the Phase 1 system. The eventual inhibition of the transferase would result from its loss of catalytic conformation due to denudation of lipid.[211] Winsnes, in his valuable early study on activation,[192] showed that not only did the activation characteristics of digitonin differ from those of Triton® X-100, but each detergent differed in the extent of its effect with different substrates. The optimal concentrations were, however, not too dissimilar.[212]

In some instances optimal concentration of detergent is markedly substrate dependent. Possibly related to digitonide formation, transferase activity in rat-liver microsomes towards estrone is sharply optimal in 1.0% digitonin, towards estradiol still increasing at 10% digitonin, and towards testosterone practically optimal over 1 to 10% digitonin.[213] In Gunn-rat-liver microsomes, no activation towards estrone occurs with digitionin.[213] Findings do not always agree, e.g., according to some,[176] cholate or Triton® X-100 did not activate rat-liver transferase towards any steroid tested, but others,[213] find these detergents good activators with steroid substrates.

The effect of surfactants upon the kinetics of crude microsomal UDPglucuronyltransferase is as complex as expected. Lueders and Kuff[172] and Mulder[139] preferred not to analyze their activation kinetically, because of the low unactivated transferase activity. Kinetic studies published do not agree. The K_m for UDPGlcUA is reported to be increased, decreased, or unchanged. It was first studied by Winsnes.[101,192] He found some ten-fold increases of apparent $K_{UDPGLcUA}$ in the mouse-, rat-, and guinea pig-liver enzyme towards 2-aminophenol and 4-nitrophenol after activation with detergents. With 4-nitrophenol in mouse and rat, interpretation was further complicated by abrupt breaks in the Lineweaver-Burk plots, suggesting negative cooperativity. Zakim and Vessey have investigated the effect of Triton® X-100 on 4-nitrophenol glucuronidation in guinea pig-liver microsomes[163] and consider it different from activation by lysophosphatide or cholate. Their bold interpretation of the kinetic data is stimulating. From calculations of the $\Delta G°$ of binding of UDPGlcUA and of UDP to the enzyme, they suggest that the enhanced affinity of UDPGlcUA for Triton®-treated, compared with fresh, transferase is due to increased affinity of the enzyme for the UDP portion of the sugar nucleotide.[187,214] $K_{UDPGlcUA}$ and K_{UDP} were lower after Triton® treatment than before. However, the K_{UDP} was not so much lowered as to result in end-product inhibition by UDP, such as seen after phospholipase A treatment. Triton® treatment abolished activation of the transferase by UDPGlcNAc but preserved the specificity of the UDPGlcUA site for its substrate.[214] These conclusions disagree somewhat with those of Winsnes, who found increased $K_{UDPGlcUA}$,[101] and increased inhibition by UDP[186] after detergent treatment. He considered end-product inhibition important.

The disagreements may originate in the wider variety of species and substrate investigated by Winsnes and his less intensive study of 4-nitrophenol glucuronidation in guinea pig liver, the system of the Californian group. Lucier et al.[133] noted no change in apparent K_m values to UDPGlcUA or to 1-naphthol on ten-fold activation of rabbit-liver transferase with Triton® X-100, even though V_{max} increased 12 times. They considered that activation increased available active sites of the transferase rather than changed its affinity for the substrates, a conclusion reached by others.[215]

Detergents have different effects in different tissues. Cetylpyridinium chloride increased transferase activity in kidney, liver, lung, spleen, and intestinal mucosa, but digitonin had no effect on the latter.[216] It is important to use a range of detergent concentrations. An apparently negative-responding tissue may be especially sensitive to the detergent in question and require only low concentration for activation before subsequent inactivation. The detergent: protein ratio is of course important in all tissues.

F. Chaotropic Agents

Chaotropic agents solubilize much microsomal protein and can activate the transferase.[217] They increase aqueous penetration and weaken the hydrophobic bonds of the membrane, causing disaggregation.[218] The monooxygenase complex was always inactivated by chaotropes,[217] but the more potent the chaotrope, the more readily it activated the transferase.[217] Activation was followed by inhibition as concentration of chaotrope rose.[217] Activity increased three- to four-fold with 0.5 M NaSCN or 1.0 M KI. At 3M, inhibition by these chaotropes was complete. Urea or KNO_3 activated a little, but never inhibited.[217]

G. Organic Solvents

Vainio[158] found that chloroform at some 1.8 mol/ℓ increased rat-liver transferase activity to 4-nitrophenol four- to fivefold; higher concentrations inhibited. Acetone was a little less effective, and, to a still lesser extent, *n*-hexane and diethyl ether.[158] Notten and Henderson[219] found *n*-pentane a better activator than *n*-hexane, and *n*-heptane only a weak activator, with 4-nitrophenol or 2-aminophenol a substrate for the guinea pig-liver enzyme. The concentration of alkane giving maximal activation increased as chain length decreased. The activation was not additional to that brought about by Triton® X-100, caused no solubilization, and increased the apparent $K_{UDPGlcUA}$.[219] Interestingly, the relative capacity of the different alkanes to activate the transferase corresponded to their liberation of phospholipid from the microsomes.[219] When fed to guinea pigs, *n*-hexane stimulated their liver transferase, possibly by activation.[220] Carbon tetrachloride, also an in vivo stimulator, had no effect on the transferase in vitro.[216] It presumably stimulates in vivo by activation during its conversion to chloroform. These in vivo effects are probably due to membrane perturbation in the cell. Similarly, exposure of the animal to carbon disulfide stimulates the transferase in vivo,[221,222] lowers Phase 1 enzymes, and damages intracellular membranes.[221,222]

Aliphatic alcohols[4,157,158] inhibit the transferase prior to activation, inhibition increasing with length of alkyl chain and, apparently,[157] being competitive with the aglycon. Methanol, ethanol, *tert*-butanol and pentanol are known to form glucuronides.[63] Very recently[222a] alkyl ketones have been shown to markedly activate crude or purified transferase from Wistar and Gunn rat liver, abolishing the genetic defect of the latter towards 2-aminophenol apparently in the same manner as does diethylnitrosamine (Section II.J. below).

H. Ethylenediaminetetracetic Acid (EDTA) and Chelating Agents

EDTA can possibly bind inhibitory ions and inhibits breakdown of UDPGlcUA by UDPGlcUA pyrophosphatase.[223] Apparent "activation" of transferase can arise by the latter action of EDTA, but when the pyrophosphatase was already inhibited by ATP, then EDTA was seen to inhibit rat-liver transferase,[223] as it did without added ATP in guinea pig-liver preparations[223] in which the pyrophosphatase is naturally low. This inhibition of transferase could arise by EDTA removing divalent metal ions not only needed for the transferase[224] but also for microsomal nucleoside diphosphatase. Inhibition of the latter enzyme allows accumulated UDP to inhibit the transferase. EDTA, inhibiting rat-liver pyrophosphatase, increased observed transferase activity.[225] These pyrophosphatase-dependent effects are further treated in Chapter 9, Section III.B. Alkaline dialysis with EDTA activated rabbit-liver transferase,[184,226] but probably from the alkaline pH (Section II.P below) for EDTA itself inhibited the enzyme. Other evidence (e.g., Reference 101), also indicates no activation of UDPglucuronyltransferase by EDTA. The reverse reaction is markedly stimulated by EDTA, probably through the almost complete inhibition of nucleoside diphosphatase allowing UDP to prime the reverse reaction.[118]

I. Sulfhydryl Reagents

UDPGlucuronyltransferase is inhibited by all types of reagents reacting with thiol groups.[227,228] This early work with 4-hydroxymercuribenzoate, 4-chloromercuribenzoate, phenylmercuric acetate, 4-chloromercuriphenyl sulfate, 2-chlorovinylarsenoxide, arsenite, pentavalent arsenicals, iodoacetamide, and N-ethylmaleimide is tabulated.[4] Glutathione or cysteine could protect against or reverse the inhibition without affecting untreated transferase. Protection was afforded by UDPGlcUA but not by the aglycon substrate.[4] In contrast, Winsnes[229] demonstrated that, with high UDPGlcUA concentrations, sulfhydryl reagents activated the enzyme. They also desensitized it to activation by UDPGlcNAc, though not by Triton® X-100. Using a wide range of substrates in mouse and rat, he concluded that the transferase active site is not dependent (as had been suggested by Goldberg[96]) on sulfhydryl groups for catalysis, but that these groups are concerned with an allosteric site, probably the one acted on by UDPGlcNAc.[229] The same conclusion was reached and extended by Zakim and Vessey.[116,230] They found that, in beef-liver microsomes, activity towards 4-nitrophenol was activated by organic mercurials reacting with what they termed Type 2-SH groups. Prior treatment of microsomes with detergent sensitized the transferase so that optimal activation occurred with less mercurial concentration. The Type 2-SH group could, therefore, be in a lipid environment.[230] As the extent of activation depends on the organic group of the mercurial, and can, with charged mercurials, completely destroy latency, they considered that the thiols involved were in a region constraining the transferase in a low-activity form or "conformation"[230] and that the constraint is due to hydrogen bonds in the neighborhood of the Type 2 SH groups.[230]

Like Winsnes,[229] they consider this region to be off the active site(s) because prior binding of either substrate will not prevent reaction with the mercurials.[230] The region need not be even on the enzyme protein, but in the adjacent constraining membrane.[230] Relatively high concentrations of organic mercurials react with another, Type 3, −SH group to produce a deactivated form of transferase.[230] N-Ethylmaleimide may react with a further (Type 1) -SH group, for pretreatment with it will not prevent subsequent activation by mercurials.[230] Dithreitol affected Types 2 and 3 groups. It reversed activation by mercurials for Type 2 and reversed inactivation for Type 3. It had no effect on the Type 1.[230] If the membrane was disturbed by phospholipase A treatment, then sulfhydryl reagents uniformly inhibited.[230] These reagents join the great variety of compounds that activate the transferase by partial disruption of the microsomal membrane. Activity towards the natural substrate pregnanediol also is stimulated by N-ethylmaleimide,[231] that towards estrone being unaffected.[232] Inhibition of the latter activity by other thiol reagents has been described,[232] with enzyme from pig kidney.

J. Diethylnitrosamine

This compound, which strongly inhibits Phase 1 enzymes,[233] was found to activate UDPglucuronyltransferase activity threefold towards 2-aminophenol and four-fold towards paracetamol when added to rat-liver homogenates and microsomes.[234] Subsequent work has extended this activation to other substrates and species; transferase activities towards 2-amino benzoic acid,[235] 4-nitrophenol,[196,236] and estriol[196] are stimulated by diethylnitrosamine, and this activation has been demonstrated in preparations from mouse and human liver.[196] Stimulation of activity towards 4-methylumbelliferone[192] and 4-nitrophenol is not great.[206,234,237] Activities towards menthol, phenolphthalein,[234] 4-nitrothiophenol,[206] and interestingly, 2-aminothiophenol,[206] remained unstimulated by the concentrations of diethylnitrosamine employed. That towards bilirubin was inhibited.[192] Winsnes[212] considered that activatability of the transferase by diethylnitrosamine only developed postnatally in rats, but his experiments were complicated by concomitant activation of the enzyme with digitonin. Burchell[196] found response to diethylnitrosamine was age-dependent, but under suitable conditions it could be demonstrated in livers of late-fetal mice; the activation was additive to that caused by previous sonication and seemed to require such membrane perturbation before being evident. Membrane perturbation potentiates the effect of diethylnitrosamine and vice versa.[192,239] This close linkage with the state of the membrane environment requires the continued presence of the compound. Preincubation with diethylnitrosamine in absence of substrate enhanced subsequent stimulation by a lower concentration of diethylnitrosamine.[237] As this stimulation was prevented by washing microsomes after preincubation,[237] no permanent alteration to the membrane occurred.

The most dramatic effect of diethylnitrosamine is its restoration of certain transferase activities to Gunn-rat homogenates or microsomes in vitro;[234] their very low activity to 2-aminophenol is stimulated to the high activated levels seen in similarly treated preparations from Wistar rats. From this phenomenon and its extension, in a lesser degree, to other substrates such as 2-aminobenzoic acid and 4-nitrophenol, Zakim's group[235,236,238] draw conclusions on the catalytic defect in the Gunn mutants (Chapter 12, Section III.B). Whether diethylnitrosamine acts on the protein itself or on the lipid constraining the protein is not yet clear. It certainly activates partially purified transferase[239] and also the apparently pure enzyme[240,241] which is resistant to inactivation by phospholipase C and contains minimal phospholipid,[242] although Lubrol® is present. Burchell,[241] therefore, considers its action intimately concerned with the protein molecule. Zakim and colleagues[239] are in favor of its affecting lipid-protein interrelationships, and hence, the protein conformation. In Gunn rat,[236] they consider it enhances binding of the UDP-moiety of UDPGlcUA to the UDPGlcUA binding site. They found, unlike previous workers[173,234] who discovered no change brought about by the compound in the apparent $K_{UDPGlcUA}$ in Wistar or Gunn rats, that diethylnitrosamine increased affinity of the transferase for UDPGlcUA in Gunn rats. Apparent $K_{aglycon}$ is unaffected in Gunn and Wistar rats.[234,236]

Hereditary defects in transferase activity are not the only ones "repaired" by diethylnitrosamine. If rats are hypophysectomized or thyroidectomized, transferase activity to 2-aminophenol falls;[237] it is restored, on addition of diethylnitrosamine to the assay,[237] to the high levels seen in a similarly treated preparation from normal rats.

Dimethylnitrosamine does not markedly activate the transferase although, like diethylnitrosamine, it induces it on administration in vivo.[233] The effect of diethylnitrosamine in vitro is now known to be reproduced by simpler ketones (Section II.G. above).

K. Trypsin

Treatment of UDPglucuronyltransferase with trypsin has already been briefly mentioned (Chapter 4, Section II), where it was noted that Finnish workers[160,168,243,244] suggested the transferase was not exposed to the cytoplasm of rat-liver microsomal membranes, for partial tryptic digestion stimulated the enzyme in fresh microsomes and inhibited it permanently in disrupted microsomes. However, a curious lack of effect of trypsin on the rat-liver enzyme in microsomes has also been reported;[156] and in guinea pig-liver microsomes, trypsin either had no effect[163] or gradually inhibited the enzyme.[161] As this contradiction affected arguments concerning the embedding of the transferase in the membrane, Wilkinson and Hallinan[162] attempted to resolve it by studies with both rat and guinea pig and "intact" and disrupted microsomes. They prepared their "intact" microsomes in 0.25 M sucrose. They found stimulation in these "intact" organelles from both species up to 165% within 40 min at 30°C with 50 µg trypsin per milligram protein. Even at lower temperatures (where no "spontaneous" activation occurred in nontrypsinized controls), stimulation could reach up to 35% with high trypsin concentrations. However, if intact microsomes were disrupted ultrasonically or with detergents, inhibitions of up to 90% replaced the stimulation. During subsequent transferase assay, the trypsin was inhibited by its inhibitor from soybean, which did not affect the transferase.[162] The authors explain the contrary results of others as follows. When trypsin was reported to inhibit the transferase, the enzyme had been studied in microsomes prepared and washed in KCl. Such preparations from guinea pig (but not rat) exhibit higher transferase activity than those prepared in sucrose[208] and might have been already activated before trypsinization. The experiments reporting lack of trypsin activity on the transferase[161,163] possibly used insufficient trypsin, according to Wilkinson and Hallinan,[162] who consider the transferase relatively trypsin-resistant among rat-liver microsomal enzymes. According to Hietanen and Vainio,[245] trypsin activates the transferase in livers of guinea pig, rat, mouse, and hamster.

L. Phospholipases and Phospholipids

Before discussing the large and often confusing literature on the effect of phospholipases on UDPglucuronyltransferase we should examine the terms "phospholipid dependence" and "phospholipid constraint".

According to Hallinan,[246] "phospholipid dependence" is an imprecise term and should be used when an enzyme requires presence of phospholipids to exhibit its maximum activity. This dependence is therefore more than a mere requirement for stability. An essential criterion would be that activity is lost on lipid depletion and restored on adding purified phospholipid. Hallinan[246] considers "phospholipid constraint" as restraint on the maximal catalytic activity of the enzyme specifically exerted by phospholipids. Removal of phospholipid might then be thought of as activating the enzyme.

By this way of thinking, phospholipid constraint could coexist with phospholipid dependence. Progressive removal of phospholipid would first activate the enzyme and then inactivate it.

It could also be argued that dependence of an enzyme on phospholipid to regulate its activity must necessarily incur "phospholipid dependence", and so phospholipid constraint would always include phospholipid dependence. Phospholipid constraint could operate structurally not only through conformation, but also through compartmentation by phospholipid barriers restricting passage of reactants, products, or modulators.

We shall, therefore, avoid these terms wherever possible and warn that different authors employ them for different situations. If used here without quotation marks, phospholipid dependence means requirement of phospholipid before any catalytic action of a pure protein can be observed in vitro, i.e., a major criterion of Fleischer et al.[247] is extended back to zero — no phospholipid, no activity.

In the case of the transferase[242] a nonionic detergent such as Lubrol 12 A9® substitutes for phospholipid in supporting catalytic activity of the pure protein in vitro. Until the detergent molecules can be removed and catalytic power shown to be lost and then restored on addition of phospholipid, we can only term UDPglucuronyltransferase "phospholipid dependent" in quotation marks.

Reviews of the effect of phospholipases on UDPglucuronyltransferase are by Zakim and Vessey[163] and Berry.[248] These authors disagree with each other's interpretations, and both should be consulted for an objective guide through the evidence.

The evidence began in 1962 when Isselbacher et al.[227] reported a four-fold increase in UDPglucuronyltransferase activity towards 4-nitrophenol in rabbit-liver microsomes treated with a crude-venom preparation of phospholipase A. They thought this "activation" due to inhibition of β-glucuronidase. Tomlinson and Yaffe[249] also noted this activation by phospholipase A towards 4-nitrophenol, but additionally reported inhibition of activity towards bilirubin by either phospholipase A or C treatment.

In 1969, Graham and Wood,[250] in the first detailed study, found that guinea pig-liver microsomes incubated with phospholipase A or C lost UDPglucuronyltransferase activity towards 4-nitrophenol. As activity was partially restored by adding back microsomal phospholipids, "phospholipid dependence" was suggested; products of phospholipase action such as lysolecithin or diglyceride did not inhibit, and albumin added to adsorb fatty acids did not prevent loss of activity. Attwood et al.[251] subsequently found lecithin the most effective phosphlipid to restore activity. No activation by phospholipase A was reported.[251] In the same year, however, Hänninen and Puukka[252] found rat-liver UDPglucuronyltransferase activity towards 4-nitrophenol to be markedly activated by phospholipases A and C, phospholipase D having no effect. They considered the activation by phospholipase C due to the phospholipase itself and not to an enzyme contaminant such as protease, for it was Ca^{++} dependent. However, the real activators, they percipiently suggested, were the products of phospholipase action, for activation also occurred after addition of lecithin, lysolecithin, or palmitic acid as micelles to untreated transferase. Moreover, phospholipase D, which did not stimulate the transferase, releases only the base from phospholipids.[252] Prolonged incubation with phospholipase A or C perturbed the membrane beyond the optimal for transferase activity.[252] They considered their disagreement with Graham and Wood,[250] who found no activation, to be due to their own use of less phospholipase C and to a purer phospholipase A. Graham and Wood's phospholipase A may have had protease activity which, in conjunction with the phospholipase, could have activated, then inactivated, the transferase.[252] Graham and Wood's[250,251] prolonged centrifugation in KCl could also have activated the guinea pig-liver enzyme so that it would not respond to activators released by phospholipase, but instead be activated by phospholipid depletion.[248]

Vessey and Zakim,[116] using beef liver microsomes incubated at 22°C, found an initially large activation of 4-nitrophenol glucuronidation by phospholipase A at V_{max}. Prolonged incubation resulted, as with Hänninen and Puukka,[252] in a slow loss of activity which they attributed to instability of the delipidated transferase because the loss continued at 37°C even with EDTA present to prevent further phospholipase A action.[116] This major disagreement — initial activation (not inactivation) and failure to reactivate with phospholipids — with the results of Graham and Wood[250] was emphasized when Vessey and Zakim[253] reported an 11-fold activation of the transferase by phospholipase A, even in guinea pig-liver microsomes which had been prepared, like those of Graham and Wood, in KCl (though, not necessarily rewashed in KCl[248]), and some activation with rat, mouse, rabbit, and man. They, therefore, considered that the transferase was not "phospholipid dependent", but was constrained by microsomal phospholipid to a conformation of low activity in vivo.[253] Treatment with phospholipase A, like other perturbing procedures, would relieve this constraint and produce a series of conformational "forms" which were less thermostable. The initial activation and subsequent inactivation were explicable without supposing compartmentation.[253]

Zakim and Vessey have recently[163] summarized their bisubstrate kinetic investigations on phospholipase A treatment of transferase activity towards 4-nitrophenol in guinea pig-liver microsomes. Kinetic studies of this activity are hampered by the downward concavity of double-reciprocal Michaelis-Menten plots when UDP is employed below 2.5 mM with the "native" untreated enzyme;[101,190] above 2.5 mM, treatment with phospholipase A decreases $K_{UDPGlcUA}$ and produces typical Michaelis-Menten kinetics.[187] If $\Delta G°$ values for binding of UDP and of UDPGlcUA are calculated from these $K_{UDPGlcUA}$ figures, and the binding energy of UDPGlcUA is assumed to be summed by the contributions of both UDP and glucuronic acid moieties, then the phospholipase A enhancement of the transferase's affinity for UDPGlcUA is due to increased affinity of the transferase for the UDP moiety.[163] Affinity for the glucuronic acid moiety is reduced.[163] UDP becomes so effective an inhibitor of the phospholipase A-treated transferase that no activation by the phospholipase appears if equimolar UDP and UDPGlcUA are present.[163] According to these workers, treatment with phospholipase A, unlike treatment with Triton® X-100[163], abolishes also the specificity of binding of UDP-sugars.[191] These results imply that phospholipase A "activation" of transferase under in vivo conditions would not stimulate glucuronidation.

Activation in vitro of the transferase by phospholipase A is partly due to increased V_{max} for both directions,[163] not only for the forward reaction as first[116] supposed. Zakim and Vessey[163] calculate the equilibrium constant for the reaction to be 6.7 for the untreated and 10.7 for the phospholipase A-treated enzyme. They believe the correspondence of these two values[163,187] proves their kinetic assumptions. Apparent $K_{4-nitrophenol}$ is reported to be increased five-fold by phospholipase A.[253]

Abolition by phospholipase A of specific activation of the transferase by UDPGlcNAc (Sections I.A. above and II.O. below) is regarded as another example of the postulated catalytic inefficiency of the phospholipase-treated enzyme in vivo.[163]

More recent work with phospholipase A has not settled this problem, but has clarified the nature of the activation. Graham and Wood[210] reported that rat-liver microsomes, unlike those from guinea pig liver, were not activated by isolation in 0.154 M KCl; from other evidence that the liver enzyme was more readily activated in guinea pig than in rat, they concluded that the membrane phospholipid structure in the two species was not identical. They then demonstrated that activation of rat-liver transferase by phospholipase A was due to the degradation products of phospholipids, i.e., lysophosphatides and unsaturated fatty acids.[255] The transferase in intact microsomes was activated by a similar concentration of these products to that liberated by phospholipase A; activation by lysophosphatidylcholine most closely resembled activation by phospholipase. Albumin protected against activation by sequestrating the lytic products. These findings confirmed most results of Hänninen and Puukka,[262] but not their activation of "intact" microsomal transferase by phosphatidycholine. Graham and Wood[255] suggested the Finnish workers' phosphatidylcholine could have been contaminated by lysophosphatidycholine. In later work[256] over 80% of the phospholipid component of KCl-prepared guinea pig-liver microsomes, was removed by phospholipase A and subsequent extraction of the products with albumin. This delipidation strongly inhibited activity towards 4-nitrophenol.[256] Activity was restored by mixtures of phosphatidylcholine and lysophosphatidylcholine. When the microsomal proteins and phospholipids of a transferase preparation were separated by gel filtration with cholate on Sephadex® G-150,[256] the enzyme became inactive, this time in the absence of any possibly lytic and inhibitory phospholipid degradation products, and activity was again substantially restored by choline glycerophosphatides.[256] Gorski and Kasper,[149,150] also found that nonenzymic removal of 96% microsomal phospholipid by gel filtration in deoxycholate almost completely inactivated rat-liver transferase towards 4-nitrophenol. Microsomal phospholipid, lecithin, or lysolecithin (dioleoyl or bovine rather than dipalmitoyl or soybean) restored up to 44% of the activity. Similar results were found with rat liver by Jansen and Arias,[257] who separated protein from phospholipid on Bio-Gel® P-30 with cholate, and partially restored activity towards bilirubin in the inactivated transferase by prolonged dialysis with phospholipid. They could not restore activity towards 4-nitrophenol this way, and Graham et al.[256] also found the rat-liver enzyme resistant to reactivation. These last authors suggest a real difference in stability between the phospholipase A-treated enzymes of the two species,[256] perhaps related to the relatively greater thermostability[253] of the guinea pig-enzyme preparation. Graham et al.[256] regard these reactivation findings as confirming their original[250] postulate that UDPglucuronyltransferase is truly phospholipid dependent, although of course detergent is present. The biphasic nature of the transferase's response to phospholipase A therefore seems established, and the immediate cause of the activation elucidated. The mechanism of this activation and succeeding deactivation depends on whether one accepts the compartmentational or the conformational viewpoint — if, as Graham et al.[161] remark, "either operates to the exclusion of the other".

Rather than phospholipase A, Berry[248] advocates phospholipase C, which liberates nonlytic diglycerides and the water-soluble phosphoryl base, even though possible interference by rat-liver microsomal diglyceride lipase would need added albumin as protectant. Winsnes[101] first conducted detailed experiments with phospholipase C, using mouse, rat, and guinea pig and both 4-nitrophenol and 2-aminophenol. Contrary to Graham and Wood,[250] but agreeing with Hanninen and Puukka,[252] he found activation with phospholipase C, though only to 1/3 detergent-activated values.[101] The bent double-reciprocal plots of the native enzyme remained after treatment, but apparent $K_{UDPGlcUA}$ and $K_{4-nitrophenol}$ values increased; the former decreased again with added lecithin, which additionally activated the enzyme.[101] Activation by phospholipase C could have been overlooked earlier because of low-substrate concentrations[101] or spontaneous activation. Berry[248] considers at least 93% total transferase activity would be lost on treatment with phospholipase C, when assayed with 0.02 mM-UDPGlcUA and 0.5 mM-2-aminophenol in mouse liver microsomes; with guinea pig and 4-nitrophenol it would be at least 87%.[248] Zakim and Vessey[163] find that phospholipase C treatment of the transferase prevents subsequent activation by UDPGlcNAc and that the resulting "form" would be inefficient in vivo. A deoxycholate-solubilized transferase preparation from human liver[258] lost 70% activity towards estriol (16α) when incubated with phospholipase C but lysolecithin restored most of this activity; phospholipids from the original preparation failed to reactivate,[258] possibly[248] from the presence of deoxycholate. Vessey and Zakim[116] also found no activation by phospholipids of phospholipase C-treated preparations.

Berry et al.[259] incubated guinea pig-liver microsomes at 5°C (to minimize thermal or "preincubation" activation) with phospholipase C preparations, measuring delipidation of the preparation (from hydrolysis of radioactive phosphatidylcholine) and transferase activity towards 4-nitrophenol (forward and reverse directions). With microsomes sonicated to "maximally express" latent transferase, extensive (> 50%) phospholipid depletion progressively inhibited the transferase activity by some 70%. In contrast, identical phospholipase treatment of latent transferase stimulated it progressively to the "inhibited activities" of the sonicates. Similar results were obtained with other perturbation procedures. Berry et al.[259] considered this supported their belief that delipidation incurs two opposing effects: (1) stimulation by removal of latency and (2) inhibition by removal of essential phospholipids. However, diglycride products themselves inhibit the transferase in intact microsomes,[259] despite earlier reports,[250] and pure phospholipase preparations were not used. Recent studies[254,260] attempting to remove these uncertainties note that extensive delipidation of guinea pig-liver microsomes by pure protease-free *Bacillus cereus* phospholipase C decreased transferase activities to 4-nitro- and 2-aminophenol up to some 85%, hydrolyzing 94% of total microsomal lipids and some 99% of their phosphatidylcholine. These activities were restored significantly (76 to 90%) by lysolecithin, but not by oleic acid, certain detergents, or albumin. After exhaustive delipidation by (the probably less pure) *Clostridium perfringens* phospholipase C, lysolecithin restored 94% of the original activity. Delipidation of intact or sonicated microsomes produced (6- to 15-fold) increases in apparent $K_{UDPGlcUA}$ and $K_{4-nitrophenol}$. Bisubstrate kinetics indicated large decreases in V_{max} on delipidation of sonicated microsomes. These kinetic results confirmed previous work with rat[252] and mouse[101] which found raised K_m values on delipidation by crude phospholipases. Glyceryl dioleate, added to test possible inhibition by diglycerides inhibited the transferase some 20% in detergent-activated microsomes; and lipase, which hydrolyzed most of the diglyceride present, restored some activity (27%) in certain phospholipase C-delipidated microsomes, optimally with albumin present. However in other highly delipidated samples, only 3% activity was restored by lipase treatment, even though it hydrolyzed as much diglyceride as before. Possibly, the residual phospholipid is too tightly bound by the transferase for the diglyceride to compete with it.[254] The authors concluded that inhibition by diglycerides does not significantly contribute to inactivation by delipidation with phospholipase C, but that transferase activity towards 4-nitrophenol and 2-aminophenol requires phospholipids, a requirement not obvious until after extensive delipidation by the *B. cereus* enzyme. As 1-lysolecithin supports this requirement and is released by phospholipase A treatment, studies with phospholipase A on phospholipid dependence[250,251] or on specific constraint by phospholipids[163] need reappraisal.[254]

In other, very recent, work, "total"[261] or 98%[261a] removal of phospholipid lowered, to 33%[261] or 0.6%[261a] the original activity towards 4-nitrophenol. Lysolecithins or lecithins restored this activity, to give an enzyme with broadly similar properties,[261a] but with its detailed kinetic behavior depending[261] on the acyl chain in the phospholipid added. Regulatory properties as existing in microsomes could not be restored.[261]

The transferase is in vitro not phospholipid dependent in the sense that only phospholipids will stimulate a largely phospholipid-depleted transferase preparation. Burchell and Hallinan[242] labeled rat-liver phospholipid with [³H]-choline in vivo, purified the enzyme to apparent homogeneity, and found minimal phospholipid content, at most some 0.7 mol phospholipid per mole protein. This contrasts with Gorski and Kasper's "47% by weight" of phospholipid in their purified enzyme,[150] but is consistent with the low phospholipid content found by Bock et al.[262] in their partially-purified preparation (0.01 mg phospholipid per mg protein, compared with 0.34 mg in the microsome). Although it could be stimulated 40 to 100% by the addition of dispersed phospholipid, the enzyme of Burchell and Hallinan[242] was not affected by phospholipase C, either because it had no phospholipid or because the little present was too tightly bound. The reason for it possessing activity with such low or absent phospholipid is probably its support by the detergent used in purification, Lubrol® 12 A9.[242] This detergent, unlike many, restores activity to microsomes whose transferase activity to 4-nitrophenol has been 65% inhibited following the hydrolysis of 95% of its phospholipid by phospholipase C.[242] Interestingly, Lubrol® does not support activity to bilirubin; phospholipid must be added to the purified enzyme to demonstrate this activity.[327b]

It is now clear that apparently pure transferase protein can operate with extremely small amounts of true phospholipid present, if any, and its catalysis may be supported for some substrates by bound detergent acting as a phospholipid substitute.

In vivo, dependence on phospholipids appears likely. Their role in constraint, conformational or compartmentational, remains unresolved, and although microsomal lipids may help to align lipid-soluble transferase substrates for optimal reaction,[170] the phospholipids do not seem to stimulate glucuronidation by dissolving the aglycon because $K_{UDPGlcUA}$ increases on delipidation as well as $K_{aglycon}$, and V_{max} falls.[258]

"Reactivation" by phospholipids is an important technique during purification of the transferase (Chapter 6, Section I). When phospholipids restore activity to a delipidated enzyme, the new environment may, as Graham et al.[161] emphasize, differ from that of the native membrane. Kinetic studies could provide criteria on the "artificiality" of such reconstituted environments.

Also, partial delipidation could inactivate transferase, not by removing an essential phospholipid X from the membrane, but by moving X further from the enzyme; subsequent reactivation by phospholipid Y would mean, not that Y is an essential phospholipid, but that Y enters the membrane and permits X to return to the transferase.[161]

There remains the question of the phospholipid requirement of the transferase activity towards steroids. Zakim and Vessey report that the rate of glucuronidation of testosterone in guinea pig liver is, unlike that of 4-nitrophenol, unaffected by phospholipase A[263] or C,[163] although specificity for UDPGlcUA is lost; as they also claim no effect of UDPGlcNAc on this transferase activity, a finding not confirmed by others (T. Hallinan, personal communication), and no effect of detergents, also unconfirmed in rats.[213] It may be premature to conclude[163] that the "steroid transferases" are not phospholipid dependent, particularly as steroids may be oriented in the membrane for catalytic efficiency.[170] For example, while phospholipase C scarcely affected UDPglucuronyltransferase activity towards 17α-estradiol in rabbit-liver microsomes under conditions inactivating UDP-N-acetylglucosaminyltransferase,[264] crude phospholipids partially restored this activity to UDPglucuronyltransferase inactivated by snake venom or 6 *M* urea.[264]

M. Fatty Acids

Hänninen and Puukka[252,265] showed that all the even-numbered fatty acids from caprylic to behenic acids stimulated transferase activity in vitro, C_{14} and C_{16} acids (the best) raising it four- to fivefold. With increasing unsaturation in the C_{18} series, less fatty acid per microsomal protein was required for optimal effect. Apparent $K_{UDPGlcUA}$ was raised some threefold. No solubilization of the enzyme occurred.[265] As the effect was not additional to that phospholipase A or digitonin it appeared due to membrane perturbation.[265] Washing microsomes failed to reverse activation by fatty acids, which were probably incorporated into, or bound onto, the membrane.[265] Although the concentration of free fatty acids needed for activations was higher than could exist in intact membranes, these compounds could contribute to the activation seen on storage or "aging"[265] (Section II.C. above).

Graham and Wood,[255] however, noted that while the saturated fatty acids palmitic and stearic (up to 0.3 µmol added per gram microsomal protein, the concentration reported optimally stimulating[265]) failed to activate transferase towards 4-nitrophenol in rat-liver microsomes, the unsaturated linoleic and arachidonic acids did so. Lipid peroxidation is not involved for similar results were obtained when fatty acids dispersed were prepared under nitrogen.[255]

Inhibition by fatty acids of UDPglucuronyltransferase activity towards bilirubin is of clinical importance (see Chapter 11, Section IV).[266,267,270] Inhibition increased with degree of unsaturation, γ-linoleic acid proving the best inhibitor of the C_{18} acids for the microsomal enzyme.[268] Even-numbered saturated acids, from butyric to arachidic, did not inhibit in slices or in microsomes. Curiously, no initial activation was reported[268] with unsaturated acids, even down to 0.01 mM. Either the fatty acid to microsomal protein ratio was too high and inactivation obscured any initial activation, or transferase activity towards bilirubin (like Phase 1 enzyme activity[269]) is more susceptible than other transferase activities to the lytic action of fatty acids. Parallel studies with bilirubin and 4-nitrophenol should decide whether the distinction is procedural or due to differences in properties of the two transferase activities.

N. Lipid Peroxides

Aerobic incubation of rat-liver microsomes peroxidizes membrane lipids, probably during oxidative cleavage of long-chain unsaturated linoleic, arachidonic, and docosahexanoic acids at the β-position of the phospholipids.[271] Phase 1 enzymes were inhibited by lipid peroxidation in microsomes[272] but transferase activity towards 4-nitrophenol was stimulated three-fold initially, followed by a fall which was not investigated long enough to show the presumably progressive inactivation to below original values.[272] Activation was accompanied by little morphological change, but by decreased turbidity of the suspension, possibly from loss of the longer-chain unsaturated fatty acids. At its onset much phospholipid phosphorus went into solution. Lipid peroxides probably contribute to activation of microsomal UDPglucuronyltransferase by fatty acids or aerobic incubation (i.e., storage); and possibly after exposure of rats to carbon disulfide.[221]

O. UDP-N-Acetylglucosamine (UDPGlcNAc)

UDPGlcNAc is a specific activator of UDPglucuronyltransferase and does not appear to perturb the membrane environment. This activation, first demonstrated by Pogell and Leloir[223] in rat liver, may have been due partly to hydrolysis of UDPGlcNAc there by the active nucleotide pyrophosphatase, thus allowing UDPGlcUA to exist at unusually high concentrations: but an additional mechanism was involved,[223] because UDPGlcNAc activated even at saturating concentrations of UDPGlcUA; and in guinea pig-liver microsomes, which lack the pyrophosphatase,[223] also, no activation occurred with the similar pyrophosphatase substrate, UDPglucose. This activation has been found with many substrates: 4-nitrophenol,[223] 2-aminophenol, phenolphthalein, 4-methylumbelliferone,[192] bilirubin,[185,273-276] indolylacylate,[277] 4-aminobenzoate,[190] 1-napthol, napththalene dihydrodiol,[278] and others. Difficulties may exist with very lipid-soluble substrates; e.g., Winsnes repeatedly[192,229] failed to find UDPGlcNAc activation towards bilirubin, and although activation has been reported with steroids,[279] "spontaneous" activation is rapid towards them[279] and could account for the lack of UDPGlcNAc stimulation at first reported for steroids.[263]

Pogell and Leloir[223] noted early that digitonin-treated transferase was not further activated by UDPGlcNAc, and activation with UDPGlcNAc, as with detergents, results in severalfold increases of apparent V.[101] As the nucleotide has no surfactant properties, its mechanism of activation, however, must differ from that of perturbants. Indeed, Winsnes[229] showed that low concentrations of N-ethylmaleimide or 4-chloromercuribenzoate abolished activation by UDPGlcNAc without affecting rates of 4-nitrophenol glucuronidation by fresh or detergent-activated transferase. Winsnes proposed that the activation by UDPGlcNAc occurred allosterically either on the enzyme[229] or by assisting transport of UDPGlcUA to the enzyme through a barrier impermeable to this substrate.[101] Berry[248] sees this as the first suggestion of the UDPglucuronic acid permease invoked by himself and Hallinan in their explanation of latency.[102] Activation with all aglycon substrates is essential for such a theory. Its opponents[263] or supporters differ regarding activation with steroid substrates[263] but, as further evidence against it, Zakim and Vessey[163] cite their kinetic studies with UDPGlcNAc. These studies, with microsomes from guinea pig and 4-nitrophenol as substrate, lead them also to the view that, in this species and for this substrate at least,[280] UDPGlcNAc is an allosteric effector of the transferase with a mechanism kinetically distinguishable from that of detergents. They suggest UDPGlcNAc, like UDPGlcUA,[190] binds to the transferase in a manner suggesting negative cooperativity and that cooperative interaction occurs between the two sugar nucleotides: UDPGlcNAc helps to bind low concentrations of UDPGlcUA to the enzyme. There is less homotropic cooperativity for the binding of effector than for binding of substrate, and (with 4-aminobenzoate as aglycon) high concentrations of UDPGlcUA assist binding of UDPGlcNAc to the enzyme.[190] This thesis has been encountered elsewhere (Sections I.A.3., II.K., II.L. above): UDPGlcNAc decreases the apparent $K_{0.5}$ of the transferase for UDPglucuronic acid,[190] and in the presence of divalent metal ions, this value approaches the overall concentration of UDPGlcUA in liver.[224,281] Unlike detergents, UDPGlcNAc would not change the specificity of the UDPGlcUA-binding site of the enzyme.[281] The activation seems reversible, and would operate allosterically, enhancing the affinity of the transferase for UDPGlcUA and decreasing its affinity for UDP. From the latter effect, it inhibits the reverse reaction of the transferase.[280] UDPGlcNAc has, therefore in the words of Zakim and Vessey,[163] "all the properties necessary for an ideal modifier of UDPglucuronyltransferase. It activates the forward reaction, inhibits the rate of the reverse reaction, limits end-product inhibition, and conserves substrate specificity." The synergism of metal ions and UDPGlcNAc at a fixed concentration of UDPGlcUA has been described.[224]

This suggestion, independently put forward by Winsnes and elaborated by Zakim and Vessey, that UDPGlcNAc is essential for the efficient operation of the transferase in vivo appears to be borne out by other work.

For example, an elegant series of papers by Bock's group suggests that only when physiological concentrations of UDPGlcUA and UDPGlcNAc are present together during incubation of microsomes does the glucuronidation rate approximate to that seen in perfused liver.[177] Galactosamine treatment of perfused rat liver markedly decreased 1-naphthol glucuronidation, but not that of bilirubin. Carbon tetrachloride treatment of rats decreased glucuronidation of both substrates in vivo. Similar results were obtained in microsomal preparations only if physiological amounts of UDPGlcNAc and UDPGlcUA were present.[179] Carbon tetrachloride treatment produced a similar fall in perfused liver and in microsomes,[180] apparently from loss of regulation by UDPGlcNAc plus increased inhibition by UDP.[180] UDPGlcNAc may also help to couple Phase 1 processes to glucuronidation in vivo.[278] In isolated rat hepatocytes, naphthalene dihydrodiol glucuronide was a major metabolite of naphthalene, but in rat-liver microsomes incubated with high concentrations of UDPGlcUA and an NADPH-regenerating system, the free dihydrodiol far exceeded its glucuronide unless UDPGlcNAc was also added.[278]

Berry[281a] has recently argued that UDPGlcNAc is not an allosteric effector, but is concerned with a UDPGlcUA permease. He found that low concentrations of Brij-35®, which did not notably affect latency of the transferase, markedly antagonized the UDP-N-acetylglucosamine response. He suggested that the site of the latter was less deeply embedded than the transferase, as befits a permease. He quoted preliminary experiments in which labeled N-ethylmaleimide was offered to rat-liver microsomes (removing all but 16% of their UDPGlcNAc-stimulated transferase activity) and the purified active enzyme then isolated from them. Less label was bound (8% of theoretical) than would be needed for one −SH group of the enzyme, suggesting that the UDPGlcNAc receptor was separable from the purified active transferase.

The function of UDPGlcNAc as modulator of transferase efficiency in vivo is, therefore, an acceptable hypothesis at present. Species other than common laboratory animals are not well documented; its "negligible effect" in pigs, cattle, and horses[282] could arise from use of enzyme preparations already partly activated by storage.

P. Ions

1. Hydrogen and Hydroxyl Ions

UDPGlucuronyltransferase exhibits activation at hydrogen or hydroxyl ion concentrations outside as well as within the usual physiological range. Preincubation for 30 min of bovine-liver microsomes above pH 8.0, before subsequent transferase assay towards 4-nitrophenol at pH 7.1, showed an activation peaking at pH 9.8 to 10.5.[116] As this peak was similar for glucose-6-phosphatase, the membrane environment rather than the protein may have been affected.[116] The earlier reported activation of the enzyme by dialysis at pH 9[226] was probably due to a similar effect. Both these papers and a subsequent report[283] record irreversible activations, and the reversible activations noted by Howland and Bohm[284] are of interest. Studying glucuronidation of 2-aminophenol by rat-liver microsomes, they found peaks of activity at pH 5.5 and 9.5 (the latter confirming earlier work[285]), as well as at pH 7.2. With 4-nitrophenol, a peak at pH 5.4 was noted. These peaks could rise three to four times above the physiological "optimal pH" activities and were freely reversible. As the substrates did not notably change their ionization over the pH range, Howland and Bohm suggested[284] that the peaks may be due to changes in the ionization of amino acids around the active center or to membrane changes, altering constraint; treatment with Lubrol®[285] or pretreatment with 3-methylcholanthrene, processes likely to change membrane composition, markedly changed the peak characteristics.[284]

Because of alterations in constraint during incubation, the pH optima published may be expected to vary somewhat even for the same substrate. Dutton[4] lists earlier reports ranging from pH 6.6 to 8.2. Specific instances are recorded elsewhere in the present review.

2. Metal Ions

EDTA is discussed in Section II.H. above. Metal ions activate UDPGlcUA pyrophosphatase and nucleotide diphosphatase. These two microsomal enzymes respectively decrease or increase observed glucuronidation by breaking down donor substrate or inhibitory product. Magnesium ion activated the transferase towards 2-aminophenol[286,287] and 4-nitrophenol.[287] Mg^{++}, Mn^{++}, and Co^{++} activated transferase[133] towards 4-nitrophenol and 1-naphthol in rat-liver microsomes fresh or already activated by Triton® X-100, showing that the ions affected the enzyme itself rather than membrane constraint, as was also noted for 4-nitrophenol and 2-aminophenol.[287] Zn^{++} inhibits;[133] $10mM$ Zn^{++} has been used specifically to inhibit the transferase in presence of β-glucuronidase.[182] Ca^{++} was almost as active as Mg^{++} in stimulating transferase activity towards pregnanediol[131] and xenobiotic phenols[288] in rat liver and towards estrone in pig kidney;[232] although with many substrates, including phenols[4] and bilirubin,[289] Ca^{++} seems inhibitory at low concentrations. Using microsomes prepared with EDTA medium, Zakim et al.[281] confirmed activation by Mg^{++}, Mn^{++}, and Co^{++}, but found little or no inhibition by Zn^{++} and Cu^{++}; Fe^{++} had virtually no effect. Degree of activation depended on aglycon and presence of UDPGlcNAc. Fe^{++}, Cd^{++}, Hg^{++}, and Co^{++} at $10mM$ completely inhibited activity to estrone in pig kidney.[232] With 4-nitrophenol as substrate, divalent metal ions increased activity at V_{max} and were needed for stimulation by UDPGlcNAc. With 2-aminophenol, the ions increased activity at V_{max}, but were not essential for stimulation by the sugar nucleotide. With 2-aminobenzoate, they decreased apparent $K_{UDPGlcUA}$ and were needed for stimulation by UDPGlcNAc.[281] These differences may reflect the membrane environment, but they convinced the authors that divalent metal ions are essential for the regulatory functions of UDPGlcNAc in vivo.[281]

Berry[248] considers much of the stimulatory effect of divalent metal ions due to their activation of nucleotide diphosphatase inside the microsomal vesicle, which prevents build-up of product UDP inhibiting the transferase.

This possibility is suggested, though not in a compartmentational sense, to explain the differential effect of Mn^{++} on the forward and reverse reactions 4-nitrophenol in guinea pig-liver microsomes.[290] The ion inhibits the reverse reaction, possibly by complexing with UDP, after which the transferase possesses less affinity for the complex than for free UDP. The UDPGlcUA Mn^{++} complex is, however, as readily bound to the enzyme as UDPGlcUA.[290] Activation of the forward reaction may thus occur by limiting the UDP-dependent hydrolysis of the glucuronide and accelerating removal of the product UDP.[290]

Magnesium ion, found by some to be stimulatory for transferase action to bilirubin,[291] may operate to some extent by increasing bilirubin solubility at subsaturating concentrations of this aglycon.[276] At saturating concentrations of bilirubin, it has no effect.[276]

Microsomes prepared in 0.154 M KCl possess higher specific-transferase activity than if prepared in 0.25 M sucrose[208] and prepared this way from guinea pig may be considered as already activated.

Q. Reactants and Products
1. UDPGlucuronic Acid
"Bent" Lineweaver-Burk plots of transferase activity towards 4-nitrophenol with varying concentrations of UDPGlcUA were first described by Winsnes,[101,192] who proposed negative cooperativity as an explanation;[101] Vessey et al.[190] suggested sequential binding of the sugar nucleotide, in which its first binding makes its subsequent binding more difficult. An alternative explanation of such "inhibition" by the substrate is saturation of a UDPGlcUA transport site on the membrane.[186]

2. Aglycon Substrates
"Bent" Lineweaver-Burk plots suggesting activation by aglycon were noted with varying concentrations of 4-nitrophenol[101,192,206] and 2-aminophenol.[117] Activation of 4-nitrophenol glucuronidation in microsomes or homogenates by 2-aminophenol,[117,292] and vice versa,[117] have been reported. When high (over 0.6 mM) concentrations of aglycon, especially phenolic aglycon, are used, this possibility of activation by substrate must be borne in mind. Its mechanism is obscure and awaits studies with the pure protein.

Sanchez and Tephly[293] reported that bilirubin and its glucuronide markedly stimulated rat-liver-transferase activity towards morphine and 4-nitrophenol, this activation not being due to contamination with bile acids nor to activation by the alkaline solution used to dissolve the bilirubin. At equimolar concentrations, bilirubin was a better activator than deoxycholate.[293] Its properties appeared similar to those of detergents (bilirubin at 0.3 mM giving a comparable effect to that of 0.25% Triton® X-100), and they suggested it acted by perturbation of membranes.[293] Supporting this, it activated only "native" microsomes, inhibited already activated preparations, and did not activate the semipurified preparation.[294] On injection, it doubled the excretion of morphine glucuronide, and they suggested it to be a "physiological" activator.[293] Yeary et al.[295] confirmed this increase of morphine glucuronide excretion, but found that, despite bilirubin's activation of 4-nitrophenol glucuronidation in vitro, intraperitoneal injection of bilirubin did not change the excretion of 4-nitrophenyl glucuronide. Whatever way bilirubin might act, these properties could invalidate much over-confident interpretation of in vitro kinetic studies with it.

Competitive inhibition by aglycon substrates, long known,[4] has been used as evidence concerning heterogeneity of the transferase (see Reference 203; Chapter 6, Section II.B). It may reflect saturation of transport sites or of aglycon pool sequestrated in the membrane,[170] but seems unrelated to activation or inhibition by perturbation.

3. Aglycon Glucuronides

Early reports[296,297] suggested that product glucuronides influenced UDPglucuronyltransferase. Digitonin-treated rat-liver-enzyme activity towards 4-nitrophenol was inhibited by high concentrations of the glucuronides of phenolphthalein and 1-naphthol. At high concentrations of UDPGlcUA inhibition was competitive with donor substrate and noncompetitive with aglycon. This suggested[298] that glucuronides are released at the UDPGlcUA site and that the enzyme binds the conjugated aglycon. Sulfate conjugates also inhibited,[299,300] possibly similarly. Neither glucuronate nor sulfate ion inhibited, indicating the requirement of a lipophilic aglycon in the conjugate.[299] Phenolphthalein glucuronide noncompetitively inhibited transferase activity towards bilirubin,[289] phenolphthalein, 2-aminophenol, 4-nitrophenol, 4-methylumbelliferone,[301] and harmol.[300] Mulder[301] found glucuronides of 2-aminophenol, 4-nitrophenol, 8-hydroxyquinoline, and 4-methylumbelliferone only marginally inhibitory to transferase activities at 10 mM. Hydrolysis to free aglycons was negligible,[300] so inhibition was from the glucuronide itself. Inhibition of excretion of harmol glucuronide (or sulfate) following injection was considered due to competition for biliary secretion.[300]

This work employed disrupted microsomes. With intact microsomes from guinea pig liver and low UDPGlcUA levels, activation by glucuronides can be demonstrated. Glucuronides of 1-naphthol, 2-naphthol, methylumbelliferone, and to a lesser extent, of phenol and estriol, activated the transferase towards 4-nitrophenol.[302] Activation possibly occurs through glucuronides allosterically increasing the affinity of the enzyme for UDPGlcUA.[302] At high 4-nitrophenol levels, activation by 1-naphthyl glucuronide was not affected by 1-naphthol itself, the conjugate probably acting at a site separate from that glucuronidating 1-naphthol. Activation is inhibited by glucosides, which without added glucuronide do not affect activity. Aglycon glucuronides may therefore bind at two sites on the enzyme. They can inhibit by binding at the aglycon site and activate by binding at the allosteric site.[302] The noncompetitive inhibition noted earlier,[298] possibly arose from simultaneous binding at both sites.[302]

Berry[248,281a] prefers a compartmentational explanation, pointing out that disrupted microsomes were usually employed, including the competitive studies[298] with UDPGlcUA and that activation occurred only at low concentrations of aglycon (the figure[307] illustrating "autoactivation" of 1-naphthyl glucuronidation by 1-naphthyl glucuronide being erroneous). He suggests either that glucuronides stimulate transport of UDPGlcUA or that they react with product UDP to produce UDPGlcUA *via* the reverse reaction. The first possibility appeared unlikely from the antagonistic effect of divalent metal ions on the activation:[248] glucuronides inhibit transferase activated by UDPGlcNAc and Mg^{++}, but stimulate it when added EDTA removes Mg^{++}.[248] For the second possibility, added glucuronide reacts with the UDP no longer destroyed by the nucleoside diphosphatase (inactive without Mg^{++}) and thereby regenerates UDPGlcUA which is used to stimulate glucuronidation. The latter explanation was rejected[302] from an isotopic experiment which may[248] only demonstrate that regeneration is not the exclusive mode of activation by glucuronides and does not allow for consumption of inhibitory UDP.

These results complicate interpretation of earlier kinetic studies. Activation of transferase activity towards substrate X by accumulation of Y glucuronide may obscure any demonstrable competitive inhibition between X and Y, kinetic "evidence" concerning heterogeneity of the enzyme could, therefore, originate in different activations by different aglycon glucuronides.

Glucuronides and sulfates can pass into the intact cell. Recently, Norling et al.[302a,302b] have confirmed this with isolated rat hepatocytes and labeled glucuronides of 4-nitrophenol, 2-naphthol, and phenolphthalein. 4-Nitrophenyl glucuronide was released from the cell on subsequent disruption, but phenolphthalein glucuronide remained bound to the fragments. Glucuronides of 2-naphthol and 4-nitrophenol had no effect on harmol sulfation or glucuronidation by the cells, but phenolphthalein glucuronide increased their sulfation and decreased their glucuronidation. As the free aglycons all depressed harmol conjugation in the cells, no hydrolysis of the glucuronides can be occurring within the cell. Whether transport through the membrane incurs transient hydrolysis and then reconjugation is not yet known.

4. UDP, UTP, and UMP

Adlard and Lathe[273] first drew attention to the inhibition of transferase activity (towards bilirubin) by the product UDP. Breuer's group (e.g., Reference 303) followed with observations on its inhibition of detergent-solubilized activity towards steroids. Subsequent studies showed that with 4-nitrophenol[117] or bilirubin[289] as substrate, UDP inhibited competitively towards UDPGlcUA and noncompetitively towards aglycon. This finding was confirmed for 4-nitrophenol, but not 2-aminophenol, by Winsnes,[186] who noted that nonspecific activation of the 4-nitrophenol enzyme increased its inhibition by UDP. He suggested that glucuronidation of 4-nitrophenol by an "activated" enzyme would be strongly inhibited by the cellular UDP whereas UDPGlcNAc would relieve that inhibition,[186] a concept developed (Section II.O) by Zakim and Vessey in their regulatory theory. Winsnes[186] noted that inhibition by nucleotides other than UMP, UDP, and UTP was minimal and confined to pyrimidine derivatives. There is no evidence of membrane perturbation by the nucleotides.

R. UDPSugars Other Than UDPUronic Acids and UDP-*N*-Acetylglucosamine

No activation or inhibition by UDPglucose of various transferase activities, activated or latent, was detected in rat liver by Hallinan or Burchell (personal communications, 1977), but Zakim et al.[191] consider that pretreatment with phospholipase A removes specificity from UDPGlcUA-binding site of guinea pig-liver transferase (towards 4-nitrophenol) allowing competitive inhibition by UDPglucose, UDPgalactose, UDPmannose, UDPxylose, and UDPGlcNAc; similar inhibition, similarly interpreted, occurs in untreated enzymeb elow 16°C.[198] Unlike UDPGlcNAc, the other UDPsugars, did not seem to bind allosterically: divalent metal ions affected the action of UDPGlcNAc but not of the other UDPsugars.[191] Inhibition by UDPGlcNAc was additive to that of the other UDPsugars.[191]

UDPXylose and UDPgalactose, as well as D-galactose and D-glucosamine, inhibited overall glucuronidation in rat-liver slices.[305] They possibly acted by inhibiting UDPglucose dehydrogenase and pyrophosphorylase (Chapter 9, Section II), for the transferase itself was unaffected, and UDPGlcUA added to the medium restored glucuronidation.[306] This again infers that the (damaged?) cell membrane is permeable to sugar nucleotides, as does the stimulation of bilirubin clearance in perfused liver by UDPGlcUA but not by glucuronate.[307]

S. ATP, AMP

ATP increased activation by UDPGlcNAc of 4-nitrophenol glucuronidation in the rat.[223] ATP alone produced a smaller activation, and AMP smaller still.[223] ATP and

AMP could have been acting as sacrificial substrates for UDPGlcUA or UDPGlcNAc, for they inhibit nucleotide pyrophosphatase[225,304,308] and exert no effect in guinea pig-liver preparations which are low in that enzyme.[223] Winsnes, using higher levels of UDPGlcUA found either no[192] or little[186] activation by ATP, whereas Adlard and Lathe[273] reported an antagonistic effect by ATP on UDPGlcNAc-activation of bilirubin glucuronidation. However, they also noted an activation with ATP alone which, being reproducible with EDTA, could be due to chelation by ATP of inhibitory calcium ions;[273] however, Mg^{++} was also present, and has greater affinity than Ca^{++} for ATP. Adlard and Lathe[273] considered that ATP (or EDTA) would not "activate" by inhibiting nucleotide pyrophosphatase because UDPglucose and UDPGlcNAc, which are also competitive substrates for that enzyme, did not reproduce its activation. However, the interpretation may be complicated by superimposed inhibition of the transferase from these UDPsugars binding at its UDPGlcUA site.[248]

Schröter[309,310] studied stimulation by ATP of transferase activity to 4-nitrophenol in newborn and adult animals, and considers it not due to inhibition of nucleotide pyrophosphatase, but either to ATP assisting UDPGlcUA access through the microsomal membrane or to allosteric effects on the transferase molecule. Stored microsomal transferase exhibited progressively less stimulation by ATP. Indeed, with a semipurified transferase preparation, high concentrations of ATP inhibited activity towards estrogens.[303]

Whereas ATP would seem to act in part by sparing UDPGlcUA, it may, with homogenates employing assays (as for 4-nitrophenol) which measure disappearance of aglycon, spuriously "activate" by encouraging formation of a sulfate or a phosphate conjugate. In assays measuring amount of glucuronide formed, this diversion of aglycon could result in observed "inhibition" of transferase. Conjugate identification is needed if a marked effect of ATP is reported.

T. Glucarate and Glucarolactone

As discussed earlier[4], the only activating effect glucarate or glucarolactone may exhibit is due to its inhibition of β-glucuronidase action which, though small, is detectable even above pH 7.0 on long incubations (Chapter 9, Section IV). Glucarolactone is added when studying the reverse reaction (Chapter 17).

III. ADDITIONAL NOTES

Section I.A: In rat liver, latency of the transferase is associated with microsomes, which were shown to be vesiculated, and not with nuclear membranes, which were shown not to be vesiculated.[163a,327a] Latency may therefore be an artifact of the vesiculation resulting from homogenization.[327a] Preliminary immunohistochemical work cited in Reference 327b suggests, however, that the catalytic site of the transferase is not accessible to the cytoplasmic side of the endoplasmic reticulum in the cell, but is accessible to the cytoplasmic side of the nuclear membrane; latency would therefore be a property of the transferase in the unhomogenized endoplasmic reticulum.

Section II.H: Further information on inhibition of the transferase by EDTA is given in Reference 170a.

Section II.L: Removal of over 95% of phospholipid by passing a transferase preparation through hydroxyapatite removed activity to estrone and 4-nitrophenol.[327c] Reactivation with phosphatidycholine and lysophosphatidylcholine was good; reactivation with phosphatidylethanolamine was poor.[327c] This and other recent work on delipidation and reconstitution of the transferase are reviewed in Reference 327b.

Section II.N: Lipid peroxides seem the major agents of transferase inactivation by membrane perturbation of microsomes.[327d]

Section II.O: A brief discussion of current views on the UDPNAcG-stimulation of the transferase, from the point of view of both the permease and the catalytic subunit[187] hypotheses, is in Reference 327b.

Section II.P.2: Magnesium ion alters the gross lipid structure of microsomal membranes, increasing the amount of a nonlamellar phase which could function as a nonspecific permease for cofactors.[327a] Stimulation of native transferase by Mg^{2+} could therefore reflect transport of nucleotides.

Section II.Q.4, II.R, and II.S: Two significant papers have appeared.[170a,170b] UDP is not normally observed with rat or rabbit liver microsomes during transferase action, being hydrolysed immediately on release from the microsomal surface.[170a] EDTA is required before UDP is detectable.[170a] UTP and UDP at physiological concentrations (0.2 to 0.4mM) are strong inhibitors of the microsomal transferase when UDPGlcUA is at 0.4mM, and ATP and UDPglucose are weak inhibitors.[170b] Perturbation of the microsomes to make them leaky, weakens the strong inhibitors and potentiates the weak inhibitors.[170b] This is consistent with active transport of UTP and UDP through the microsomal membrane.[170b] β-γ-Methylene-UTP, which is not hydrolysable by nucleoside diphosphatase, is a poor inhibitor of the transferase, suggesting that prior hydrolysis of UTP to UDP is needed before entry of the nucleotide into the microsomes and inhibition of the transferase there.[170b]

Chapter 6

PURIFICATION OF UDPGLUCURONYLTRANSFERASE AND ITS HETEROGENEITY

I. SOLUBILIZATION AND PURIFICATION OF UDPGLUCURONYLTRANSFERASE

Solubilization of the transferase is necessary for its purification, and purification is necessary before we can establish its requirement for phospholipid, classify its heterogeneity, and investigate its physical and chemical structure. The preceding chapters will have suggested, however, that the transferase is so intimately involved in the endoplasmic reticulum enzyme complex that the properties of the pure protein could have little direct physiological relevance, especially if reactivation with phospholipids has been necessary after purification.

Razin's[328] criteria for solubilization were (1) nonsedimentability at 100,000×g for 1 hr, (2) inclusion in Sepharose® 4-B, and (3) no vesicles visible on electron microscopy. Criterion (1) is difficult to apply to a membrane-bound enzyme, (2) has been achieved, and (3), recently achieved during purification,[239,240] need not apply if the physiologically regulated enzyme requires extensive membrane organization.

Solubilization cannot be usefully discussed apart from purification. Preparations of an arguably pure transferase have now been achieved,[241] and studies on its properties are beginning. We briefly summarize chronologically the various approaches made; early work is more fully recorded elsewhere.[4]

In the earliest work, freezing and thawing or butanol-extraction[329] produced some "solubilization" into the high-speed supernatant, but too rapidly inactivated the enzyme. More successful attempts involved digitonin treatment,[223] sonication, or incubation with snake venom.[227,563] Mowat and Arias[315] transferred up to 40% of the transferase activity of an EDTA-dialyzed guinea pig-liver particulate fraction into the high-speed supernatant (100,000 × g for 45 min) following ultrasonication. There was no solubilization, or retention on Sephadex® G-200, and vesicles of 80 to 200 nm diameter were visible.[315] The vesicles had probably merely been reduced in size by the treatment for no detergents were employed.[315] Insufficiency of gravitational criteria for solubilization is therefore obvious. The transferase preparation was active towards 4-nitrophenol, 2-aminobenzoic acid, and 4-methylumbelliferone. Further steps, involving ammonium sulfate-precipitation and acetone, methanol, and ether inactivated the enzyme.[315] Isselbacher et al.[227] had been more successful with ammonium sulfate fractionation, treatment with calcium phosphate gel, and final elution from a DEAE column of their venom-treated enzyme. They claimed 30-fold purification before losing activity, quoted as some 30% of the venom treated supernatant. Noting that venom treatment raised specific activity, they wrongly believed that separation from unsolubilized microsomal β-glucuronidase was responsible. Activation during purification has frequently resulted in over-optimistic interpretations of the increases in specific activity observed.

Howland et al.[288] activated rat-liver microsomal transferase with the nonionic detergent Lubrol® and extracted 25% activity towards 4-nitrophenol and 15% towards 2-aminophenol at 105,000×g for 60 min. This preparation[288] exhibited heterogeneity on sucrose-density-gradient centrifugation, activity towards 2-aminophenol appearing in fractions of density less than 1.056, and towards 4-nitrophenol activity in more dense fractions. Published data reveal no clear separation. The "solubilized" enzyme was much more stable at 4°C or −20°C than in microsomes.

Freeze-dried human-liver microsomes were treated with deoxycholate[258] and activity eluted from Sepharose® 4-B towards the 16α-group of estriol in low yield at six-fold purification. The impure eluate contained phospholipids and had a molecular weight around 2 million.

Activity towards the 3α-OH group of estradiol was extracted also[330] from pig intestine using sodium dodecyl sulfate (SDS) to "solubilize" the enzyme, followed by centrifugation at 230,000×g twice for 30 min. After passage through Biogel® A-5M, several peaks of activity were noted, but in 0.05% SDS only two, of molecular weight 750,000 and 410,000. Sucrose-density-gradient centrifugation without SDS gave many peaks, but with SDS, again only two, of 58,000 and 147,000 daltons. The former peak gave a molecular weight of 50,000 on SDS-gel electrophoresis. However, activity was virtually lost in SDS. Conclusions were that different aggregates could occur, and that the minimal molecular weight could be 50,000 to 58,000, interesting in view of the similar figure recently found.

Labow et al.[264,331] have separated various rabbit-liver steroid UDPglycosyltransferases. UDPGlucuronyltransferase activity towards the 3-OH group of 17α-estradiol was solubilized into the 105,000×g supernatant by Triton® X-100 after preliminary trypsinization and after ultrafiltration it passed through Sephadex® G-200, but was retained on Sepharose® 6-B. Preparations chromatographed in 6M urea on the Sepharose® and subsequently partially reactivated with phospholipid, gave two separate peaks, presumably from a fragmented vesicle. This is a warning of the difficulties of interpreting such peaks with an impure enzyme.

Digitonin (10%) was used[137] to solubilize transferase activity from rat liver towards bilirubin. Although nonsedimentable at 100,000×g, 9 hr at 320,000×g concentrated this "soluble" enzyme, with no "formed elements" on electron microscopy. Though retention occurred on Sephadex® G-200 (molecular weight is 150,000), the column was not equilibrated with detergent as required.[328] By analytical ultracentrifugation, molecular weight was around 135,000. As activity was eluted with the protein peak which[137,248] represented 6.6% total liver nitrogen, the claim of a "homogeneous", transferase preparation[137] was premature.

Hänninen et al.[332] used prior induction of transferase by 4-methylcholanthrene to boost its level in rat-liver microsomes before trypsinization, extraction with digitonin, centrifugation and ultrafiltration and elution Sepharose® 4B. Activity followed flavoproteins and UDPGlcUA pyrophosphatase, but preceded hemoproteins and bulk protein, accompanying most of the phospholipid. Purification was almost 100-fold, but,[332] discounting activation, only 20-fold or so. A subsequent paper describes reactivation with different elutions of activities towards various substrates, but no separation and little effective purification.[333]

Lucier et al.[334] solubilized transferase from control rats and rats treated with 2,3,7,8-tetrachlorodibenzo-p-dioxin (TCDD) using deoxycholate (0.4 µg/mg) which had to be removed immediately after centrifuging. Ammonium sulfate precipitation followed by elution from Sephadex® G-200 gave activities towards 4-nitrophenol, estrone, and testosterone in the void volume. Some purification (10-fold above the solubilized enzyme) was noted. When Sepharose® 4-B was used, two activity peaks were obtained, identical for 4-nitrophenol and testosterone. As centrifugation of the eluates brought activity from peak I into the pellet, only peak II seemed free of endoplasmic reticulum fragments.[334] Little significant purification was achieved.

Graham et al.[161] removed 85% of phospholipid phosphorus was removed from guinea pig-liver microsomes with phospholipase A and the readily collected sediment was reactivated towards 4-nitrophenol with a mixture of phosphatidylcholine and lysophosphatidylcholine. This method avoided the difficulty of sedimenting lipid-rich fragments.

Reactivation was also employed by Jansen and Arias[257] for the rat-liver enzyme. After treatment with deoxycholate, the 100,000×g supernatant was mixed with albumin and put through BioGel® P-30 with buffer at pH 8.1, deoxycholate, EDTA, and glycerol. Albumin and glycerol preserved the enzyme for subsequent reactivation. Some 92% phospholipid was removed and activity towards bilirubin disappeared until regained through dialysis (not mixture) with lecithin, microsomal lipid, or membranes (phosphatidylserine or phosphatidylethanolamine could not replace lecithin). Activity to 4-nitrophenol was not regained. As it can be regained following delipidation,[150,161] it may have been lost in this procedure.

Glycerol, dithiothreitol and the nonionic detergent Emulgen® 911 were used by del Villar et al.[294] to solubilize transferase activity from rat liver towards 4-nitrophenol and morphine. The supernatant (100,000×g for 4 hr) was put through Sephadex® LH-20 to remove detergent and the active fractions (in the void volume) were placed on a DEAE-cellulose column with buffer, glycerol, and dithiothreitol. Three peaks were eluted with KCl, one for 4-nitrophenol, one for morphine, and an intermediate active for both, possibly due to incomplete "solubilization" yielding fragments possessing both activities. Subsequent[335] work with rabbit liver and Emulgen® 911 yielded only two peaks, for 4-nitrophenol and morphine, respectively. Purification was only some 15-fold, but was the clearest separation of existing activities reported. However, results are expressed as specific activity; if expressed as absolute activity (units/mℓ), the activity to 4-nitrophenol coincides with major elution of protein, and the peaks seem less dramatic.

Bock et al.[336] solubilized liver transferase from phenobarbital- and methylcholanthrene-treated Sprague-Dawley rats with cholate and performed ammonium sulfate fractionation. In a fuller later purification[262] the ammonium sulfate (50 to 70%) precipitate was dissolved in glycerol, Brij® 58, and dithioerythritol and applied to a BioGel-A® column. Transferase activity to 1-naphthol was eluted just after the void volume and placed on a DEAE-cellulose column, equilibrated with buffer at pH 7.7, and eluted with a linear KCl gradient in buffer. Active fractions were dialyzed at pH 6.5, placed on a CM-cellulose column, washed and eluted at pH 6.5 with 0.2 M potassium phosphate. The enzyme was then concentrated by dialysis against polyethelene glycol 20,000 and subjected to isolelectric focusing at 4° C. The last step destroyed activity. Transferase activity to 1-naphthol eluted as a single peak from DEAE-cellulose coinciding with bulk protein. Activity to morphine was eluted here and at a second peak which, with low protein content, had a much higher specific activity. When the activity to 1-naphthol was purified further, two isoelectric points were found at pH 6.5 and 7.0. SDS-Gel electrophoresis of the activity to 1-naphthol eluted at pH 7.0 showed two bands with a molecular weight of 48,000 and 52,000. Activity to morphine showed several bands of molecular weight above 56,000. Bock et al. suggested the two transferase activities stem from proteins containing different polypeptides.[262]

Work by Bock et al.[337] just available, improves this procedure. The two differentially inducible transferase activities separated at the DEAE-cellulose step were purified by chromatography on UDP-hexanolamine agarose. The eluted activities (towards 1-naphthol and morphine, respectively) each accompanied a single, broad protein band on SDS-polyacrylamide gel electrophoresis. They possessed a molecular weight of 54,000 and 56,000, respectively.

The work of Gorski and Kasper[150] marked, with that of Bock et al.[262,337] and of Burchell,[239-241] a great advance. Gorski and Kasper [150] used Lubrol WX® to solubilize liver transferase activity to 4-nitrophenol from phenobarbital-treated Sprague-Dawley rats and put it through urea extraction, a DEAE-cellulose flow-through at pH 6.0, and importantly, affinity chromatography on UDP-hexanolamine Sepharose® 4-B with elution by 25 mM-EDTA followed by 5 mM UDPGlcUA. Traces of contaminating proteins were removed through a DEAE-agarose column at pH 7.5 with 3 m*M* UDPGlcUA. The substrate was removed by Sephadex® G-25 chromatography, and the enzyme concentrated on DEAE-agarose at pH 7.5. The product exhibited only one band on subsequent electrophoresis in SDS-polyacrylamide gel, giving a calculated molecular weight of 59,000. Phospholipid coeluted, suggesting the enzyme to be a lipoprotein.[150]

This active lipoprotein complex exhibited both size and charge heterogeneity by gel filtration and electrofocusing, a changing phospholipid (not polypeptide) content accounting for the heterogeneity. Isoelectric focusing in polyacrylamide gels showed two activity peaks and when performed in sucrose gradients yielded a doublet of molecular weight of 59,000 and 52,600. The latter could not be separated from the transferase protein, and both proteins may be part of the same phospholipid complex in the membrane.[150] The smaller protein could be a transferase specific for a substrate other than 4-nitrophenol or a modulator of transferase action.[150] It could be a degradation product.

The crude solubilized enzyme had a molecular weight of 200,000 which, with a phospholipid composition of some 47% by weight, could contain two polypeptide chains.[150] Complete chemical removal of phospholipid destroyed activity, restored by microsomal lipids or synthetic phospholipids.[150] The authors unfortunately gave no details of the later work.

The amino acid composition of the purified enzyme was published and resembles that reported by Burchell.[241] The results of Burchell, of Bock's group, and of Gorski and Kasper agree strikingly, suggesting the enzyme is ready for chemical investigation as a protein and for the basic study of its heterogeneity and regulation by phospholipids.

Yuasa,[337a] in work too recently received to review fully, purified transferase from rabbit-liver microsomes 45-fold, with cholate solubilization and passage over ω-aminohexyl-Sepharose® 4-B, DEAE-cellulose and hydroxyapatite. SDS treatment gave several electrophoretic bands, that considered to be transferase had a molecular weight of 60,000 and, interestingly, stained for covalently linked sugars.

Burchell[239-241] assayed Wistar rat liver transferase activity towards three different substrates at each purification stage while analyzing it for polypeptide content by SDS-polyacrylamide gel electrophoresis. Detergent-treatment and ion-exchange chromatography was followed by affinity chromatography.

The detergent treatment, based on that of Howland et al.,[288] used a greater (1%) concentration of Lubrol® 12 A-9 and yielded over 90% of the (activated) activity towards 2-aminophenol and 4-nitrophenol and 87% of that towards bilirubin in the high-speed (105,000×g for 1 hr) supernatant; 82% microsomal protein accompanied the activity, which was retarded on detergent-equilibrated Sephadex® G-200 and Sepharose® 6-B, gave no visible vesicles on electron microscopy[239] and thus satisfied Razin's critieria for solubility. Activity, concentrated by 25 to 60% saturated ammonium sulfate, was chromatographed three times successively by ion-exchange on DEAE-cellulose, CM-cellulose, and then DEAE-Sephadex®. Most cytochrome P-450 was removed by binding to the CM-cellulose, and all NADPH-cytochrome P-450 reductase by binding to the DEAE-Sephadex®. Three bands were visible on SDS-polyacrylamide gel electrophoresis of molecular weights 58,000, 50,000, and 15,000.[239] The latter, a minor band, was believed to be cytochrome b_5. One of the first two was believed to be epoxide hydratase. Activity towards 2-aminophenol and 4-nitrophenol had increased by 43-fold and 46-fold, respectively.

The final step,[240,241] utilized a UDP-hexanolamine agarose column. Bulk protein was eluted by Lubrol®-buffer and transferase activity was subsequently eluted by 5 m*M*-UDPGlcUA in Lubrol®-buffer. The further 40-fold purification gave a relative purification from the 10,000×g (homogenate) supernatant of 908 and a yield of 1.8% from the Lubrol®-solubilized preparation. No epoxide hydratase was detected, and on gel electrophoresis in 0.1% SDS only one band remained, of 57,000 daltons.[240,241]

This molecular weight, for UDPglucuronyltransferase or its subunit is consistent with the result of Gorski and Kasper,[150] who found a major polypeptide of 59,000 daltons; with that of Bock et al.[262] who found for their two transferase peaks weights of 48,000 and 52,000, and later, on fuller purification,[337] of 54,000 and 57,000; and with the crude SDS preparations of Vollrath et al.[330] which gave a minimal weight of 50,000 to 58,000 daltons. Antiserum raised against the pure enzyme produced a single sharp precipitation line after Ouchterlony double diffusion analysis.[241] Burchell[241] found no obvious difference in gel pattern when preparing pure enzyme from phenobarbital-treated and normal rats by this method, so no evidence for a separate species of phenobarbital-induced transferase (or its subunit) is obvious; the phenobarbital-stimulated enzyme of Bock et al.[337] had a molecular weight of 57,000, and the methylcholanthrene-stimulated enzyme (of different substrate specificity) exhibited a molecular weight of 54,000.

Work on the properties of the purified enzyme has begun. First, phospholipid content is minimal. Burchell's preparation[242] (Chapter 5, Section II.L.) has apparently even less than that of Bock et al.[262,337] Secondly, treatment of the pure enzyme with phospholipases has little effect. Phospholipase A or C decreased activity to morphine or 4-nitrophenol by only 30%,[337] and pure phospholipase C had no effect at all on Burchell's preparation[242] (which may, therefore, be the purer). The detergent inevitably present may support activity in place of the phospholipid.[242] Its removal incurs precipitation and inactivation of the enzyme. Thirdly, certain properties persist from the crude "activated" transferase preparation. Examples are a similar $K_{UDPGlcUA}$ (5.4 mM), although as low levels of UDPGlcUA have not yet been used, persistence of the bent double-reciprocal plots is unknown. No activation by UDPGlcNAc occurs, suggesting that this activation requires the membrane. Activation by diethylnitrosamine persists at a similar level to that in activated crude enzyme,[241] and so appears connected with the protein itself.

The pure enzyme appeared electrophoretically and immunologically similar, from normal Wistar, phenobarbital-treated Wistar,[241] and Gunn[323] rat liver. It did not accept UDPglucose or UDPgalacturonic acid as donor substrates.[241] Its acceptor substrate specificity is considered later (Section II.C. below).

Work, too recent to review,[261,261a] studies the phospholipid-free purified enzyme, with kinetic properties, as expected (Section II. B.8 below), depending on the molecular structure of the phospholipids employed to reconstitute its activity.

One of the major reasons for purifying the transferase is to investigate its heterogeneity. This can now be discussed.

II. HETEROGENEITY OF UDPGLUCURONYLTRANSFERASE

A. Types of Heterogeneity Possible

Of all the questions concerning UDPglucuronyltransferase, none has produced so much discussion, inconclusive experiment, and premature satisfaction, as that of its degree of heterogeneity. Is there one transferase for all aglycons? If not, what are the boundaries of specificity between the various transferases. The author's remark in 1966 that "clarification cannot be long delayed"[4] was unduly optimistic. Only recently has progress been made.

It is important to define "heterogeneity", particularly for a membrane-bound enzyme. Mulder's[338] tentative tabling of types of heterogeneity was developed by Dutton and Burchell.[203] They suggested four main categories: (1) sequential, (2) conformational, (3) accessional, and (4) artifactual.

Sequential heterogeneity would arise from changes in the amino acid composition or sequence in the polypeptide chains. It would ensure different enzyme proteins and its existence, only provable by sequence determinations, would be strongly suggested by the demonstration of isoenzymes purely of protein composition. This has not been achieved for the transferase at the time of writing, but is now a practical possibility.

Conformational heterogeneity would result from changes in tertiary or quaternary structure. Such changes would not arise without sequential change unless the microenvironment dictates them, as by alteration in neighboring lipids. This environmental modification could alter the shape or charge of part of the transferase molecule, and thereby, alter its catalytic efficiency by modifying active and/or allosteric sites. This phenomenon was proposed to explain the removal of conformational constraint (see previous chapter) by "activation" procedures. It could occur in vivo if change in neighboring proteins redistributed the lipid population of the membrane next to the transferase. Polymerization or depolymerization of any transferase subunits could also be dictated by the environment. How far these conformational changes would change the observed specificity of the transferase is doubtful. Perhaps, at the most, they might encourage or discourage glucuronidation of certain aglycons by an enzyme already accepting a wide range of aglycons, and to that extent would help to obscure the boundaries of its specificity.

Gorski and Kasper[150] considered the heterogeneity in their largely purified transferase preparation on isoelectric focusing due to phospholipid differences, the polypeptide composition of the three forms being identical. However, other purified transferases appear to have very little phospholipid, and the heterogeneity of one is believed due[247,337] to polypeptide, not phospholipid, differences.

Accessional heterogeneity is a related phenomenon by which changes in the membrane structure differentially limit access of substrates, or modulators of substrate catalysis, to the enzyme. Enzyme properties are unaltered, but barriers are put up or dismantled as the neighboring proteins and lipids undergo change. This type of heterogeneity is, of course, consistent with the "compartmentation" theory of transferase latency, and its likelihood is subject to the same qualifications as conformational heterogeneity.

Artifactual heterogeneity has no existence in vivo and is due to misleading experimental design or observation. Examples may have been embarrassingly common, e.g., selective inhibition due to selective inhibition of assay procedures where these differ widely between substrates, differential solubility or pH requirements for different substrates, and differential susceptibility of substrates to the reverse reaction.

Before discussing how far each of the major experimental approaches can yield evidence of the above categories of heterogeneity, we should mention two additional related points.

First, it should be clear from the above classification that different transferase *proteins* (category (1) above) and different transferase *specificities* (categories (1), (2), or (3) above) should be distinguished from each other. The latter is the practical, physiological, property. We might have a single protein A in environments x and y giving catalytic entities of transferase activity Ax and Ay which are, respectively, specific towards substrates X and Y, or we might have two proteins A and B which in vivo would operate as Ax and By, again, respectively, specific towards substrates X and Y. Delipidation (if lipids had determined the immediate environment) would in the second case give us two distinct proteins A and B, and the transferase is clearly heterogeneous. But if only one protein A is found, we would have in vivo the two enzymes Ax and Ay, and the transferase is again heterogeneous if it is regarded as the catalytic entity and not as the apoprotein. This, or similar, "functional heterogeneity" arising from a single protein A must be kept in mind throughout the following discussion. The membrane environment of the transferase plays a large part in its reported properties and has been responsible, or has been argued to be responsible, for a large part of the "evidence" for or against heterogeneity.

The second point concerns use of endogenous and xenobiotic substrates. It should be remembered (Chapter 1)[4,11,339] that any different transferase(s) eventually isolated will have been evolutionarily selected to serve "natural" substrates — endogenous substrates and those routinely encountered by the species in its ecological niche. Newly designed drugs or laboratory xenobiotics are thus unlikely to trace the boundaries of transferase specificity. Overlap probably will occur. Two transferases developed for two different natural substrates may well, to varying degrees, accept a particular xenobiotic. A third may not.

An analogy occurs in sulfate conjugation. Estrone is not a substrate for the phenol sulfotransferases of guinea pig liver,[340] but the related estrogens equilin and equilenin are substrates;[341] some phenol sulfotransferases may function as estrone sulfotransferases, and some estrone sulfotransferase(s) function as general phenol sulfotransferases.[341]

B. Experimental Approaches to the Problem of Heterogeneity of UDPglucuronyltransferase

1. Amino Acid Sequencing

This has not been reported at the time of press, but results may soon appear.

2. Physical Separation

This would include separation of enzymes A and B towards substrates X and Y by chromatography, electrophoresis, centrifugation, etc. It need not utilize the pure enzyme. In fact, because of "functional heterogeneity", a nondelipidated preparation should be more relevant to specificities in vivo. Separation is the only valid evidence for functional heterogeneity and is the best we have so far obtained for any form of heterogeneity. However, activities towards both X and Y must be recovered, and each must be free of the other, as apparently found by Bock et al.[337] If only X activity is recovered, then a separate protein for Y might have been lost, but there might equally well be only one protein A still retaining environment x (and thus accepting X), but which has lost environment y (and thus no longer accepting Y). For example, Burchell[241] suggests that activity for bilirubin and testosterone may still exist in his pure transferase, but the phospholipid membrane necessary for the orientation of these substrates to the active site[170] has been lost. Successful separation of unpurified activities must be followed by purification and sequencing before we can distinguish between the Ax-Ay and the A-B situation. Where purification involves relipidation with phospholipid micelles before activity is demonstrable, the imposition of the new microenvironment may have artificially changed specificity of the catalytic entity.

3. Species Difference

Such evidence is not very useful. Rat liver has activities towards 2-aminophenol and bilirubin, chick liver (of the two) only towards 2-aminophenol. This indicates that either the protein or its environmental membrane differs genetically between rat and chick, which is scarcely surprising. However, reports of grouped omissions, e.g., that rat has activities for bilirubin, chloramphenicol, and 4-methylumbelliferone, whereas chick has none of these,[342] may be worth recording in the context of heterogeneity.

4. Strain or Tissue Difference

Such evidence is somewhat stronger than from species difference, and with careful controls is genetically informative. For example, the Gunn strain of rat lacks liver activity towards bilirubin and possesses only very low activity towards 2-aminophenol. With a less sensitive assay the latter activity could indeed have been assumed absent. However, when certain ketones are added to its purified liver transferase in vitro, the Gunn rat is seen to possess activity towards 2-aminophenol at the same high level as Wistar rat. The protein of the Gunn rat enzyme is genetically defective rather than absent. The difference between strains with this aglycon is not so evident in intact liver cells.[173] Therefore, a degree of artifactual heterogeneity may also be involved.

5. Development

It has long been believed, and recently proved, that transferase activities to various substrates develop at different rates (Chapter 10). As activation characteristics of the mammalian liver enzyme change during development (Chapter 10, Section III.B.4) constraint may progressively change, altering the relative activities of a single protein A to X and Y. Controls must cover the effect of activation procedure on age of tissue and protein present, and (if "microsomes" are employed) on the age-dependent harvest of endoplasmic reticulum fragments at any one centrifugal speed (Chapter 10). As apparent K_m values may change on development, assays must be at V_{max} for all substrates. Then it is a clear case of functional heterogeneity if activity to X develops before activity to Y. The separate development of groups of transferase activities (i.e., "clusters"), their appearance presumably related to protein or membrane synthesis, has now been reported (Chapter 10, Section III.E.1). Each cluster of activities could arise from simultaneous appearance of several catalytic entities, not just of one.

6. Induction, or Loss of Activity, in Intact Cells

This can be brought about in vivo or in cultured cells by age, disease, and administered hormones or xenobiotics. Usually, one activity is changed relative to another, e.g., in culture, activity towards estriol declined but towards 2-aminophenol increased.[197] Phenobarbital administration to rats increased liver transferase activity towards bilirubin over that towards 4-nitrophenol, whereas 4-methylcholanthrene produced the opposite effect.[336] Such evidence is not very reliable as a criterion of heterogeneity. Differential changes in the membrane can be occasioned by administration of different xenobiotics,[324] and probably also by the other treatments. Species differences may be profound — methylcholanthrene, but not phenobarbital, inducing the activity to bilirubin in the guinea pig,[336] contrary to the effect in the rat.[336] We may thus be following changes in microenvironment and observing a case of Ax and Ay. If so, then, as these effects can be followed through to a semipurified enzyme,[334,336] they must concern tightly bound membrane components. In rat,[337] they seem to concern different proteins.

7. Activation or Inhibition

This may be brought about by physical factors, molecules, or ions, including, of course, H^+. For example, activity towards X is increased by Triton® X-100, but not that towards Y; activity towards Y, but not towards X, is inactivated rapidly by heat or by trypsin, etc. Such information is one of the most quoted and least reliable for claiming heterogeneity. The effects probably arise from change in constraint; how far these reflect preexisting "physiological" forms of the transferase-membrane complex is uncertain. Ionic changes (as H^+ for acidic substrate or Mg^{++} for bilirubin solubility), will also differentially affect substrate availability in the assay cocktail and so lead to "artifactual heterogeneity".

8. Kinetics

Because of, rather than despite, increasing sophistication, severe qualifications remain with this approach. Kinetic studies on purified preparations conducted with due regard for humility can tell us much about the mechanism at the active and allosteric sites on the protein, and when carried out on both microsomal and purified preparations together, something about constraints imposed by the in vivo environment. Conducted on crude preparations alone they may provide stimulating hypotheses to be tested subsequently by more appropriate techniques, but are also likely to encumber the literature with misleading data on heterogeneity.

Mulder[338] and others,[203] emphasizing the dangers of undue "kineticizing", quote work describing shifts in optimal pH and kinetic parameters when membrane-bound enzymes are solubilized and how membrane microenvironments affect behavior to substrates. An enzyme can change in substrate specificity once immobilized, partly due to substrate access through the charged immobilizing environment. Anomalies due to such kinetic compartmentation have been detailed.[344-347] As lipid-soluble substrates of the transferase may be differentially concentrated in the microsomal lipids,[170] and as substrates[293] and products[302] can both activate the enzyme (Chapter 5, Section II.Q.), detailed interpretation of experiments on competitive inhibition becomes difficult, even assuming noncompartmentation of the enzyme. Once compartmentation is admitted, it becomes largely unprofitable, even with stepped levels of donor and acceptor substrates, unactivated microsomes, short linear reaction times, and a careful eye on the reverse reaction.

Although certain kinetic features persist through the first stages of purification,[285] the concavity in Lineweaver-Burk plots decreases,[349] suggesting[310] that membrane effects were initially responsible for these deviations. No concavity was observed with the immobilized enzyme,[349] but comparison was not made with the free transferase. However, pH optima and $K_{4-nitrophenol}$ were similar in solubilized and agarose-bound transferase preparations (Chapter 7, Section II.E.1), and now that the virtually phospholipid-free enzyme is available,[242] more information should be forthcoming. Burchell[241] found the $K_{UDPGlcUA}$ of his pure transferase to differ little from that of the crude solubilized enzyme.

C. Examples of Heterogeneity of UDPGlucuronyltransferase
1. General

Not surprisingly, considering the above somewhat deprecatory account of the preparations and techniques, there are as yet (early 1978) few examples of the heterogeneity of UDPglucuronyltransferase within a single species worth discussing, and probably only one scientifically convincing. As category (1), sequential heterogeneity, the only one provable chemically, has not yet been examined, we must rely on the other approaches for information on the degree of heterogeneity currently probable. Even here we are restricted by the relatively infrequent use of endogenous substrates.

Some current evidence is shown in Table 3, extended and modified from that of Dutton and Burchell.[203] It is an incomplete index, and no individual appraisal can be possible here for most entries. The original papers should be judged from the critical remarks offered above. Dutton[4] reviews the earlier (pre-1965) work fairly fully.

We shall examine the more likely evidence, almost all of which comes from approach (2) of Section II.B.2 above, separation, and ascertain how far it is supported by the other approaches. The evidence suggests that:

1. **Heterogeneity** seems very probable between activities towards
>   L-morphine and 4-nitrophenol;
>   L-morphine and 1-naphthol;
>   bilirubin and 4-nitrophenol;
>   and estrone and 4-nitrophenol.

TABLE 3

Some reports concerning heterogeneity of UDPglucuronyltransferase[a]

Major substrate groups	References and criteria
Simple substrates *v* polymeric substrates (glycans, etc.)	Heterogeneity (see Reference 4)
O-glucuronides *v* *N*-glucuronides	227B*, 350D, 351C, 352F
O-glucuronides 4- & *N*-glucuronides	563G
O-glucuronides & *S*-glucuronides	206 EFGH
'Ether' glucuronides *v* 'ester' glucuronides	No obvious division
Endogenous substrates *v* xenobiotic substrates	No obvious division
Steroid substrates *v* nonsteroid substrates	No obvious division
Bulky molecules *v* small planar molecules	145aEF, 336F

Endogenous substrates	
Estriol & 17β-estradiol	112DFGH, 303BH*
Steroid 16α *v* 17α groups	353GH, 354GH, 355H
16α & 17α groups	
Steroid 16α *v* 3β groups	357GH, 358C
16α *v* 3α groups	213H, 359H, 360C
17β *v* 3α groups	361H
Estrone *v* testosterone	138F
Estriol *v* bilirubin	see 182C
Estrone *v* bilirubin	138H
Testosterone, estriol *v* bilirubin	213C
Tetrahydrocorticosterone *v* bilirubin	362D
Pregnanediol *v* bilirubin, estrone and some other steroids	231EH
Bilirubin *v* bilirubin monoglucuronide	125D, 136D, 363H

Xenobiotic substrates	
4-Nitrophenol *v* morphine	294B*, 335BH*, 364EFH, 365AD
4-Nitrophenol *v* 2-aminophenol	114C, 186H, 192G, 234DG, 288BGH* 292H, 336D, 367H, 368D
4-Nitrophenol & 2-aminophenol	241B*, 287H
4-Nitrophenol *v* 4-methylumbelliferone	369C
4-Nitrophenol & 4-methylumbelliferone	326DG
4-Nitrophenol *v* phenolphthalein	130D, 370E
4-Nitrophenol & 1-naphthol	113 FGH, 241B*, 336F
2-Aminophenol & 1-naphthol	241B*
2-Aminophenol *v* 4-methylumbelliferone	192G, 342C
2-Aminophenol *v* phenolphthalein	234G, 368D, 370E, 371C
2-Aminophenol *v* morphine	372D
1-Naphthol *v* morphine	262BF, 337BF
Morphine levallorphan, nalorphine, naloxone, naltrexone	335H
Morphine or 4-nitrophenol *v* salicylamide, salicylate, 3- or 4-acetylaminophenol, 4-aminophenol, 4-aminobenzoate, phenolphthalein, resorcinol [also estrone]	335H
Phenolphthalein *v* 1- or 2-naphthol, 4-acetamidophenol, morphine	373C

[a]   See Section III., Additional Notes at the end of the chapter.

TABLE 3 (continued)

Some reports concerning heterogeneity of UDPglucuronyltransferase[a]

Major substrate groups		References and criteria
4-Acetamidophenol	v 4-nitrophenol, phenol-phthalein, menthol [also bilirubin]	234G
Hydroxyphenylhydan-toin	v 4-nitrophenol, 4-methy-lumbelliferone	374H
Diethylstilbestrol	v diethylstilbestrol glucu-ronide	375H

Endogenous and xenobiotic substrates together

Bilirubin v 2-aminophenol	234G, 276H, 369C, 371C, 376F, 377E
Bilirubin v 4-nitrophenol	135D, 184G, 249E, 257B, 276H, 336F, 366D, 378D, 379G, 380F, 381F, 382F, 383CDF, (others in 4)
Bilirubin & 4-nitrophenol	384H
Bilirubin v 4-nitrophenol, 2-aminophenol	385D
Bilirubin v 1-naphthol	240B*, 241B*, 179BG, 336F
Bilirubin v chloramphenicol	336F
Bilirubin v 4-aminobenzoate	276H
Bilirubin v 4-methylumbelliferone	382F
Bilirubin v diphenylacetate	386D
Bilirubin v morphine	372D
Bilirubin, progesterone, testosterone, chloramphenicol, bunamiodyl, tyropanoate, iodopanoate, diphenylacetate v 4-nitrophenol, 2-amino-phenol, morphine	C(see 371)
Estrone v 4-nitrophenol	138H, 387B*
Estrogens, testosterone v 4-nitrophenol	263GH, 281H, 335F, 356D, 388FE
Estriol v 2-aminophenol	197F
Estrone, estradiol, di-ethylstilbestrol, testos-terone v 4-nitrohpenol, 4-methy-lumbelliferone, 1-naphthol	389DEF
Bilirubin, estradiol, tes-tosterone, chloram-phenicol, morphine, phenolphthalein v 2-aminophenol, 2-ami-nobenzoate, 4-methy-lumbelliferone, 4-ni-trophenol, 1-naphthol, serotonin [see also Chapter 10 Section III.E. for further addi-tions to these 2 groups]	390EF, 391EF
Tetrahydrocortisone v 4-nitrophenol	362D
Tetrahydrocortisone & salicylate	392H
Bilirubin, serotonin, chloramphenicol 4-methylumbelliferone, harmol v estriol, 2-aminophenol, 4-nitrophenol	C (see 342)

Note: Only quoted are direct comparisons made in one paper. 'v': evidence for heterogeneity towards two substrates or groups of substrates; '&': evidence for homogeneity in such cases; *: referred to further in text of this Section. Criteria for evidence: B, physical separation; C, species difference; D, strain or tissue difference or subcellular location; E, developmental; F, differential activity changes in intact organism, tissue or cells; G, differential activity changes in broken cells in vitro; H, kinetic.

2. **Homogeneity** seems very probable between activities towards
2-aminophenol and 4-nitrophenol; and
1-naphthol and 4-nitrophenol.

2. L-Morphine and 4-Nitrophenol

These two activities were separated from rat[294] and rabbit[335] liver microsomes using chromatography on DEAE-cellulose after Emulgen solubilization. The final preparations from rat liver, eluted by buffered KCl gradient, were not very pure, but consisted of three peaks: (1) one for morphine eluted first (Peak I), (2) one for both morphine and 4-nitrophenol (Peak II), and (3) one for 4-nitrophenol (Peak III).[294] Peaks I and III, with no overlap, convincingly separated activities towards morphine and 4-nitrophenol. How far this demonstrated heterogeneity of the transferase is debatable because of the relative crudity of the preparations, but evidence included the following: (1) 4-nitrophenol did not inhibit morphine glucuronidation by fractions from Peak I, nor morphine that of 4-nitrophenol glucuronidation by those from Peak III, and (2) as apparent Michaelis constants for UDPGlcUA, morphine, and 4-nitrophenol approximated to those found with microsomes, the constraint of the transferase had not been greatly changed, and the heterogeneity may truly reflect conditions in the endoplasmic reticulum and not be an artifact of preparation. Peak II may exist because: (1) it represents another separate transferase activity with broad specificity, accepting both substrates, or (2) it represents a protein-membrane fragment in which the morphine and 4-nitrophenol transferases of Peaks I and III have not been completely dissociated;[335] (2) is more likely.[294] The procedure may have merely isolated three types of vesicle, each containing the same transferase, but separable because of different degrees of membrane change; as a result of this change, one (Peak I) has lost activity towards 4-nitrophenol, another (Peak III) has lost activity towards morphine, and one (Peak II) remains relatively undamaged and possesses both these activities. Reappearance (from zero), during purification, of the activity to bilirubin following addition of lipids[257] should inspire caution.

The authors subsequently found, with rabbit liver[335] only two peaks, for 4-nitrophenol and morphine, respectively. Kinetic experiments on these semipurified preparations indicated no mutual competition for acceptor substrate, but competitive inhibition for the morphine enzyme existed between morphine and the narcotic antagonists naloxane, nalorphine, levallorphan, and naltrexone. These compounds did not affect the 4-nitrophenol-glucuronidating enzyme. Morphine 3-*O*-glucuronide and 4-nitrophenyl glucuronide acted specifically as product inhibitors for their enzymes.[335] Interestingly, the kinetics were similar in both crude and separated preparations, indicating either that they were closely associated with the enzyme's active center or that separation had not greatly changed the enzyme's membrane environment.

Evidence from differential induction, development, and kinetic parameters (Table 3) is consistent with separate transferases to morphine and 4-nitrophenol, although by itself inconclusive.

3. 2-Aminophenol and 4-Nitrophenol

These two activities were separated by zonal centrifugation on linear sucrose gradient after Lubrol®-solubilization of rat-liver microsomes.[288] Activity towards 2-aminophenol was found in fractions of density less than 1.056 and that towards 4-nitrophenol only in those of greater density. There was considerable overlap, both activities appearing with bulk protein, in the "4-nitrophenol fractions". When absolute, not specific, activities are calculated per fraction. The authors' data do not support a separation and they rely more on absence of mutual inhibition, on differential kinetics, and on activation by ions. Similar kinetic differences have been quoted by others (see Table 3), but their significance is disputed,[138,139] activation by substrate and/or glucuronide, and the physical difference of these two substrates being quoted. Burchell,[240,241] found that activities towards 2-aminophenol and 4-nitrophenol copurified, maintaining approximately the same ratio of specific activities over five purification stages from an ammonium sulfate precipitate through to the pure transferase preparation. This suggests a single enzyme for the two substrates. Although the final preparation also catalyzed glucuronidation of 1-naphthol, morphine, and 2-aminobenzoate,[241] activity to morphine did not copurify and may originate from a closely similar concurrently purified transferase. This is especially likely with morphine (Sections II.C.2 above and II.C.5 below).

Supporting evidence for homogeneity with 2-aminophenol and 4-nitrophenol comes from developmental and induction studies (Table 3).

4. 1-Naphthol and 4-Nitrophenol

Burchell[240,241] found that activities also towards these two substrates copurified in the same proportion through the final stages. Homogeneity is consistent with work from developmental and induction studies (Table 3).

5. 1-Naphthol and L-Morphine

Bock et al.[262] separated activity towards morphine from that towards 1-naphthol. Two peaks were eluted, the second containing the activity to morphine, the first towards 1-naphthol, but also a little towards morphine. Very recently,[337] these two activities have been completely separated from rat livers differentially "induced" with phenobarbital and 3-methylcholanthrene. The final preparations seemed as pure as any yet obtained. No "alternate substrate" inhibition could be demonstrated with rat or human liver.[337] However, the sources of the two enzymes are differently treated sets of animals. Separation of the two enzymes from one set of animals has yet to be demonstrated, and in that respect proof of heterogeneity is incomplete.

Consistent with two enzymes were their findings[262] that prior treatment of rats with phenobarbital gave different relative specific activities towards the two substrates in their respective eluates different to those found after pretreatment with 3-methylcholanthrene. The probability of separate enzymes for these two substrates is also borne out from developmental and other studies[390,391] (Table 3) and by the likely homogeneity of activities towards 1-naphthol and 4-nitrophenol (Section II.C.4 above) and towards morphine and 4-nitrophenol (Section II.C.2 above).

6. Bilirubin and 4-Nitrophenol

Transferase solubilized from rat-liver microsomes with deoxycholate and chromatographed on Bio-Gel® P30, lost activity to these two substrates.[257] Dialysis with lecithin or microsomal lipids restored activity to bilirubin, but not to 4-nitrophenol. Differential loss of activity also occurred during purification of Lubrol®-solubilized transferase,[239,240] when activity to bilirubin progressively disappeared, whereas that to 4-nitrophenol, 1-naphthol, and 2-aminophenol copurified.

Until both activities can be separately recovered, the evidence for heterogeneity from separation remains very indirect, but is supported by observations from other approaches (Table 3).

Jansen[363] found, kinetically, no obvious heterogeneity in cat liver between activities towards bilirubin monoglucuronide and 4-nitrophenol, but as the diglucuronidation step has recently[175] in rat been attributed to another enzyme than the transferase, this observation may need reinterpretation.

7. Estrone and 17β-Estradiol

Homogeneity for the semipurified pig intestinal activity towards the 3-position of 17β-estradiol with that towards estrone is based on competitive kinetics and similarity of kinetic parameters, and not by copurification through several stages. It is supported by further circumstantial evidence from elsewhere[112] (Table 3).

8. Estrone and 4-Nitrophenol

These activities have been recovered separately after column chromatography.[387] Supporting evidence is listed in Table 3. Separation was achieved from rabbit liver, using Emulgen® 911, and DEAE cellulose column chromatography.[387] More recent work[391a] uses isoelectric focusing on dextran slab gels. Peak A contained transferase activity to estrone only, Peak B that towards 4-nitrophenol and a very little towards estrone. The latter "impurity", 1/1000 of the activity to 4-nitrophenol, accompanied the 4-nitrophenol activity through purification. Peak A exhibited a pI of 7.97 ± 0.19, Peak B, 6.74 ± 0.37. Delipidation of Peak A did not affect the pI value, whereas Peak B was largely delipidated during preparation. Reconstitution with phospholipid, following delipidation on hydroxyapatite, increased specific activity of both peaks, providing further evidence that two transferases had been separated. Kinetic evidence supported this conclusion; 1-naphthol inhibited the "4-nitrophenol activity", but not the "estrone activity". Peak A showed only two major bands on gel electrophoresis, and its relatively poor (15-fold) purification from microsomes probably relfected its great lability.[391a] Peak B, purified to "near homogeneity", had been purified 76-fold from microsomes.

9. N -Glucuronides and O -Glucuronides

Venom solubilized activity from rabbit-liver microsomes towards phenolic, alcoholic, and acidic substrates, but that towards aniline remained at its original value in the residue. No inhibitor of aniline glucuronide formation was found in the supernatant. This early separation requires modern reexamination. Heterogeneity is consistent with indirect evidence (Table 3).

D. Practical Importance of the Problem of Heterogeneity of UDPGlucuronyltransferase

UDPGlcUA is glucuronyl donor to a wide range of aglycons which arise from both endogenous and xenobiotic sources, and which are linked in the conjugate through O-, S-, and probably N-atoms. It is important to know if one single UDPglucuronyltransferase is responsible for catalysis in all these instances, and if not, how many enzymes are involved and what their specificities are. The answer would help us to understand the nature of the catalytic mechanism and assist certain practical problems.

These problems concern prediction of the degree of glucuronidation for any one species, strain, age, or substrate by extrapolation from information already known. For example, low UDPglucuronyltransferase activity occurs in the newborn, particularly if premature, because the enzyme(s) does not develop fully until after birth (Chapter 10); and in certain rare hereditary conditions, such as the Crigler-Najjar syndrome, the enzyme activity to most substrates remains low throughout life. In these instances, jaundice may develop as a result of insufficient transferase to catalyze formation of bilirubin glucuronide or by drugs competing with bilirubin for glucuronidation at the few active enzyme sites available. We need to know what other substrates are poorly conjugated at birth and whether they are conjugated by the transferase accepting bilirubin. Further, if activity to one drug develops early or is induced by a xenobiotic, then will activity to another drug, or to bilirubin, appear at the same time? Do all transferase activities decrease together in disease, or is there some phasing-out of less-essential activities, and if so, what are they and what types of compound would they have conjugated?

Testing for these effects in man with the appropriate aglycon are not usually practicable, especially during commercial drug development. Scarcity of human tissues, or ethics, forces the use of other species. The difficulty of measurement of certain aglycons encourages the use of other compounds as substrates. We are, therefore, frequently faced with assessing the validity of the extrapolation to substrate *B* in species

X of results found there with substrate A, substrate B in species X of results found with substrate B in species Y, and even to substrate B in species X of results found with substrate A in species Y. The validity of this procedure depends on the degree of heterogeneity of the transferase or, practically, upon the constancy of its behavior among species and substrates. Are results obtained with substrate A in species X applicable to substrate A in species Y, or to substrate B in either species X or Y?

It should be noted that here we are considering *functional* heterogeneity, the practical property. For the remainder of this section (II.D), heterogeneity is equated with functional heterogeneity.

Formerly, there were three main possibilities: (1) only one UDPglucuronyltransferase exists among mammals; (2) only one transferase exists in each organism, but differs in substrate range among different species or strains, and (3) there are many transferases in each organism, and different species or strains possess different shares.

Possibility (1) is now untenable. As Chapter 12 indicates, even among mammals, different species (and even different strains) possess quite different spectra of transferase activity. Cats have low activity towards phenols, but the argument that cat has a low level of its one transferase is rebutted by the evidence that cats can form glucuronides of other substrates quite well. No single transferase is shared by all mammals, and extrapolation from one species to another is unjustifiable.

Possibility (2) is now, from the evidence just discussed, also untenable for "functional" heterogeneity. Possibility (3), therefore, appears the most reasonable explanation. This likelihood of clear functional heterogeneity among species and substrates forces us to plan future investigation to identify the known and then predict the unknown groups of xenobiotic and endogenous "activities" of transferase which rise and fall together, as is already being done in perinatal studies (Chapter 10).

This way, we may determine the boundaries of functional heterogeneity. These will be valid only for one species. Extrapolation to man requires parallel experiments with laboratory (or other) species and man to determine which, for that group of transferase activities, is the animal most resembling man. Smith and Caldwell[392a] discuss this animal but much remains to be done. Satisfactory information on human glucuronidation requires work with human tissues. Cultured human tissue being the most convenient, study of the transferase on culture is currently important (Chapter 15, Section II.).

III. ADDITIONAL NOTES

This chapter is effectively updated by a short stimulating review of work on purification and heterogeneity of the transferase, principally over the period from 1977 to the beginning of 1980.[327b] The major advances have come from the groups of Bock and of Burchell. Bock has linked his two separable forms of the transferase, GT_1 and GT_2, with further substrates, mostly xenobiotic, e.g., Reference 392c. Burchell has proposed four groups of transferase activity, from his separative work, A, B, C, and D; A and D approximate respectively to Bock's GT_1 and GT_2. Burchell has employed many endogenous substrates and the picture emerging is, as forecast, of considerable specificity for xenobiotic substrates and of overlapping specificity for xenobiotic substrates. Interesting examples of functional and artifactual heterogeneity are quoted.

Activity to bilirubin, estrone and testosterone can be restored during purification by addition of phospholipid; detergent is insufficient for these activities.[327b] Burchell's form A glucuronidates 2-aminobenzoate, 2-aminophenol, 3′-hydroxybenzo(a)pyrene*, N-hydroxy-2-naphthylamine*, morphine, 1-naphthol*, 4-nitrophenol*, testosterone (those marked * are glucuronidated by Bock's form GT_1); form B, bilirubin, morphine; form C, estrone, 4-nitrophenol; form D, deduced from Bock's form GT_2, chloramphenicol, 4-hydroxybiphenyl, morphine. This refers to rat liver. References are in Reference 327b.

Chapter 7

ACCEPTOR SUBSTRATES OF UDPGLUCURONYLTRANSFERASE AND THEIR ASSAY

I. RELATION OF ACCEPTOR SUBSTRATES TO AGLYCONS GLUCURONIDATED

Early work indicated that "ether"[90] and "ester"[329] O-linked glucuronides, steroid glucuronides,[329] and N-glucuronides[392b] were formed by UDPglucuronyltransferase. Since then, apart from a few aromatic amines, bilirubin monoglucuronide,[125] and possibly diethylstilbestrol monoglucuronide, there is no clear case of an aglycon known to form a glucuronide, or an aglycon glucuronide known to form a diglucuronide in vivo, not being accepted by UDPglucuronyltransferase in vitro. (The C-linked glucuronides have not as yet been investigated enzymically). Consequently, we may assume biosynthesis of all known glucuronides isolated from, or identified in, living systems utilizes UDPglucuronyltransferase activity. The number of such glucuronides is immense. In 1966, Marsh could publish[63] detailed lists of chemically and biologically synthesized glucuronides classified by chemical structure and degree of characterization with sources indicated. Since then, the task has become worthy of a separate text. A recent publication[64] is not comprehensive and omits much listed by Marsh. The present section must be confined to selected known or probable substrates. Biosynthetic production of reference glucuronides on a large scale by animal dosage is treated by Williams[5] and Marsh.[63] Biosynthesis of ^{14}C-labeled reference glucuronides in vitro using ^{14}C]-UDPglucuronic acid[393] and of ^{3}H]-UDP-^{14}C]-glucuronic acid have been described.[394] For chemical synthesis, see References 63 and 63a.

II. SOME NEW SUBSTRATES

"New" constitutes compounds reported as substrates, or as aglycons in biosynthesized glucuronides, since the last major review.[4]

A. Substrates Forming S-Glucuronides

UDPGlucuronyltransferase was required for formation of S-glucuronides, the last of the three major types of glucuronides (O, N, and S) to be studied. The mechanism[395] and properties[206] were investigated with 2-aminothiophenol, 4-nitrothiophenol, thiophenol, and diethyldithiocarbamate as substrates. S-Glucuronides of these compounds were identified by several tests after chromatographic isolation, both aglycon and uronic acid being present. Cysteine and glutathione did not, under these conditions, behave as substrates. Differences between enzyme activities synthesizing S-glucuronides and their O-glucuronide analogs occurred in development, in induction by phenobarbital or organ culture, in activation, and in kinetics,[206] but were not greater than already found between those synthesizing O-glucuronides. Mercaptobenzothiazole,[159] also a substrate for the enzyme, was studied by a nonradioactive method. These papers detail assay and identification. The antithyroid drug 6-n-propyl-2-thiouracil forms a confirmed S-glucuronide in vivo in rat and with rat-liver microsomes.[396] Thiouracil, also a substrate,[397] forms an uncharacterized S-glucuronide.

B. Substrates Forming N-Glucuronides

Little new is reported on the anomalous N-glucuronides. Biosynthetic N-glucuronides are described in Chapter 2, Section I.D, in Chapter 1, Section I.2, in Marsh's list,[63] and by Irving.[52,53] Recent *N*-glucuronides isolated include those of carbameza-pine[398] and desmethylimipramine,[399] the latter an aliphatic *N*-glucuronide apparently hydrolyzed by β-glucuronidase, although such hydrolysis is difficult.[74,75] Other ali-phatic *N*-glucuronides of mono- and didemethylated chlorpromazine have been re-ported in human blood, but their characterization as *N*-glucuronides was incomplete and their properties not studied.[400] The psychoactive compound EMD 16923, a 1-pyr-azoloethyl-4-chlorophenylpiperazine, forms in man a main metabolite characterized by mass spectrometry and impulse resonance spectrometry as a glucuronide where glu-curonic acid is linked to the *N* of the pyrazole nucleus.[73] This new type of *N*-glucuron-ide is unexpectedly stable to hydrolysis.[73] Conjugation of glucuronic acid occurs through an amido, not a glycosyl, linkage with the amino group of glycine,[401] the *N*-terminal amino acid of a glycoprotein; the mechanism is unknown.

C. Substrates Forming *C*-Glucuronides

C-glucuronides are formed in vivo (Chapter 2, Section II.D.). The C—C link, wide-spread in plant glycosides,[76] in man occurs in glucuronides of sulfinpyrazone and phen-ylbutazone.[77] After administering labeled phenylbutazone to man 40 and 12%, respec-tively, of the urinary radioactivity was due to the C(4)-glucuronides of phenylbutazone and γ-hydroxyphenylbutazone.[78] The role of the transferase and β-glucuronidase in their metabolism requires study. Probably many *C*-glucuronides remain unsuspected.

D. Substrates Forming *O*-Glucuronides

1. O-Glucuronides of Xenobiotics

a. Comparison of Substrates

Transferase activities to a series of substrates, usually chemically related, have some-times been compared. Relevance of these findings to the enzyme mechanism is dis-cussed in Chapter 8, Section III. Mulder and van Doorn[402] using their new assay method (Section V.1. below) compared activities of the rat-liver microsomal enzyme to 26 phenols, giving physicochemical data on each substrate. All these phenol deriva-tives were glucuronidated faster than phenol itself, except for the 2-,3- and 4-hydrox-ybenzoic acids, the 2-,3- and 4-aminophenols, *N*-acetyl-4-aminophenol, and 4-hydrox-ybenzenesulfonic acid, the substrates of least solubility in *n*-octanol when extracted from aqueous solution at pH 7.4.

Increasing lipid solubility by introducing an *N*-methyl group into 4-aminophenol or ethylating the carboxylic group of 4-hydroxybenzoic acid increased glucuronidation rate considerably. Bromine at C-4 of phenol increased glucuronidation markedly, chlorine less so, and fluorine hardly at all. A nitro or methyl group enhanced the reaction more in the 4-position than in the 2-position, a chlorine atom the contrary. The highest rate was found with 3-methyl-2-nitrophenol. Using the conventional assay, again 3-methyl-2-nitrophenol was the nitrophenol most rapidly glucuronidated. No activity appeared with (4-hydroxy-3-nitrophenol)-arsonic acid (not an inhibitor), and very little was found if chlorine was present at the 2- and 6-position of 4-nitrophenol.[402]

Members of a series of phenolic opiates likewise appeared better substrates for rat-gut transferase as their lipophilicity (i.e., partition in heptane and phosphate buffer at pH 7.4) increased.[403,404]

Nemoto and Gelboin[20] compared transferase activities to metabolites of benzo(a)pyrene. Among the phenols, glucuronidation of the hydroxy group at C-1, -3, -7, -9, -10, or -12 was greater than at C-2, -4, -6, or -8. The quinones were not accepted as substrates. The 4,5-oxide was conjugated, possibly via epoxide hydratase and the dihydrodiol. The 7,8-oxide was conjugated at a greater rate than the dihydro-diol formed from it by the hydratase. As the 7,8-dihydrodiol may be more readily converted to the probably carcinogenic 7,8-diol-9,10-oxide than to the glucuronide,[20] transferase activity could be rate-limiting in carcinogenesis with benzopyrene.[20]

Batt et al.[374,405] compared the enzymic glucuronidation of phenol, its various hydroxy- and nitro-derivatives, and certain drug metabolites including hydroxyphenobarbital. Bile acid substrates have been compared (Section II.D.1.c below). For comparision of substrates during development, see Chapter 10, Sections III.E.1 and IV.D.2.

b. New Xenobiotic Substrates

Among the xenobiotic compounds reported as forming *O*-glucuronides since the previous[4,63] lists are the following. Biosynthesis of the glucuronide of the first quaternary ammonium substrate, 3-hydroxyphenylethyltrimethylammonium (neostigmine) was studied in rat-liver microsomes,[215] and UDPglucuronyltransferase considered responsible for its previously noted formation in perfused liver and intact rats.[215] It was hydrolyzed by β-glucuronidase, but no specific inhibition of hydrolysis was attempted. The substrate, a cation, is a rare example of a polar substrate for a microsomal enzyme. However, at physiological pH the positive charges on the quaternary group could be internally balanced by a negative charge on a dissociated phenolic hydroxyl group, and the resulting neutral zwitterion might penetrate the lipid barrier.[215]

Harmalol, and its dehydro congener harmol, derivatives respectively of the monoamine oxidase inhibitors harmaline and harmine, have been studied as substrates of the transferase,[406,407] as have 4'-hydroxyamphetamine,[408] chloramphenicol,[336] desmethylimipramine,[409] oxazepam,[410,411] and various isoflavones. Among the numerous xenobiotic glucuronides isolated from tissues are those of 5,6-dihydro-5,6-dihydroxycarbaryl,[412] hydro- and dihydromorphine,[413] pyridinethione,[414] 2-methyl-14-naphthohydroquinone (menadione),[415] cannabadiol and hydroxycannabadiols,[416] methocarbamol metabolites,[417] and metabolites of the narcotic analgesic triannular benzomorphan derivatives, cyclazocine, ketocyclazocine, volazocine, and pentazocine.[418]

Although 4-propylthiouracil forms an *S*-glucuronide (Section II.A above), 4-propyluracil itself, despite its *C*-4 hydroxy group, did not form a glucuronide in vivo or in vitro.[396]

2. New Endogenous Substrates and Glucuronides

a. Phenolic Amines and Related Compounds

Phenolic amines are excreted as glucuronides in urine,[419-421] and tyramine is probably a substrate of the transferase.[422] Human urinary 2-phenylethylamine glucuronide[423] increased in the manic phase of manic depression and decreased in depression.[423] Concerning catechol glucuronides, 3-*O*-methyladrenaline, 3-*O*-methylnoradrenaline, 3-methoxytyramine, and 4-hydroxy-3-methoxy-phenethanol are substrates,[406] but not homovanillic (4-hydroxy-3-methoxyphenylacetic) acid;[406] see also References 424 and . 424a.

Purification and properties of L-epinephrine glucuronide are described[425] and a method[426] for serotonin (5-hydroxytryptamine) has been quantitated. Enzymic formation and hydrolysis of the glucuronide of 3-hydroxyanthranilic acid is reported[427] and isolation of that of indolylacrylic acid.[277]

b. Bile Salts

Bile salts are substrates, as well as activators, of the transferase. With cases of human hepatobiliary disease, when the mono-, di-, and trihydroxy bile acids were high in plasma, Back et al.[428] using chromatography and hydrolysis including treatment with β-glucuronidase, found in the urine glucuronides of cholic (3,7,12-trihydroxy-5-cholan-24-oic) acid, chenodeoxycholic acid (the 3,7-dihydroxy analog), deoxycholic acid (the 3,12-dihydroxy analog), lithocholic acid (the 3-hydroxy analog), and 3-hydroxy-5-cholan-24-oic acid. All were minor constituents.[428] Back[429] later found chenodeoxycholic glucuronide in plasma from patients with intrahepatic cholestasis. The β-

D-glucuronides of mono-, di-, and trihydroxy bile acids, chemically synthesized and characterized[430] and seem identical with the conjugates found in vivo, but Back[431] assumes that the chenodeoxycholate glucuronide, for example, was originally present as a glycine or taurine conjugate. These glucuronides are stable to alkaline hydrolysis.[431] Lithocholate, chenodeoxycholate, and cholate were shown substrates of UDPglucuronyltransferase, and confirmed present in cholestatic urine as up to some 25% of the total excretion of bile salts.[432] No bile salt glucuronides were found in normal urine.[432] Clinical aspects are discussed in Chapter 15, Section VIII.

The transferase activity behaved similarly to most other transferase activities reported.[433] It was inducible with phenobarbital, and despite its substrate being a membrane perturbant and present at 1.0 mM, it was activated by Triton® X-100. Activity was greatest towards lithocholate and least to cholate.[433]

c. Simple Lipids and Steroids

Other recently reported endogenous substrates include retinol and retinoic acids,[434] and various hydroxy derivatives of testosterone and androstenedione.[435] Enol-glucuronides[436] are now commonly identified in human tissue fluids, e.g., of progesterone[437] and of androstenedione.[438] Although steroids with the 17-hydroxy group are better substrates than those such as pregnenolone and corticosterone,[213] steroids conjugated with glucuronic acid at other hydroxy groups exist; e.g., aldosterone 18-glucuronide.[439]

d. Complex Lipid as Substrate

Stern Tietz[107,441] and Shaw and Pieringer[440] independently demonstrated UDPGlcUA-dependent formation of glucuronosyl diacylglycerol by Pseudomonad preparations. Glucuronate, α-glucuronic acid 1-phosphate, and sugar nucleotides such as UDPgalacturonic acid, UDPglucose, etc. were without effect.[108] Dipalmitoyl [^{14}C]-glycerol donated glycerol,[440] but the in vitro system required exogenous diacyglycerols containing ester-linked unsaturated and shorter-chain fatty acids.[108] The relationship of this UDPglucuronyltransferase to the vertebrate enzyme has not yet been examined. Its power of glucuronyl transferase to phenolic types of substrate appears negligible (B. Burchell, unpublished results). It occurs in the 34,800×g supernatant and particles from disrupted *Pseudomonas diminuta* ATCC 11568, is very heat labile (80% being lost after 5 min at 37°C), is activated by Triton® X-100 and has an optimum pH of 7.1.[108] It may be located near the surface of the membrane, being readily "solubilized";[108] a site convenient for both its hydrophobic and its hydrophilic substrates. This interesting enzyme may shed light on the hypothetical lipid glucuronic acid-donor in mammalian tissues and the observed (Chapter 3, Section II.B) lipid-soluble glucuronic acid-acceptor there.

3. Substrates Other Than Bilirubin Forming Di- or Mixed Glucuronides

Although many substrates possess more than one group conjugable with glucuronic acid, double conjugates are comparatively rare. They are of special interest in that the transferase performing the first glucuronidation may not be the same as that adding the second glucuronic acid to the now quite polar substrate. It may not even be a UDPglucuronyltransferase.[125] Glucuronic acid is usually added first to mixed glucosides.[436] Detailed enzymic work has only concerned bilirubin, an important aglycon treated below (Section III). The following outlines work with other substrates.

a. Diglucuronides Other Than of Bilirubin

Diglucuronides (probably at C-3 and C-16) of 17β-estradiol[357] and of estriol and 17-epiestriol[442] are formed by rabbit liver in vitro, and by pregnant women,[443,444] and the 4,8-diglucuronide of xanthurenic acid by rats after intraperitoneal (i.p.) or oral administration of the aglycon during vitamin B deficiency.[445] Glycyrrhetic acid, a β-D-glucuronosyl-β-D-glucosiduronate occurs in licorice root.[446]

A little diglucuronide of diethylstilbestrol was found in rat bile,[447] but not in everted sacs of rat intestine[448] or in UDPGlcUA-fortified liver or kidney homogenates from various species.[449] When the monoglucuronide was offered to hamster-liver homogenates in presence of activators known to stimulate monoglucuronidation of diethylstibestrol,[375] the diglucuronide was still not formed. The monoglucuronide, like that of bilirubin (Section III.B below), appears not a substrate of the enzyme glucuronidating the free aglycon.

A diglucuronide has been formed in two stages in vitro by utilizing two different species.[358] Mouse- and guinea pig-liver preparations glucuronidate at respectively the 16α - and the 3 - positions estriol. The 3-glucuronide, preformed by guinea pig liver, was presented to the mouse-liver enzyme, which added glucuronic acid at the 16α position. However, the guinea pig-liver enzyme would not accept the 16α-monoglucuronide.[358]

b. Mixed Glucuronides

Mixed conjugates exist with glucuronic acid and another sugar or an amino acid. Artifactual transfer during isolation or assay, as with bilirubin,[450] should be borne in mind, however.

17α-Estradiol and the 17- and 16,17-epimers of estriol form double glycosides at positions C-3 and C-17, utilizing glucuronic acid and either N-acetylglucosamine or D-glucose. Formation[436] and structure[451] of the former has been discussed. In vitro, only the 3-glucuronides (or sulfates) of the steroids concerned (possessing a phenolic ring and an α-oriented C-17 hydroxyl group) are accepted by the N-acetylglucosaminyl transferases. The free steroid is not.

Biosynthesis of double conjugates with glucuronic acid or glucose again distinguishes between aglycon and aglycon glucuronide as substrate for the enzyme adding the second sugar. Free estradiol-17α and free estriol are substrates for human-liver UDPglucosyltransferase, but their 3-glucuronides are not.[452] In sheep, the 3-glucuronides, not the free steroids, are the substrates.[453] Glucosylation of the β-glucuronides of estradiol-17α and estradiol-17β, is described.[454]

Mixed conjugates of glucuronic and sulfuric acids are frequent and possibly physiologically important (see also Chapter 16, Section III.A). Much circulating estriol in human late pregnancy is the 3-sulfate, 16(17)-glucuronide. The fate of estriol so conjugated is complex.[455] Some 25% of injected labeled estriol appears in human bile as the mixed conjugate, which is then hydrolyzed, and the free steroid reconjugated with glucuronic acid at C-3, a reaction possible for man only in the intestine. This glucuronide is reabsorbed and excreted in the urine.

Diethylstilbestrol injected into rat as the mono- or disulfate,[456] appears in bile significantly as the sulfoglucuronide, with little mono- or disulfate being excreted. Injected sulfoglucuronide is excreted unchanged.[456] Injected monosulfate appeared largely in bile as sulfoglucuronide. If disulfate were injected, only a little sulfoglucuronide appeared there.[456] As desulfation rates of mono- and disulfate were similar, one factor responsible for their different fates may be affinity of a UDPglucuronyltransferase for the monosulfate.[456] This work implies that glucuronidation is the second, not the first, stage in a double conjugation. Injected cortisone 21-sulfate was rapidly excreted, by rat as the 3-glucuronide 21-sulfate, gain suggesting acceptance by the transferase of an already sulfated aglycon.[1189] The in vitro formation of a sulfoglucuronide with tritiated 17β-estradiol as substrate is reported;[457] although no UDPGlcUA was added to the concentrated microsomal mix, sequential hydrolysis after the biosynthetic incubation revealed a sulfoglucuronide at 10% of the total conjugate.[457]

A mixed conjugate of xanthurenic acid, with glucuronic acid at C-8 and serine at the C-2 carboxyl, has been reported,[445] and Javitt[458] gives references to the double conjugate of the phenoltetrabromphthalein monosulfonate with glucuronic acid and glutathione. All the monosulfonate and no di- or tetrasulfonate was glucuronidated, so the free OH group of the sterically less-cluttered monosulfonate could be accepted by UDPglucuronyltransferase. An unmixed glutathione conjugate was found, but not, apparently, an unmixed glucuronide conjugate. Glucuronidation would again seem to succeed conjugation, this time with glutathione.[458] Similar diconjugates with phenol-dibromphthalein sulfonates[458] and the nonsulfonated phenoltetrachlorphthalein occur.

III. RECENT WORK WITH BILIRUBIN AS SUBSTRATE

This frequently quoted and physiologically important compound requires a section to itself. Pharmacologically, it is of interest because it shares the glucuronidation pathway with many drugs. Its own conjugation can be impaired by competition from administered xenobiotics or increased by their induction of its specific transferase activity. Authoritative recent reviews exist of the assay,[460] and of the physiological and enzymic aspects[111,461,462] of bilirubin conjugation.

A. Reasons for Conjugation of Bilirubin

In mammals, degradation of hemoglobin and other heme-containing proteins such as cytochromes predominantly removes the α-bridge of heme to yield biliverdin-IXα, which is then reduced to bilirubin IXα by NADPH and biliverdin-IX reductase.[463] Cleavage of heme elsewhere can occur forming a certain amount of the IX-β, -γ, and -δ isomers of biliverdin, and hence of bilirubin.[464,465] As pronounced differences in polarity occur between these isomers, assays designed for the IX-α isomer could miss the other three. These have now been measured in bile of man, dog, pig, and rat,[464] where they occur as 5% of the total bilirubin-IX isomers and only unconjugated. When the four isomers were injected separately into Wistar rats, all were rapidly excreted at equal rates;[465] complete conjugation of bilirubin IX-α occurred, with partial conjugation of the γ (some 50%), β (16%), and δ (16%) isomers. When injected into Gunn-strain rats which lack transferase activity towards bilirubin-IX-α, only the non-α isomers were excreted. They were excreted unconjugated and at rates comparable with rates in Wistar rats.[465]

Conjugation, therefore, appears necessary before the IX-α isomer is excreted. This may be due to the ready formation of multiple internal hydrogen bonds in the IX-α isomer, shielding the polar character of the carboxyl groups at the NH—CO groups of the "outer" pyrrolenone rings. These bonds cannot be formed with the non-α-isomers.[465] Conjugation of the IX-α ring would stretch its "ridge-tile" structure[16,17] and break the hydrogen bonds, accounting for the water solubility of the conjugate and for its bilary excretion.[16]

Transferase activity accepting the IX-α isomer may accept the other three isomers, as the Gunn rat appears equally unable to conjugate all four.[465] Also, the β-, γ-, and δ-isomers were conjugated on one specific propionic acid group, which suggests an enzymic conjugation.[465]

B. Mono- and Diglucuronides of Bilirubin

Early work on bilirubin conjugation has been frequently reviewed (e.g., Reference 466). Independently, Billing et al.,[467] Schmid,[468] and Talafant[469] provided evidence that the "direct-acting" bile pigment in the classical Malloy and Evelyn diazotization assay was the glucuronide of bilirubin. The glucuronic acid was linked to the propionic acid groups.[461] The existence of a monoglucuronide was long debated. It was held by some to be a complex of free bilirubin and diglucuronide.[470,471] Although dipyrrole exchange can result in monoglucuronide formation in vitro from bilirubin and its diglucuronide,[363,472] this mechanism is not considered to contribute significantly to the demonstrable formation of the monoglucuronide in vivo.[473,474]

The diglucuronide and monoglucuronide vary between species in their proportions in bile (see Reference 475). The diglucuronide is the predominant bilirubin conjugate in rat and human bile,[467,476-478] but only the monoglucuronide was formed when bilirubin was incubated with rat- or human-liver microsomes and UDPglucuronic acid.[185,274,479] However, some workers[136,275] found both mono- and diconjugates. Assay interpretation, difficult with bilirubin, may have been responsible for the differences, but the report that "smooth" microsomes from rat formed the monoglucuronide, whereas "rough" or "total" microsomes formed the diglucuronide,[136] is interesting because Jansen[363,480] could separate the two postulated steps and confirm their sequential participation in diglucuronide formation. He found rat-liver microsomes preferentially made monoconjugate. As cat bile largely contained diglucuronide, he incubated monoconjugate with UDPGlcUA and solubilized transferase from cat liver, to form diconjugate. At that time,[480] he postulated two enzymes or two active sites on the transferase, from the differing solubilities of the two substrates,[111] the different "microsomal" origin[136] and stability of the "enzymes" responsible, and their different pH optima in cat-liver microsomes. Overall formation of the diglucuronide exhibited an optimum of pH 8.1, that of the second step pH 6.5. 4-Nitrophenol inhibited the second step competitively, not the first. Optimal pH of nonenzymic conversion of diglucuronide from two molecules of monoglucuronide was also 6.5,[472] but was not responsible for the second step studied in cat-liver microsomes because it produced ring structures $III\alpha$ and $XIII\alpha$, not the natural $IX\alpha$ from enzymic incubation.

Fevery et al.[462] suggested that an unknown factor present in the rough endoplasmic reticulum was necessary for diconjugate formation and was deficient in those species, e.g., ox, pig, sheep, rabbit, guinea pig, mouse, chicken,[475] whose bile contains proportionately less diglucuronide.

Recent work by Jansen et al.[125] has provided evidence of quite different mechanisms for the two stages of bilirubin glucuronidation in rat, where the unknown "factor" appears to be an enzyme distinct from UDPglucuronyltransferase. These authors confirmed earlier suggestions of the differential subcellular distribution of the two stages.[136] The first stage was in the centrifugal fractions P_3 and P_4. The second stage, although found in these fractions, was concentrated in fraction N_2, characterized by high 5-nucleotidase and phosphodiesterase I activities. Electron microscopy showed P_4 to contain predominantly ribosome-studded vesicles and N_2 to contain only smooth membranes without ribosomes.[125]

Optimal pH value for the second stage was 6.7, that for the first, 8.1.[125] The term "stage", not used by the authors, is misleading, for in formation of diglucuronide a second molecule of glucuronic acid is not added to every molecule of monoglucuronide. Stoichiometry indicated that 2 mol of bilirubin-monoglucuronide were converted to 1 mol diglucuronide and 1 mol free bilirubin; free bilirubin was extracted in confirmation.[125] Products of Stage 2 were the naturally occurring $IX-\alpha$ isomeric form, and the process would, therefore, seem enzymic, as supported by its occurrence in a specific subcellular fraction of a specific tissue, its enhancement on washing the fraction, and its thermolability.[113] The enzyme is not UDPglucuronyltransferase for UDPGlcUA does not enhance the reaction rate,[125] is even slightly inhibitory,[125] and UDPGlcUA labeled in the glucuronyl moiety did not transfer label to the bilirubin diglucuronide formed in its presence.[125] Nor is the enzyme likely to be β-glucuronidase, for its subcellular distribution is different and it is not inhibited by glucaro-1,4 lactone.[125]

Jansen et al.[125] believe that the N_2 fraction is a mixture of bile canalicular, contiguous, and blood-sinusoidal surface membranes. Membranes of erythrocytes, for example, do not contain the Stage 2 enzyme.[125] The diglucuronide appears to be formed just before excretion into the bile. The monoglucuronide is found in blood when its rate of formation is thought to exceed the liver cell's excretory capacity, the diglucuronide only when the common bile duct is obstructed. The authors conclude that Stage 2 occurs in the plasma-membrane fraction of rat liver and that biliary excretion of bilirubin and the formation of its diglucuronide are related.[125]

In discussion,[481] Lathe points out that bilirubin monoglucuronide is not one molecule, but two asymmetric molecules, of which only one might be specific; and, whereas Arias[481] and Jansen et al.[481a] note that if bilirubin monoglucuronide is infused into Gunn rats or their isolated liver, some 19% diglucuronide is excreted (infused bilirubin forming no conjugate whatever), Javitt[481] points out that bilirubin monoglucuronide is a different substrate to bilirubin, and Gunn rat might possess UDPglucuronyltransferase activity to the monoglucuronide, as it does to phenylacetic acid.

This new enzyme is quoted[481a] as now being obtained pure, and work is under way to determine whether it is a transferase or an isomerase. If a transferase, UDPGlcUA is not required, and neither 4-nitrophenol nor phenolphthalein appears an acceptor (P. L. M. Jansen, personal communication, 1978). If an isomerase, it directs isomerization to the IXα form, not to the random pattern seen in vitro.[472] However, Jansen et al.[481a] point out that cat liver does need UDPGlcUA to form the diglucuronide from monoglucuronide, forming it mole for mole. Gunn-rat liver needs 2 mol of monoglucuronide to form 1 mol of the diglucuronide. The situation in Wister-rat liver is difficult to ascertain. The plasma-membrane transferase may only be important, or exist, in rat and certain other animals.

Recent work by Fevery et al.[482] throws doubt on this mechanism in man. When UDPglucuronyltransferase to bilirubin is lower than normal (e.g., in Gilbert's syndrome), the proportion of bilirubin diconjugate in the bile falls from 75 to 50%; and where the enzyme is virtually absent (e.g., in Crigler-Najjar syndrome), conjugated bilirubin exists solely as monoglucuronide.

Further, in hemolytic jaundice, the proportion of diglucuronide remained normal when the transferase was normal, and fell when transferase fell.[482] In Wistar rats, clofibrate or phenobarbital treatment raised liver UDPglucuronyltransferase activity to bilirubin and at the same time increased the proportion of diglucuronide over monoglucuronide in the bile (Fevery, Kutz and Steenbergen, unpublished results; J. Fevery, personal communication). Early work[275] suggested diglucuronidation proportionately increasing with neonatal age. All this suggests a direct relationship between measured transferase activity and the excretion of bilirubin diglucuronide.

Chowdhury et al.[483] report that E_2 at normal levels converts some 17% mono- to diglucuronide. Fevery (personal communication) notes that this level of conversion is found (N. Blanckaert, unpublished results) in vitro (10 to 12%) as well as in rat (10 to 20%). The problem is obviously far from being settled.

Another outstanding problem is the predominant secretion of diglucuronide into bile by some species (e.g., rat), whereas the liver microsomes synthesized only monoglucuronide.[475] Even with "physiological" concentratios of UDPGlcUA and the endogenous effector UDPGlcNAc, and with untreated microsomes, the monoglucuronide is still the conjugate formed (M. T. Campbell, unpublished results). To bridge this whole-animal-microsomal gap, Campbell and Dutton[483b] studied the conjugation of bilirubin by rat-liver slices. Virtually only monoglucuronide appeared in the medium, as with microsomes; diglucuronide was formed, but not secreted, existing within the slice at levels approaching that found in bile. A factor concerned with secretion from the slice seems to be responsible. The problem invites liver perfusion.

Wolkoff et al.[483a] believe that in rat the intrahepatic pool is mostly bilirubin monoglucuronide, the diglucuronide in bile being formed from the canalicular pool. As assay in bile is subject to overestimation of monoconjugates by methods utilizing ethyl anthranilate and overestimation of free bilirubin by solvent-partition methods,[484] considerable caution is required.

The mixed glucuronyl-glucosyl or -xylosyl conjugates of bilirubin in dog bile[485] suggests that in dog, at least, sugars are added consecutively to the aglycon, and this could equally happen if both sugars were glucuronic acid, i.e., in sequential diglucuronidation by UDPglucuronyltransferase.

C. Properties of Bilirubin UDPglucuronyltransferase

Reports on the properties of the "enzyme glucuronidating bilirubin" must from evidence of the preceding section be somewhat suspect. Dependent on species, the microsomal fraction will form either the monoglucuronide (e.g., man and rat) or both mono- and diglucuronide (e.g., cat); and the former has little direct relevance to the situation in vivo.

Transferase activity towards bilirubin will not, therefore, be separately discussed further. References occur throughout this work and a summarized presentation of findings up to 1975[462] and a more detailed review[111] are available. Relevant papers are referred to there, including several extensive studies,[185,276,289,321] to be evaluated in light of the above remarks and of the critical review of assay methods by Heirwegh et al.[460] A more rapid modification of the classical procedure of van Roy and Heirwegh[313,460] has been published,[485] which separates the bilirubin isomers and conjugates from bile and diazotizes, extracts, and chromatographs them within 6 to 8 hr. Transferase assay for bilirubin is described in Chapter 17.

D. Other Conjugates of Bilirubin

1. Simple Conjugates

Although glucuronidation is the major pathway, significant conjugation of bilirubin with D-glucose and D-xylose occurs. The ratio of conjugation with glucuronic acid, glucose, and xylose is 1.00:0.05:0.03 in human bile and 1.00:0.04:0.07 in activated human-liver microsomes exposed to saturating amounts of the UDP-sugar. Further details are given by Fevery et al.[462] and in Chapter 16, Section II.

2. Mixed Conjugates

Because of sensitivity of bilirubin and its conjugates to light, oxygen, and unphysiological pH values, most investigations employed diazotization. Separations achieved have been of azopigments, allowing opportunities for manipulative artifacts. One recent controversy is settled, that of the physiological significance of several mixed conjugates of bilirubin postulated from study of the azopigments.

Heirwegh et al.[186] noted that a number of complex conjugates appeared in human T-tube bile in addition to glucoside and xyloside conjugates. At the same time, Kuenzle provoked controversy by[487-489] presenting chemical and physical evidence of conjugates from human T-tube bile of bilirubin with aldobiouronic acid, with pseudoaldobiouronic acid, and with a hexosyluronic acid containing a branched-chain hexuronic acid. He even suggested[490] that bilirubin was never conjugated solely with glucuronic acid, and so-called "bilirubin glucuronide" was an artifact. Lathe[461] and Fevery et al.[476] pointed out that Kuenzle's recovery was only some 5% of his starting material and that its source was abnormal, being stored, post-operative, T-tube bile from cholecystomectomized patients. Fevery et al.[476] compared normal human bile and bile from patients with mechanical biliary obstruction and also the bile from normal rats and from rats with experimentally obstructed bile ducts. In both species, the bulk of bile pigment was diglucuronide, with monoglucuronide and a small amount of xyloside and glucoside. In both species, however, obstruction of the duct changed this pattern so that a heterogeneous collection of complex azopigments appeared.[476] Noir[491] also reported unidentified pigments in cholestatic biles which resisted β-glucuronidase, although containing hexuronic acid. Kuenzle[492] recently briefly reported further work on his original preparation, and discussed his interpretation.[493]

Further work by the Leuven group[450] involving sophisticated chemical procedures with fresh normal bile, stored normal bile (37°C under N_2 for 1 to 6 hr), and cholestatic bile of man and rat, showed that fresh bile yielded homogeneous azoderivatives, but those from the stored and cholestatic biles separated into four main bands on thin-layer chromatography (TLC). These bands appeared due to sequential migration of the 1-acyl-aglycon to the 2-, 3-, and 4-positions of glucuronic acid. The conversions also occurred in media buffered at pH 7 to 9, and took place with isolated bilirubin glucuronides as well as with their azopigment derivatives.[450,493a]

The "branched-chain" structure of Kuenzle appeared to be incompletely silylated forms of glucuronolactone and glucuronic acid.[493b] Kuenzle's own samples were used in much of this work.[493b]

Gordon et al.,[478] using a slightly different isolation procedure, also failed to find any of the "Kuenzle" conjugates in normal human bile taken from duodenal juice.

Many of the complex bilirubin conjugates of cholestatic bile are therefore formed nonenzymically after excretion into the bile duct, and possibly, even before such excretion; extreme care is needed during work with bilirubin derivatives at every stage.

In dog bile, but not in human or monkey bile, diconjugates of bilirubin IXα with glucuronic acid and glucose, and with glucuronic acid and xylose, have recently been reported in small amounts.[485]

E. Recent Work with Biliverdin as Substrate

Biliverdin exists conjugated in green bile, and its separation and investigation have been described.[494] Green post-mortem human bile contains both mono- and diglucuronides of biliverdin.[494] Glucuronic acid, identified chromatographically by the naphthoresorcinol test, appeared conjugated through an alkali-labile ester linkage.[494] The same compounds were readily obtained by oxidation of the bilirubin glucuronides.[494] There is no report that biliverdin is a substrate of UDPglucuronyltransferase; it seems not, in chick and mouse liver (J. E. A. Leakey, unpublished results). Biliverdin glucuronides in mammalian green biles may, therefore, arise through storage in the presence of oxidative conditions in vivo or in vitro,[494a] although a greenish pigment exists in fresh rabbit bile,[475] identified as biliverdin glucuronide.[495] Rabbit should be studied for a biliverdin UDPglucuronyltransferase. Biliverdin glucuronide occurs in snake bile. According to some,[496] biliverdin represents 94% of biliary bile pigment in the chicken as aginst 6% from bilirubin, but others[495] find that conjugates there and in turkey bile resemble bile acid complexes of sodium biliverdinate.

IV. ADDITIONAL NOTES

Section III.B: The bilirubin glucuronoside glucuronyltransferase (EC 2.4.1.95) of Jansen et al.,[125] has now been purified 2000-fold and shown to be a tetramer of four 35,000-dalton subunits.[495a] Its relevance in vivo is still strongly contested, e.g., Reference 495b.

Chapter 8

ENZYMIC MECHANISM OF GLUCURONYL TRANSFERENCE BY UDPGLUCURONYLTRANSFERASE

I. INVERSION OF GLYCOSYL LINK

UDPGlcUA has an α-glycosidic link, but only β-glucuronides are formed by UDP-glucuronyltransferase. Inversion of the link must occur during the reaction. Axelrod et al.[329b] pointed out that, as amino-, hydroxyl-, and carboxyl-groups were all acceptors in the reaction, glucuronides could be formed by nucleophilic substitution in which the electron-donating aglycon displaced UDP from the opposite side of the electrophilic anomeric carbon atom of UDPGlcUA. This makes inversion of the α-link in the nucleotide to a β-link in the glucuronide understandable. A similar inversion takes place in the formation of glucosides by UDPglucosyltransferase.

Bedford et al.[496a] have studied the glycosylating reactivity of UDPGlcUA and UDPglucose. They find no spontaneous reaction and consider that special attributes of the enzyme constrain the substrate to attack exclusively the C-1 of the glycoyl moiety at physiological pH.

II. REVERSIBILITY OF REACTION

Irreversibility with 2-aminophenol,[93] has been confirmed,[497] that with phenolphthalein[497] has not (G.J. Wishart, unpublished work), and that with tetrahydrocortisone is doubtful.[498] Analogy with the cycling of PAP $\rightleftharpoons$ PAPS in which nitrophenyl sulfates (possessing group potentials approaching that of PAPS) acted as sulfate donors to PAP, suggested[4,114] that nitrophenyl glucuronides would be better substrates for a reverse reaction. This was demonstrated in bovine-liver microsomes,[116-118] using 4-nitrophenyl glucuronide and UDP as substrates, with glucarolactone present to minimize β-glucuronidase activity, and extended to 1-naphthyl glucuronide[177] and estrone glucuronide.[232] A condition for demonstrating the reverse reaction in vitro is the addition of UDP before liberation of the aglycon from its glucuronide can be observed (see Chapter 17). Despite earlier reports,[116] both forward and reverse reactions are activated by membrane perturbants,[118,177] but constraint on the reverse reaction is greater,[163] which could minimize loss in vivo of the glucuronide already formed; for the physiological role of the reverse reaction and its inherent "transglucuronylation", see Chapter 3, Section II.C. Even under optimal conditions in vitro the reverse pathway is only some 10% of the forward reaction with 1-naphthol as aglycon.[177] With estrone as substrate, but not necessarily optimally, only some 0.4% of the forward reaction was reversible.[232]

III. NATURE OF THE REACTION AT THE ENZYME

From previous chapters, work so far claiming to study the reaction at the enzyme active center must be viewed with caution, if not scepticism. With unsolubilized preparations results are susceptible, if compartmentation is assumed, to factors deriving from diffusion or from permease action, especially with the nonpolar substrates; if compartmentation is denied, they are susceptible to factors deriving from sequestration of the more lipid-soluble substrates in the microsomal lipids. Use of the "solubilized" enzyme is still unreliable, for the remaining degree of membrane, phospholipid, or lipoprotein attachment is unknown and must vary from procedure to procedure. Only work on the purified enzyme can pronounce on the chemical mechanism responsible for catalysis. We may then see how far phenomena such as the bent Lineweaver-Burk plots and substrate and product inhibition and/or activation are a feature of the active center itself and not of its nonenzymic environment. Publication of such work appears (in early 1978) imminent.

Understandably, much has been published on the reaction mechanism using crude preparations, and may be briefly reviewed.

Thiol groups are not now thought part of the active site of the transferase, but are involved in its constraint, existing elsewhere on the enzyme or on an adjacent protein (Chapter 5, Section II.I.).

Concave Lineweaver-Burk plots for fresh UDPglucuronyltransferase at low ($<$ 2.5 mM) UDPGlcUA concentrations were noted with 4-nitrophenol,[101,190,192] 2-aminobenzoic acid,[190] 4-nitrothiophenol,[206] tetrahydrocortisol,[318] and in some conditions, with 2-aminophenol.[212] One explanation could be negative cooperativity, the binding of the first nucleotide molecule hindering the subsequent binding of others, but variation with the different aglycons used is difficult to account for. This variation also lessens the possibility of saturation of a transport process for UDPGlcUA in the membrane, first proposed by Winsnes[101,186] and developed by others.[102]

Bisubstrate kinetic analysis is essential,[163] but complicated by positive and negative modulation of both products and substrates,[169] and earlier work[117,289] did not sufficiently allow for this. One study[289] used bilirubin, an unfortunate choice for kinetics because of its characteristics as an activator,[293] its probable sequestration in the microsomes,[170] and its diglucuronidation. Despite further commitment to extrapolation, this study nevertheless considered the enzymic mechanism to be Bi-Bi sequential, not Ping-Pong, and ordered, not random.[289] Work with 4-nitrophenol[117] suggested a rapid-equilibrium random-order reaction, an ordered mechanism of the Theorell-Chance type being excluded by isotope experiments possibly too ingenious for the conditions employed. The analysis of results obtained with probably partially activated microsomes from pig kidney (estrone as substrate)[232] suggested that an iso-Theorell-Chance mechanism took place. Estrone was added first, followed by UDPGlcUA, with the release of estrone glucuronide and then of UDP. A central complex of enzyme-substrates and enzyme-products was not likely to exist.[232] For both forward and reverse reactions, it was concluded[232] that increasing concentrations of one substrate favored the binding of the other to the enzyme. As apparent K_m values for the products of the forward reaction were much larger than for the reactants, and as V_{max} for the reverse was a small fraction of that for the forward reaction, the reverse reaction with this substrate may be of minor importance physiologically. However, the assay did not try to reproduce in vivo concentrations.

Isotope exchange was observed between labeled 4-nitrophenol and certain glucuronides separable in the aqueous phase, especially 4-nitrophenyl glucuronide itself;[348] it occurred on incubation with microsomes or partly purified enzyme, UDPGlcUA being apparently absent and β-glucuronidase being inhibited by glucarolactone. One explanation[244,348] is a "substitution" enzyme reaction. However, more evidence is needed.

Separate binding sites for UDPGlcUA and for aglycon at the active center and the possibility of allosteric sites have been discussed (Chapter 5) and are reviewed from one point of view by Zakim and Vessey.[163] There is evidence, for example, that UDPGlcNAc, which does not compete with UDPGlcUA in the forward reaction, inhibits the reverse reaction competitively towards UDP by an "allosteric" effect.[280]

If conjugates containing the negatively charged glucuronic or sulfuric acid residues are bound to the UDPGlcUA site, then, because disulfate has a higher affinity for the enzyme than monosulfate, the UDPGlcUA binding site must be positively charged.[299] Moreover, the lipophilic aglycon moiety is needed as well as the glucuronic acid moiety because inorganic sulfate[90,299] or glucuronate[90,311] do not compete. Competition of the product glucuronide with UDPGlcUA[298] suggests either[244] that the glucuronide is synthesized at, and released from, the UDPGlcUA binding site, or that the two compounds hinder each other's transport to or from the enzyme.

Much evidence suggests that glucuronyl donor and acceptor do not compete with each other.[96,287,298] Affinity of the transferase for the aglycon is lessened, but not abolished, by the transfer of the glucuronyl residue to the aglycon.[298]

Multiple binding sites for 2-aminophenol and 4-nitrophenol on the rat-liver transferase protein are suggested by a series of activity optima at high and low pH values[284] (Chapter 5, Section II.P), but environmental effects cannot be ruled out.

Several approaches have studied relative affinities of structurally similar aglycon substrates, but with little progress. Both electron-releasing and electron-withdrawing ring substituents increased glucuronidation rate of the phenolic ring.[402] Bulky substituents at the 2- or 6-position hinder conjugation at the C-1 hydroxyl, probably sterically; substitution at C-2 does not necessarily cause steric hindrance. Phenols poorly soluble in n-octanol at pH 7.4 are poor substrates or not substrates.[402] Details are given in Chapter 7, Section II.D.1. An N-alkyl group may be important in binding a compound to the aglycon site of the transferase responsible for glucuronidating morphine.[335,499] Stereoselectivity was not observed, and the pK_a value and lipid solubility of the aglycons appeared of secondary importance.

The effect of the phospholipid component of UDPglucuronyltransferase, or of the lipids in the surrounding membrane, in correctly aligning a lipid-soluble substrate to the active center is implied,[170] but not substantiated.

Eletr et al.[199] (see also Reference 169) reported a preliminary temperature-dependence and spin-label study on largely latent transferase activity to 4-nitrophenol, finding discontinuities in Arrhenius plots at 19°C and at 32°C, when they noted breaks in fluidity of the microsomal matrix. In disrupted microsomes, these breaks appeared only at 32°C, and treatment with phospholipase A prevented them. As apparent activation energy was higher above 19°C than below, transferase activity seemed related to fluidity of the matrix.[199] Subsequently,[198] a break in Arrhenius plots was reported at 16°C, with data suggesting that activation energy was now less when above the transition point than when below it. The authors considered that membrane lipids underwent a crystalline/liquid crystalline phase transition at 16°C. Changes in the enzyme's regulatory characteristics were also claimed at 16°C (Chapter 5, Section II.B). At high temperatures (above 26°C) where thermotropic changes in the phospholipid bilayer are not usually observed, no breaks were found;[202] membrane perturbants lowered activation energy from 84 kJ/mol for UDPGlcUA and 78 kJ/mol for 4-nitrophenol to 45 kJ for each substrate.[202]

Pechey et al.[200] studied temperature dependence of transferase activity and the associated structural changes of the membrane in more detail. Activation energy in largely latent preparations was 56 to 63 kJ/mol. Agreeing with Eletr et al.,[199] they found Arrhenius plots of nonlatent activity to 4-nitrophenol linear from 5 to 40°C, whereas largely latent preparations showed a break at 20 to 25°C associated with changes in membrane structure.[200] They did not report finding the second break at 32°C. They observed that perturbants activated latent transferase at all temperatures between 5 and 50°C, and that the thermal transition was rapidly reversible.[200] Activation energy at temperatures above the transition point was greater than below it, but perturbation of the membrane greatly reduced this difference.[200] They consider that activation energy values are characteristic of a rate-limiting permeation process. From studies with fluorescent probes,[200] they conclude that the sharp transition point reflects a thermotropic increase in substrate permeation. Phase separation or clustering of phospholipids may occur, facilitating access of substrate to the enzyme in "native" microsomes.[200] The crystalline/liquid crystalline transition previously suggested[198] is unlikely at such a high temperature in membranes with highly unsaturated phospholipids.[200] This thermotropically increased permeability to substrate in latent preparations would be small compared with that engendered by activating procedures.[200] Pechey et al.[200] consider that the activation energy values found with the activated enzyme could reflect the catalytic reaction itself.

Chapter 9

METABOLIC PATHWAYS IMMEDIATELY PRECEDING AND SUCCEEDING GLUCURONIDATION — ANABOLIC PATHWAYS

I. TISSUE UDPGLUCURONIC ACID

A. Structure of UDPGlucuronic Acid

Figure 4 gives the structure of UDPGlcUA, as deduced from behavior of the molecule isolated from mammalian liver[4,90,93] and confirmed by its biosynthesis (Section II below) and chemical synthesis.[63,97] UDPGlcUA with a β-glycosidic link has been synthesized chemically and possessed no activity with UDPglucuronyltransferase.[97] Isolation, characteristics and some properties of the α-linked form (henceforth the only one referred to here) were compiled earlier.[4] Pseudo-uridine diphosphate glucuronic acid has been chemically synthesized.[98] It exhibited no trace of activity with UDPglucuronyltransferase[98] even though pseudo-uridine diphosphate glucose participates slowly in glucosyl transfer.[98]

B. Assay of UDPGlucuronic Acid

Two approaches measure this nucleotide. One separates it chromatographically from a tissue extract and so quantitates it physically or chemically. The other estimates it enzymically as a substrate for UDPglucuronyltransferase. The latter procedure (see also Reference 4) is described in Chapter 17. Although requiring careful controls because of possible concurrent activation (e.g.., with endogenous UDPGlcNAc) or inhibition (e.g., with competing endogenous aglycons), it has been widely used[177,500-502] because of its simplicity and rapidity for comparative studies.

Chromatography of UDPGlcUA on paper has been described.[4] More recent reports use paper or t.l.c. with similar solvents[189,503] or a Dowex-1 ® formate column, followed by gas chromatography,[507-510,518] the latter using double labeling.

C. Distribution of UDPGlucuronic Acid among the Tissues

Chromatographic separation is usual, with liver appearing the major source. Several laboratories confirm the levels of the nucleotide in this tissue, (e.g., or as μmol g^{-1} wet liver), adult guinea pig, 0.4;[504,511] rat, 0.3.[177,189,504,511a] Similar levels were reported earlier.[501,512]

The role of UDPGlcUA in biosynthesis of glycans and ascorbic acid has stimulated studies of its distribution. Tissues containing UDPGlcUA and the degree of the evidence offered were listed earlier.[4] They included liver, kidney, gastrointestinal tract, skin, cartilage, placenta (trace), uterus, milk and colostrum, and certain tumors. The brain gave one positive and two negative findings, but has since been found positive in several species.[513] No contradiction has yet appeared for the other negative findings,[4] i.e., muscle and cockroach fat body. Recent additional tissues reported negative are spleen, pancreas, testis, and lung (trace) of rat, pancreas of guinea pig, pituitary gland (both lobes) of calf, and the isthmus and magnum of hen oviduct.[506] As UDPglucose and UDPGlcNAc were found present at quite high levels in these tissues,[506] the absence, or a very low level, of the nucleotide is likely. Methods of detection are insufficiently sensitive to distinguish "trace present" from "absent". As perfused lung forms little glucuronide (Chapter 14), the "trace" in lung recorded above may limit the transferase activity found there.

Species also were listed earlier.[4] It would be surprising to find any species in which glucuronic acid occurs in a simple or in a polysaccharide conjugate, without the nu-

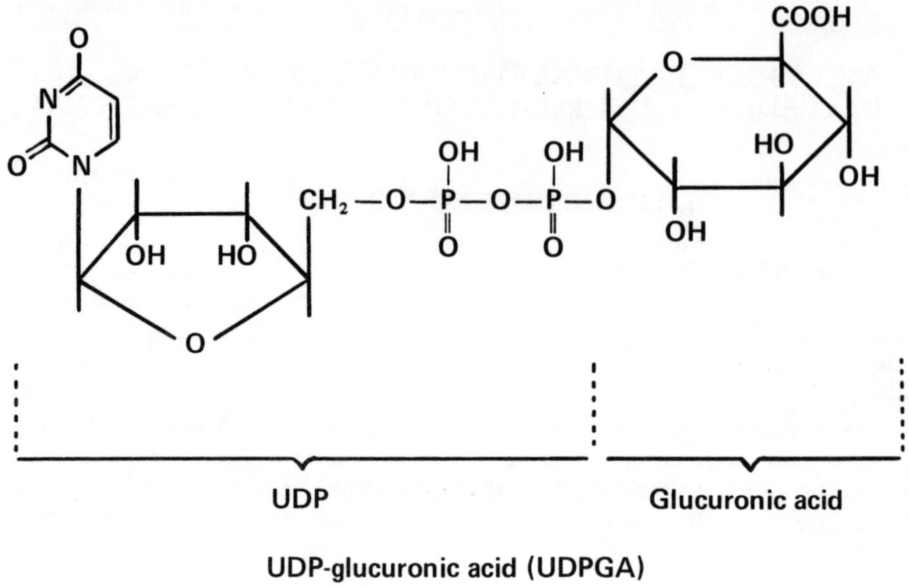

FIGURE 4.   Uridine diphosphate α-D-glucuronic acid.

cleotide being present. Plants and bacteria, therefore, also occurred in the list,[4] and we need not detail species investigated then or subsequently. Zhivkov et al.[506] published a well-documented comparative list. UDPGlcUA levels in liver varied from 0.41 μmol g⁻¹ wet weight in guinea pig to 0.02 in carp, but no species pattern could be observed except that birds, fish, and amphibia examined possessed less (0.02 to 0.12) than mammals (0.12 to 0.41),[506] a trend supporting observations[502] with the enzymic assay of UDPGlcUA.

Low levels of tissue UDPGlcUA could indicate rapid metabolic utilization rather than lack of importance. For example, rat liver has less of this nucleotide than guinea pig liver, yet turnover experiments[504,505] suggest more rapid synthesis in rat liver, consistent with greater activity of the hydrolyzing enzyme UDPGlcUA pyrophosphatase in rat than in guinea pig liver (Section III.B.2 below).

How far these measured levels per gram of tissue reflect availability of UDPGlcUA for glucuronidation is debatable. Not only latency of the transferase, but also physical and metabolic compartmentation of the nucleotide must be considered. However, homogenates of hepatoma cells, whose transferase is not latent in vivo, exhibited the same rate of glucuronidation as intact cells when supplied with UDPGlcUA at the concentration calculated from the liver extracts.[514] This suggests that these levels are physiologically valid and any compartmentation of UDPGlcUA not seriously limiting.

II. MECHANISMS OF BIOSYNTHESIS OF UDPGLUCURONIC ACID

A. General

Pathways of biosynthesis of UDPGlcUA (Figure 5) will be discussed separately. General factors affecting their activity, e.g., age, pretreatment with drugs, etc., are best treated under the effect of these factors on overall glucuronidation. UDPGlcUA production and fate in plant tissues is a specialized aspect not wholly relevant, being largely concerned with polysaccharide metabolism. Recent papers by Dalessandro and Northcote[515,516] give useful references to that subject. The most significant pathway of UDPGlcUA formation for drug metabolism in vertebrates appears to be that involving the oxidation of UDPglucose by UDPglucose dehydrogenase and NAD.

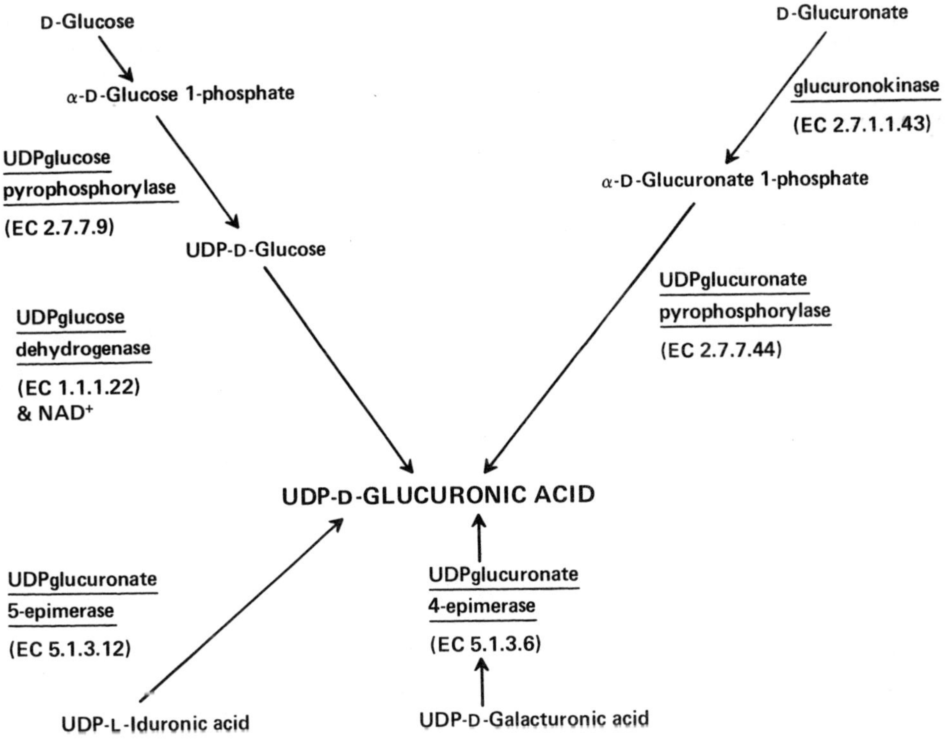

FIGURE 5.   Biosynthesis of UDPglucuronic acid.

B. Biosynthesis of UDPGlucuronic Acid by UDPGlucose Dehydrogenase

1. Enzymic Mechanism of UDPGlucose Dehydrogenase (EC 1.1.1.22, UDPglucose: NAD⁺ 6-Oxidoreductase)

Strominger et al.[517] found that microsomes and high-speed supernatant from calf-liver homogenate synthesized glucuronides in the presence of the aglycon, UDPglucose and NAD. The supernatant itself formed, from UDPglucose and NAD, an intermediate active in microsomal glucuronyl transference, shown to be UDPGlcUA. For each mole of UDPglucose, 1.95 mol of NAD were reduced. No reversal could be demonstrated:

$$\text{UDPglucose} + 2\text{NAD}^+ \rightarrow \text{UDPglucuronic acid} + 2\text{NADH}$$

Deamino and acetylpyridine analogs could replace NAD, but NADP could not.[96,517] For each 2 mol of NAD reduced, 0.7 to 0.9 mol UDPGlcUA were obtained.[517]

When UDPGlcUA was formed from UDPglucose in $H_2{}^{18}O$, the UDPGlcUA contained ^{18}O, but UDPxylose prepared from it did not; ^{18}O from H_2O is therefore incorporated into the carboxyl group of UDPGlcUA.

UDPGlucose dehydrogenase is unusual in catalyzing a four-electron oxidation. This oxidation was suggested[519] to be the sum of two reactions, the first oxidizing UDPglucose to a compound X-UDPglucose at the expense of one NAD⁺, the second oxidizing X-UDPglucose to UDPGlcUA at the expense of the second NAD⁺, and each of them being the common two-electron type. Early workers, finding no evidence of an aldehydic intermediate[517,520] or of a lactone, suggested that any intermediate was bound to the enzyme.[517]

Nelsestuen and Kirkwood[521] enzymically synthesized the possible intermediate

UDPglucose-6-aldehyde and demonstrated it as a substrate for the purified enzyme from bovine liver. They suggest the two-step mechanism to be reversible in the first step and demonstrated the irreversibility of the second step, probably rate limiting.[521] More recent work[522-524] suggests two reversible oxidation steps, with the hydrolysis of a thioester intermediate conferring irreversibility. Each step may occur on a different enzyme subunit.

Both hydrogen atoms on C-6 are transferred to NAD⁺.[518] From experiments with the 5-fluoro-6-azo[525] and 5,6-dihydrouracil[526] analogs of UDPglucose, the uridine moiety does not appear to participate.

Other work with substrate analogs,[527] suggests that two sites, the C-3 hydroxyl of the D-glucose and the −NH of the pyrimidine, are essential for specificity, for their modification incurs loss of activity. Mannose or galactose derivatives are not substrates because their C-2 or C-4 substituents block the correct enzyme-substrate interactions.[527] The thio analog of UDPglucose is a potent inhibitor.[528]

Recent work on the enzymic mechanism[529] concludes that the enzyme subunits are related in couples, and the glucosyl moiety of UDPglucose is flanked on one side by NAD⁺ and on the other by an essential thiol group.

A highly purified enzyme has been prepared at 80% yield after sequential chromatography.[530] The enzyme from bovine liver has a molecular weight of some 300,000,[531] with six subunits of 52,000.[532] Amino-acid analysis and peptide mapping[533] revealed only one type of N-terminal amino acid, methionine, and indicated that the subunits are similar if not identical. Thiol groups are present on UDPglucose dehydrogenase,[534,535] and on the hexameric enzyme, the two rapidly reacting groups are protected from 5,5′-dithiobis-(2-nitrobenzoate) by NAD⁺, UDPglucose, and UDPxylose.[533] Six thiol groups maintain catalytic activity.[533] UDPXylose, shown to be a specific inhibitor,[536] exerts a cooperative homotropic effect at the UDPglucose-binding site.[533,537,538] Allosteric binding of UDPxylose may also occur,[539] and the type of interaction depends on relative concentrations of the ligands.[539]

Control of UDPglucose dehydrogenase by UDPxylose and other factors during the biosynthesis of connective tissue has been discussed[503,509,540] and the enzyme mechanism is further explored.[524] These references report other sources of information. For assay, see Chapter 17.

2. Distribution of UDPGlucose Dehydrogenase

Histochemical demonstration is referred to in Chapter 17.

Distribution of UDPglucose dehydrogenase follows that of UDPGlcUA, but not necessarily quantitatively because of the frequent occurrence of the specific inhibitor UDPxylose. Generally, the amount of glucuronic acid in tissue polymers varies inversely to concentration of UDPxylose,[503] but sometimes, as in sheep nasal septum,[509] the UDPxylose present is sufficient theoretically to have prevented the turnover of UDPGlcUA observed in the tissue. However, compartmentation of UDPxylose seems unlikely. This and the associated problem of inhibition by the product UDPGlcUA is discussed by Gainey and Phelps.[502]

Intracellularly, distribution of the dehydrogenase appears entirely cytoplasmic in rat liver,[502] and work with other tissues supports this.[4]

UDPGlucose dehydrogenase has been found in all mammalian and avian tissues known to contain UDPGlcUA or to incorporate glucuronic acid into simple conjugates or glycosaminoglycans. Early sources are given by Dutton,[4] and include human placenta.[541,542] The same principle holds for amphibia, plants, and bacteria.[4] The problem in fish,[4] where UDPglucose (in one instance) and the dehydrogenase (in several others) were not found, requires investigation. So far, insects do not appear to be a source.

The rate of synthesis of UDPGlcUA, UDPglucose, and UDPGlcNAc in various mammalian tissues, the former presumably by the UDPglucose dehydrogenase pathway, has been examined by chromatography of radioactive UDPsugars after injection of 1-^{14}C-glucose.[505] In all rat tissues examined, synthesis of UDPglucose exceeded that of the other two UDPsugars.[505] In liver and kidney, but not intestine, UDPGlcUA was formed faster than UDPGlcNAc.[505] In guinea pig tissues, with UDPGlcUA higher than in rat, its rate of formation was lower, indicating its more rapid turnover in rat than in guinea pig,[505] as suggested by rat's more active UDPGlcUA pyrophosphatase (Section III.B.2).

3. Preceding Step in this Pathway of UDPGlucuronic Acid Biosynthesis

This preceding step utilizes the cytoplasmic enzyme UDPglucose pyrophosphorylase (EC 2.7.7.9) catalyzing:

$$\text{Glucose 1-phosphate} + \text{UTP} \rightleftharpoons \text{UDPglucose} + \text{pyrophosphate}$$

Early work on the enzyme's role in glucuronidation has been outlined,[4] and reviewed.[543]

Linkage of the pathway with glucuronidation has been demonstrated in vitro[95] with glycogen and UTP, muscle phosphorylase to produce glucose-1-phosphate, *Zwischenferment* to produce UDPglucose, calf-liver UDPglucose dehydrogenase to produce UDPGlcUA, and subsequent incubation with liver microsomes and 2-aminophenol to produce 2 aminophenyl glucuronide.

C. Biosynthesis of UDPGlucuronic Acid by UDPGlucuronic Acid Pyrophosphorylase (EC 2.7.7.44, UTP: α-1 phospho-D-glucuronate uridyltransferase)

1. Mechanism

Pyrophosphorolysis of UDPGlcUA to UTP (and presumably glucuronic acid 1-phosphate) occurred in a preparation from mung-bean seeds.[544] Reversible formation of the nucleotide by this soluble pyrophosphorylase was then reported:[545]

$$\text{UTP} + \text{α-glucuronic acid 1-phosphate} \rightleftharpoons \text{UDPglucuronic}$$

$$\text{acid} + \text{pyrophosphate}$$

Product UDPGlcUA was isolated and identified,[545] radioactivity only appearing in it when the α reactant not the β, was labeled.[545] UTP, but not ATP, together with α-glucuronic acid 1-phosphate forms UDPGlcUA in mung-bean leaves also.[546] Roberts[547] partially purified the enzyme from barley seedlings and lists its properties.

2. Occurrence

UDPGlcUA is not synthesized from glucuronic acid 1-phosphate in mammals, or not significantly. Early evidence[4] indicated that glucuronate is not a direct precursor, and synthesis of UDPGlcUA from α- or β-glucuronic acid 1-phosphate and UTP was not demonstrable in mammalian liver[93,548,549] or cartilage.[509] Reports suggesting stimulation of glucuronidation by α-glucuronic acid 1-phosphate in fractionated liver homogenates[550] or by ethyl glucuronate in slices[551] are difficult to interpret and have not been confirmed. In chick embryo epiphysial cartilage, however, UDPGlcUA was reportedly formed from α-glucuronic acid 1-phosphate and, presumably, UTP.[552]

In fish, formation from α-glucuronic acid 1-phosphate could account for presence of UDPGlcUA in certain fish not possessing UDPglucose or UDPglucose dehydrogenase.[4] Liver of the salmon *Oncorhyncus kisutch* yielded UDPGlcUA heavily labeled from administered C^{14}-myoinositol, UDPglucose being virtually unlabeled;[553]

UDPGlcUA may have been formed here by pyrophosphorolysis from glucuronic acid 1-phosphate, which originates from inositol via free glucuronic acid,[553] but this fish is claimed to synthesize UDPGlcUA by UDPglucose dehydrogenase.[554] Possibly both routes of UDPGlcUA synthesis exist to different extents in fish.[4]

The enzyme is found in plants,[545-547] the carbon source often being myoinositol (see References 543 and 555). A recent review is by Loewus et al.[556] In some plants, e.g., barley seedlings, where UDPglucose dehydrogenase is very low, this route (myoinositol → glucuronate → α-glucuronate 1-phosphate → UDPglucuronic acid) may be the major one.[547]

3. Preceding Reactions

α-Glucuronic acid 1-phosphate itself is formed in plants from glucuronate and ATP[557] by glucuronokinase (EC 2.7.1.1.43, ATP: D-glucuronate 1-phosphotransferase). In mammals, α-glucuronic acid 1-phosphate is produced from UDPGlcUA in sheep nasal cartilage,[509]rat skin,[558] rat liver,[223] and rat kidney,[559] but as noted above (Section II.C.2), is not apparently utilized there for UDPGlcUA synthesis.

D. Biosynthesis of UDPGlucuronic Acid by UDPGlucuronate 5-Epimerase (EC 5.1.3.12)

This is the reverse, and presumably less important, direction of the reaction producing UDP-L-iduronic acid:

$$\text{UDP-D-glucuronic acid} \rightleftharpoons \text{UDP-L-iduronic acid}$$

NAD⁺ is necessary,[560] NADP⁺ less effective, and NAD.H inhibits.[560] The enzyme was originally found in rabbit skin.[560] No evidence of its role in supplying UDPGlcUA for conjugation of simple molecules has been encountered since the original work was reviewed earlier.[4] Conversion of D-glucuronic acid into L-iduronic acid during biosynthesis of heparin by microsomes from mouse mastocytoma incurs loss of the C-5 hydrogen.[561]

E. Biosynthesis of UDPGlucuronic Acid by UDPGlucuronate 4-Epimerase (EC 5.1.3.6)

This is the reverse direction of the reaction producing UDPgalacturonic acid:

$$\text{UDP-D-glucuronic acid} \rightleftharpoons \text{UDP-D-galacturonic acid}$$

NAD⁺ is necessary, at least in bacteria.[562] The epimerase is also found in various plants,[545,564,565] but like the 5-epimerase, its role in preceding biosynthesis of simple glucuronides is unknown and probably not significant.

III. BREAKDOWN OF UDPGLUCURONIC ACID

A. General

The metabolic fates of UDPGlcUA are summarized in Figure 6. We will here consider only certain breakdown pathways.

Polysaccharide synthesis is not relevant; the transferase involved appears quite distinct.[597]

Formation of L-iduronic and D-galacturonic acids (Sections II.D and II.E above) has no obvious bearing on simple glucuronidation. Supply of UDPGlcUA for polysaccharides or for the epimeric acids could be restricted by dosage with glucuronidogenic drugs, but no evidence has been found.

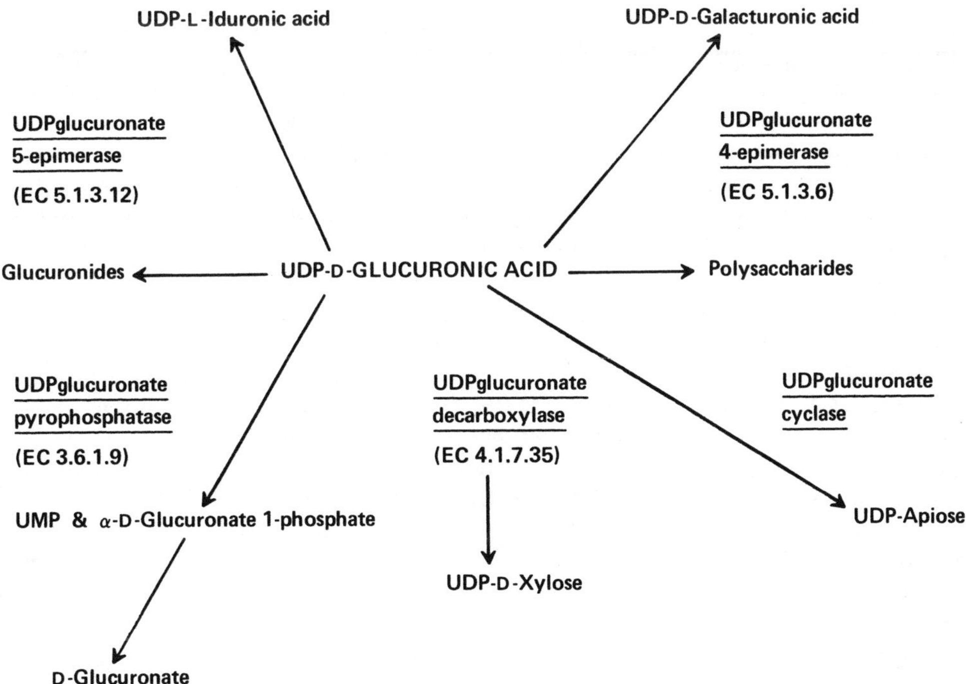

FIGURE 6.   Breakdown of UDPglucuronic acid.

Glucuronyl transference is treated elsewhere in this book. We now discuss the three remaining pathways of UDPGlcUA transformation.

B. Action of UDPGlucuronic Acid Pyrophosphatase (EC 3.6.1.9, Unspecific Nucleotide Nucleotidohydrolase)

1. Specificity

There is probably only one nucleotide pyrophosphatase, hydrolyzing all nucleotide pyrophosphates added as substrates.[304,566-569] When acting on UDPGlcUA it has been termed UDPGlcUA pyrophosphatase,[225,304,570] a name convenient to retain here. Its destruction of NAD^+, NADH, $NADP^+$ and NADPH[569] suggests interference with many pathways including NADPH-dependent Phase 1 drug metabolism.[571]

2. Occurrence

Mammalian UDPGlcUA pyrophosphatase was reported in early studies from rat kidney,[569,572] skin,[558] liver,[559] and intestine.[555] More active in rat liver than in that from man, monkey, or rabbit, it is very low in guinea pig liver;[225,570] its activity in rat tissues requires high UDPGlcUA concentrations for transferase assay there.[225,573,574] Higher turnover of UDPGlcUA in rat than in guinea pig tissues is confirmed (Section II.B.2).[505] It exists in microsomes,[304,575] nuclei,[566] and plasma membranes.[576-578] The partly purified enzymes from microsomes and nuclei may, from kinetic studies, be identical, possibly derived from plasma membrane.[568] Separation of these fractions is difficult, and most of the activity in rat liver preparations seems microsomal, largely in rough microsomes.[577,578]

3. Assay

Assay of UDPglucuronic acid pyrophosphatase involves colorimetric determination of glucuronic acid after removal of nucleotides by absorption on charcoal (e.g., Reference 304), or measurement of unhydrolyzed UDPGlcUA by added UDPglucuronyl-transferase.[225,572] A radioactive assay[579] uses UDPGlcUA labeled with ^{14}C in the glucuronyl residue and involves paper-chromatographic separation for 16 hr; unchanged UDPGlcUA, α-glucuronic acid 1-phosphate, and glucuronic acid (and also glucuronide if required) are located by color reagents and UV light. A more rapid and sensitive method[569] employs purified alkaline phosphatase to hydrolyze the phosphate esters (UMP and α-glucuronic acid 1-phosphate) produced by pyrophosphatase action of dilute microsomal suspensions. Previous attempts were troubled with pyrophosphatase activity of crude alkaline phosphatase and with endogenous microsomal phosphate released during incubation.

4. Properties

Activity of UDPGlcUA pyrophosphatase in vitro depends on incubation buffer, being highest with Tris-HCl and diethanolamine.[569] Of nucleotides inhibiting the enzyme,[566] UDP and UMP are particularly effective.[304,580] UDPGlcUA itself inhibits further breakdown of UMP,[581] possibly a mechanism to conserve tissue UDPGlcUA. Citrate[579] and EDTA are powerful inhibitors. EDTA at 10 mM was claimed to completely,[225,568] or 75%,[580] inhibit the pyrophosphatase. During transferase assay, rat liver pyrophosphatase consumes 7 to 9 times more UDPGlcUA than does the activated or latent transferase,[571,579,580] and 10 mM EDTA increased the apparent UDPglucuronyltransferase ability in rat-liver microsomes,[225] though insufficient to abolish the species difference seen between in vitro glucuronidations of rat and hamster.[317]

The pyrophosphatase, activatable by membrane perturbants,[569,571,580] is solubilized by various procedures.[580] Solubilized from rat liver and purified to a homogeneous protein of 137,000 daltons, it possessed pyrophosphatase and alkaline phosphodiesterase activity,[578] as suggested earlier,[576] and as also shown with mouse liver enzyme.[567] Surfactants can substitute for lipids as stabilizers.[567,578]

5. Succeeding Step: The Hydrolysis of α-Glucuronic Acid 1-Phosphate

α-Glucuronic acid 1-phosphate is hydrolyzed by one of the phosphoric monoester hydrolases (EC 3.1.3):

$$\alpha\text{-glucuronate 1-phosphate} \rightarrow \text{glucuronate} + \text{phosphate}$$

It exists in kidney microsomes,[572] but at first appeared doubtful in liver.[580] Recent evidence[579,580] shows that in rat-liver microsomes 80 to 90% of the increase in labeled glucuronic acid accompanies similar decrease of labeled α-glucuronic acid 1-phosphate, and so in vitro at least, it breaks down UDPGlcUA; and is probably rate limiting in the production of glucuronate by liver.[579,580]

The phosphatase is inhibited by chelators and probably also by phosphate.[579,580] Chromatography readily separates reactants and products.[579]

C. Action of UDPGlucuronate Decarboxylase (EC 4.1.1.35, UDP-Glucuronate Carboxylyase)

The reaction

$$\text{UDP-D-glucuronic acid} \rightarrow \text{UDP-D-xylose} + CO_2$$

is catalyzed by an enzyme first noted in particulate and soluble fractions of mung bean[582] or wheatgerm[583] extracts; UDP-D-arabinose is also formed.[582] The enzyme exists in animal tissues.[584,585] UDPxylose is formed in certain plants also by an enzyme[586] closely associated with UDPGlcUA cyclase.

D. Action of UDPGlucuronic Acid Cyclase (EC unclassified)

UDPGlcUA can be converted into UDPapiose [uridine 5-(α-D-apio-D-furanosyl pyrophosphate)]:

$$\text{UDPglucuronic acid} \rightarrow \text{UDPapiose} + CO_2$$

by a NAD$^+$-dependent enzyme, UDPcyclase, in parsley (*Apium petroselinum*) and duckweed *(Lemna minor)*.[586-588] The enzyme has been isolated[394,589] and its mechanism involves a hydride shift from C-4 of UDPGlcUA to C-3' of UDPapiose.[590] The structure of UDPapiose was only recently described.[394] It has not been recorded from animals.

E. Action of UDPGlucuronic Acid 4, 5-Epimerases

See Sections II. D and E above.

F. Role of UDPGlucuronic Acid Metabolism in Limiting Glucuronidation

From the above sections glucuronidation by mammalian tissues depends not only on UDPglucuronyltransferase, but also on their receiving NAD$^+$ and UDPglucose, on UDPglucose dehydrogenase forming UDPGlcUA, and on the relative activity of pyrophosphatase in breaking down UDPGlcUA formed.

Glucuronidation capacity normally exceeds basal requirements, for glucuronidation is immediately increased three- to fourfold by increase of aglycon load, even with no increase in enzyme activities.[543] Either the fall in UDPGlcUA concentration due to glucuronide synthesis stimulates flow of glucose through UDPglucose dehydrogenase, or UDPGlcUA is normally produced in excess, and the increasing aglycon load diverts the nucleotide to the transferase from the pyrophosphatase.[543] In the latter case, free glucuronate would be diminished, and flow along the "glucuronic acid pathway" of glucose metabolism would consequently be lessened.[543] This problem of compensation has not been much investigated since early observations[311] suggested that, in fact, the flow through the glucuronic acid pathway was not necessarily diminished under these circumstances. The pathway itself has been much investigated[220,591] (Chapter 13, Section II.K).

The role of UDPglucose, UDPglucose dehydrogenase, and the pyrophosphatase in limiting glucuronidation in specific conditions is discussed under the conditions themselves (e.g., age, disease, drug administration) in subsequent chapters, but it should be noted that effective limitation in vivo has not been unequivocally established. Evidence for limitation of glucuronidation by UDPGlcUA supply in vivo is:

1. Fall of glucuronidation on depletion of carbohydrate, noted long ago (see Reference 4 and 5).
2. From pharmacokinetics, Levy[31,36] has proposed the saturation of the glucuronidation system with increasing doses of salicylamide.
3. UDPGlcUA supply appears limiting for glucuronidation in cultured MH_1C_1 hepatoma cells,[592] but in cultured human skin epithelial cells, with less transferase activity, it appeared adequate;[385] conclusions were based on comparison of glucuronidation in whole cells and in homogenates prepared from them.
4. The level of UDPGlcUA in perfused[177] and fresh[188] liver of some 0.3 mM appears to limit glucuronidation; for glucuronidation in perfused liver was almost that of native microsomes offered 0.3 mM UDPGlcUA, a concentration clearly limiting their in vitro UDPglucuronyltransferase activity. This concentration of UDPGlcUA is approximately the apparent K_m value of the transferase for UDPGlcUA[133,336] and is maintained under a variety of stress conditions[179,180] in-

cluding perfusion with insulin, though some find [512] insulin treatment of intact animals raises liver UDPGlcUA.

Interesting recent work[511a] indicates a drop in rat liver UDPGlcUA level when hepatocytes are offered a transferase substrate (4-methylumbelliferone), followed by a slow increase. Rate of increase is low compared with the glucuronidation rate, and synthesis of UDPGlcUA may be well below maximal in absence of a transferase substrate; presence of the substrate calls out its increased synthesis.

However, in sliced tissue or isolated cells carbohydrate is required for maximal UDPGlcUA generation; whereas liver slices or cells from fed animals are rich in glycogen, those of intestinal mucosa usually need glucose in the medium for effective glucuronidation,[574] and inhibition of UDPglucose dehydrogenase by rotenone in isolated rat hepatocytes [166] or by UDPxylose in liver slices[306] almost abolished their glucuronidation. Certain aglycons may be preferentially allotted UDPGlcUA when the nucleotide is restricted: e.g., injection of galactosamine lowered liver UDPGlcUA by almost one half, causing glucuronidation of 1-naphthol to fall in the liver on perfusion, but not that of bilirubin[179] (1-naphthol was increasingly excreted as sulfate under these conditions, but bilirubin forms negligible sulfate).

Limitation in vitro is of course apparent in broken cell preparations also, if UDPGlcUA concentration is too low because of faulty experimental design or because of unexpected increase in nucleotide pyrophosphatase activity (e.g., employing a species (rat) or tissue (kidney or intestine) rich in pyrophosphatase). NAD$^+$ supply largely determined activity of UDPglucose dehydrogenase and hence of glucuronidation in an artificially reconstituted broken cell system.[95] NADH inhibits the dehydrogenase[96] and increase in NADH/NAD$^+$ ratio could lower its activity.[511a]   Evidence suggests that change in nucleotide status (as after fasting or administration of nicotinamide or ethanol) alters flux along this pathway,[543] but direct linkage between UDPglucose and UDPGlcUA levels is not evident: e.g., administration of D-glucosamine markedly lowers liver UDPglucose but raises liver UDPGlcUA.[593]

IV. BREAKDOWN OF GLUCURONIDE

A. General

β-D-Glucuronides are hydrolyzed by β-glucuronidase (EC 3.2.1.21, β-D-glucuronide glucuronohydrolase) present in all animal tissues examined.[74,594] Its role in breaking down, in vitro or in vivo, glucuronide already formed by UDPglucuronyltransferase has been long discussed and is not settled, though it does not seem major. Nonenzymic breakdown of glucuronide in vivo seems limited to labile conjugates such as N-glucuronides (Chapter 2, Section II.C), usually unstable physiologically (as in blood or bladder), or to others such as "ester" glucuronides (Chapter 2, Section II.A.4) easily hydrolyzed during incautious separation and assay.

We shall treat the enzymic pathway only; in effect, that due to β-glucuronidase. Breakdown of glucuronide by reversal of UDPglucuronyltransferase action is minimal physiologically (Chapter 8, Section II). Hydrolysis by β-glucuronidase during development is discussed in Chapter 10, Sections III.B and III.D.

B. Properties of β-Glucuronidase

This enzyme has been extensively documented. For early work see References 75, 594.

It possesses hydrolase and some transferase properties (Chapter 3, Section III.A). The former is considered here. Hydrolysis only occurs at the β-link, as with simple β-glucuronides, including the (synthetic) UDP-β-glucuronic acid[595] and β-glucuronic acid

1-phosphate;[596] certain other compounds are also hydrolyzed by mammalian β-glucuronidase.[74,597] The glucuronosyl-*O*-bond breaks during this hydrolysis.[598] D-Glucuronic acid linked through C-atoms other than its C-1 is not liberated.[450]

It accepts all known β-glucuronides as substrates, with the possible[63a,71] exception of *N*-glucuronides (*N-O*-glucuronides are accepted,[74] and certain *S*-glucuronides.[599,601] *E. coli* β-glucuronidase failed to hydrolyze thio-β-glucuronides of aliphatic thiols and thiophenol.[599] Bovine liver β-glucuronidase hydrolyzed the thioglucuronide of 2-benzothiazole;[600] and thio-β-glucuronides of diethyldithiocarbamate, 2-aminothiophenol and 4-nitrothiophenol were hydrolyzed by specific (i.e., glucarolactone-inhibited) β-glucuronidase activity from rat preputial gland.[395] As thio-β-glucuronides of 2-benzothiazole and 4-nitrophenol are hydrolyzed by rat liver β-glucuronidase, whereas that of thiophenol is not,[601] an induced shortening of the C−S bond towards C−O dimensions may exist in the hydrolyzable conjugates.[601]

Mammalian β-glucuronidases have a pH optimum usually 4.5 to 5.0, but sometimes as low as 3.5 and as high as 6.0;[74] molluscan enzymes often possess a more, and those from bacteria a less, acid optimum. Figures from early work are suspect because optimal conditions were not attained.

Many purified preparations of β-glucuronidase possess some β-galacturonidase activity, possibly inherent in the enzyme from both limpet and mammal.[602]

β-Glucuronidase is a glycoprotein[603] and has been highly purified,[74,604] recently by employing an antibody column.[604,605] It has been obtained crystalline and its isomeric forms investigated.

In mouse liver, the lysosomal form is a tetramer of mol wt 280,000.[609] The microsomal form, when solubilized with Triton X-100®, consists of several complexes, M_1 to M_4, of higher molecular weight, composed of a tetramer core complexed with from 1 to 4 chains of the glycoprotein egasyn (mol wt 64,000) which stabilizes the binding of the enzyme to the endoplasmic reticulum membrane.[610]

In rat liver,[604] the microsomal enzyme is a tetramer of 290,000 daltons and the lysosomal enzyme is electrophoretically distinct, but catalytically and immunologically identical, with indistinguishably different molecular dimensions. The two forms appear to be charge isomers.[604] The microsomal and lysosomal mouse-liver enzymes are derived from a single structural gene on chromosome 5.[606]

Genetic considerations have been treated generally[606] and for man.[607,608] The human gene for β-glucuronidase is on chromosome 7, and its deficiency leads to mucopolysaccharidosis Type VII.[608]

C. Occurrence of β-Glucuronidase

The enzyme occurs in animals, plants, and bacteria, and lists have been published.[74,594] Its source is important for hydrolyzing glucuronides (Section IV.D below). Its intracellular distribution is of interest. It occurs in lysosomes, but unlike most lysosomal enzymes, also in microsomes,[611,612] its quaternary structures being distinct at these two sites (see preceding section). Whether microsomal β-glucuronidase serves solely as precursor of the lysosomal enzyme or whether it has a distinct physiological function is not yet clear. Many workers have described multiple forms of β-glucuronidase, some restricted to certain subcellular sites (for early references see Reference 613). A study of these forms in lysosomes, microsomes, and high-speed supernatant, concluded that those in the latter arose from leakage of organelles, probably lysosomes, during preparation.[613] β-Glucuronidase activity occurs equally in lysosomes and crude microsomes and washing the latter left some 76% bound.[613] The washed microsomal enzyme was not latent, unlike lysosomal β-glucuronidase, and therefore judged free of contaminating lysosomes;[613] its incubation with lysosomes converted it into the lysosomal form.[613] Recent results[614] suggest that in rats treated with a specific elevator of plasma glucuronidase activity (di-isopropyl phosphorofluoridate), microsomal β-glucuronidase is a precursor of plasma glucuronidase, and probably, of the lysosomal enzyme, for activity fell in microsomes and rose concurrently in plasma.[614]

Consideration of polymeric structure earlier suggested that microsomal and lysosomal enzymes were both derived from one precursor.[610] Because mouse-kidney Golgi fraction contains high β-glucuronidase activity more sensitive to androgen stimulation than the microsomal enzyme and electrophoretically resembling the lysosomal enzyme, the Golgi complex may distribute the precursor after ribosomal synthesis to both microsomal and lysosomal sites.[615]

Supporting this idea, β-glucuronidase in phenobarbital-treated mice remained constant, but egasyn (Section IV.B above) almost doubled, and the enzyme redistributed itself to increase its proportion in microsomes.[616] Availability of egasyn, therefore, could determine distribution of β-glucuronidase, agreeing with conclusions[606] from genetic studies that glucuronidase in the proximal-tubule epithelial cell of mouse kidney leaves the ribosome as a polypeptide which then either enters the lysosome or takes on egasyn and binds to the endoplasmic reticulum.

D. Role of β-Glucuronidase in Hydrolyzing Preformed Glucuronides
1. Role In Vitro

At neutral pH values, or at values above 7 where UDPglucuronyltransferase is normally assayed, hydrolytic activity of β-glucuronidase appears negligible according to most reports[4,225] even following activation or induction of microsomal enzymes (e.g., Reference 192 and subsequent chapters). Glucaro-1,4-lactone has little effect on glucuronidation in vitro; it might sometimes slightly increase it.[277,282] Microsomal β-glucuronidase activity towards 4-nitrophenyl glucuronide is essentially lost as the pH increases from 7.0 to 8.5.[55] When both transferase and hydrolase were assayed under identical conditions using rat-liver microsomes at pH 7.5, the same aglycon (4-nitrophenol) free and conjugated, and various modulators, β-glucuronidase was found active at pH 7.5 (up to 20% the activity at pH 4.5, but depending markedly on substrate and species; rat was particularly active) and glucarolactone slightly increased the observed transferase activity.[282]

2. Role In Vivo
a. Role of Intracellular β-Glucuronidase

The work quoted above[282] concluded that a conjugation-deconjugation-reconjugation cycle could operate in the endoplasmic reticulum with transferase and β-glucuronidase participating. Much literature concerns the effect on glucuronidation of administered glucarolactone,[543] but interrelationships are unprofitably complex. Glucarolactone administered to mouse inhibits liver and kidney β-glucuronidase for several hours,[617] but there is no evidence yet that glucuronidation there is increased.

β-Glucuronidase activity is inversely proportional to observed transferase activity in many tissues and physiological states, and after treatment with certain drugs,[618,619] suggesting that increased hydrolytic activity may decrease observed glucuronidation. However, there are well-documented instances where a rise in transferase activity is accompanied by no change, or even by a rise, in β-glucuronidase: e.g., membrane perturbation,[105,192,210] organ culture,[171] disease,[382] and administration of phenobarbital.[171,620] Separate estimation of washed microsomal β-glucuronidase activity, however, was not usually performed in these studies.

Hydrolysis by β-glucuronidase of endogenous glucuronides has often been invoked to explain increased urinary glucuronate and its metabolites after administration of certain drugs,[223,570] especially those (such as aminopyrine and barbital) which inhibit UDPGlcUA pyrophosphatase.[580]

b. Role of β-Glucuronidase in Bile and Intestine

β-Glucuronidase occurs in bile, intestinal juice, and salivary secretions,[74] arising from sloughed cells or bacterial infection; its often high pH optimum suggests the latter origin. In bile, it may initiate formation of stones of calcium bilirubinate.[74]

In the intestinal lumen it is very active,[74,594,621] originating mainly from sloughed intestinal epithelial cells and in the lower reaches also from the microflora. As many glucuronides of both endogenous and foreign compounds are excreted in bile, the action of β-glucuronidase in promoting liberation, and possible reabsorption and enterophepatic circulation, of aglycons, is physiologically and pharmacologically important. Enterohepatic circulation is treated in Chapter 14, Section IV.F.2, but the specific role of β-glucuronidase in the process is considered here.

Therapeutic administration of the β-glucuronidase inhibitor, glucaro-1,4-lactone, has been proposed to lessen liberation of toxic aglycons from their glucuronides.[622,623] Glucarolactone is rapidly absorbed from the gut,[617] and its administration inhibited the enzyme in bacteria of human bile.[624]

The prolonged effect of morphine has been attributed to its enterohepatic circulation after biliary excretion.[625] Addition of glucarolactone to an intraduodenal infusion of stilbestrol monoglucuronide markedly decreased absorption of free stilbestrol into blood,[626] consistent with inhibition of β-glucuronidase in duodenal contents. The period of depression of locomotor activity caused by phenobarbital or progesterone administered to male rats i.p. was shortened after oral pretreatment with glucarolactone,[627] which given alone had no effect on locomotor activity. Both depressants are excreted by rats as biliary glucuronides; zoxazolamine (not forming a glucuronide with a pharmacologically active aglycon); when likewise injected, continued its paralytic action unchanged during oral treatment with glucarolactone.[627] The β-glucuronidase in luminal contents throughout the intestinal tract was inhibited within 6 hr in rats treated this way with glucarolactone, and the enzyme from (mouse) intestinal contents and epithelial cells was inhibited by glucarolactone at pH values of in the tract.[627] As orally administered glucarolactone therefore enhances elimination of compounds excreted in the bile as glucuronides, the enterohepatic circulation of their aglycons must largely arise from the action of luminal β-glucuronidase. Considering the relatively high environmental pH values (even though bacterial β-glucuronidase, with its higher pH optimum, participates) hydrolysis must imply high concentration of enzyme and long period of incubation.

c. Role of β-Glucuronidase in Plasma

Plasma β-glucuronidase rises in certain diseases,[74] specifically, after injection of diisopropyl phosphofluoridate,[614] and increases after administration of certain pesticides and hepatotoxic agents.[628] Its significance in hydrolyzing plasma glucuronides is not known, but probably small.

d. Role of β-Glucuronidase in Urine

Factors increasing the enzyme in urine have been discussed.[74] It arises from sloughed-off genito-urinary epithelial cells and during local infection from bacteria. When liberated aglycons are carcinogenic or otherwise toxic, inhibition of urinary β-glucuronidase in the bladder is important, and oral administration of glucarate (to give glucarolactone) has been recommended as prophylaxis for workers in the chemical industry exposed to glucuronidogenic toxins;[594,622] urine of subjects so treated contains an increased amount of β-glucuronidase inhibitors.[622] In this regard, the (1 → 4), (6 → 3) dilactone is less toxic on administration[629] than the (1 → 4) lactone. Bacterial β-glucuronidase hydrolyzes the N-O-glucuronides of N-hydroxy-2-naphthylamine,[71] but in acid urine, this and similar unstable glucuronides may also be chemically hydrolyzed to carcinogens. This harmful role of β-glucuronidase in urine is illustrated by the elegant work[18,19] quoted in Chapter 1, Section V.

E. Use of β-Glucuronidase in Characterizing β-D-Glucuronides

Many years ago Levvy[630] demonstrated that a boiled solution of glucaric acid inhibited β- but not α-glucuronidase. The specific inhibitor, glucaro-1,4-lactone,[594] is widely used in characterizing conjugates as β-glucuronides; as most commercial β-glucuronidase preparations contain other glycosidases, sulfatase, and phosphatase, this control is essential to verify that hydrolysis is due to β-glucuronidase activity. A β-galacturonide cannot be ruled out by this procedure because this link also is hydrolyzed to some extent by an enzymic activity inhibited by glucaro-1,4-lactone.[602] The lactone being unstable above neutrality,[594,622] it is less helpful with the bacterial than with the mammalian or molluscan enzyme. Recent comparisons of glucuronide hydrolysis with β-glucuronidase from mammalian liver and from bacteria use both "ether" and the more rarely studied "ester" glucuronides.[631,632] A recommended source of β-glucuronidase largely free of interfering hydrolases is rat preputial gland.[633,634]

With "ester" glucuronides, incomplete hydrolysis by β-glucuronidase can arise from the aglycon linking at positions other than C-1 of glucuronic acid. Sequential migrations of the aglycon group from C-1 to C-2, C-3, or C-4 have been reported during storage at neutral to slightly alkaline pH or in the body (e.g., bile[450]). Sequential migration with lack of hydrolysis by β-glucuronidase is also suggested for clofibrate glucuronide following incubation at 37°C in plasma for 20 min, or in slightly alkaline buffer.[634a]

V. ADDITIONAL NOTES

Section III.B: [31]P n.m.r. studies indicated no measureable breakdown by rabbit liver microsomes of UDPGlcUA within 20 min, unless Mg^{2+} was present.[170a]

Part II
Factors Affecting Glucuronidation In Vivo

From previous chapters, glucuronidation may be controlled:

1. At passage of precursor or aglycon into the cell
2. At metabolism of precursor into aglycon
3. At accessibility of enzyme to aglycon
4. At availability of UDPglucuronic acid
5. At activity of UDPglucuronyltransferase itself
6. At secretion of conjugate from the cell
7. At hydrolysis of conjugate to free aglycon

Control at any stage depends on the state of the organism — its age, genetic and hormonal complement, diet, intake of xenobiotics, response to physical changes in its environment, and health.

The following chapters treat glucuronidation during these various states of the organism and as far as possible relate it to each of the above stages.

Chapter 10

THE EFFECT OF AGE ON GLUCURONIDATION

I. GENERAL

Glucuronidation of all simple molecules examined is low or undetectable in early fetal or embryonic tissues, ranging from perfused organs to UDPGlcUA-fortified microsomes. Glucuronidation increases with age, the rate depending on substrate, tissue, strain, and species, until approximate "adult" activity is reached, usually perinatally. Several reviews exist. Dutton[4] gives early references; later reviews[182,203,635,636,636a] update successively.

Poor fetal and neonatal glucuronidation is of considerable pharmacological importance. It goes far to explain toxicity of glucuronidogenic xenobiotics to the fetus and newborn, especially the premature newborn. It also contributes to the icterogenic effect of many drugs at this age. The human neonate, like that of most animals, is slow to glucuronidate bilirubin and to excrete the conjugate. Drugs competing with bilirubin anywhere in the process (c.f., stages 1 to 7 above) may precipitate jaundice. Drugs may also decrease glucuronidation of other endogenous substrates not so clinically recognizable as bilirubin.

Relative deficiency of glucuronidation in fetus and newborn compared with that in the adult is typical of most Phase 2 processes[636] and, although to a lesser extent in the human, of Phase 1 processes also.[637-640] Glucuronidation being so common a pathway, its deficiency understandably complicates medication, e.g. inadequate allowance being made for the deficient glucuronidation and subsequent excretion of chloramphenicol by young babies[641] caused almost 25% of infants receiving chloramphenicol to die, and the more immature the infant, the greater the mortality. Longer plasma half-lives of drugs (e.g., salicylates[641-643]), when administered neonatally on a body-weight basis, also indicate less efficient clearance of xenobiotics in the newborn, and perinatal medication has recently been drastically revised to allow for "immaturity" in detoxication. A useful compilation of LD_{50} values in newborn and adult animals points out that when detoxication depends largely on excretion (as with antibiotics) LD_{50} increases progressively with maturation, but when it depends on metabolism, sharp shifts occur that cannot be predicted by any general dosage formula.[644]

In today's medicated and xenobiotic-contaminated environment, the fetus is at considerable risk. Most drugs cross the placenta relatively easily,[645] even though they seem fully ionized (e.g., salicylates,[646] quaternary ammonium compounds[647]). Drugs may influence utero-placental-umbilical vascular systems so that placental perfusion, directly related to drug exchange between mother and fetus, is increased.[648] As Yaffe and Juchau[640] point out, the question is no longer whether a drug crosses the placenta or not, but at what rate it does cross.

As will be gathered from the evidence below, the principal cause of low fetal and neonatal glucuronidation is lack of UDPglucuronyltransferase activity, probably due to lack of the enzyme, not to its inhibition or excessive latency. Relative lack of UDPGlcUA usually, but not always, accompanies lack of transferase. Defective uptake of aglycon and/or defective secretion of conjugate may contribute, but do not seem the primary cause.[649] Recent reports on the problem are encouraging. Development of UDPglucuronyltransferase, and glucuronidation as a cellular process, can be induced precociously by xenobiotics and, for many substrates, by hormones. We are beginning to understand the natural mechanisms of its onset.

Extrapolation of animal studies to man, always quantitatively hazardous, is more so in developmental studies. Even qualitatively there may be great differences, as in possession by midterm human fetal liver of an apparently functional Phase 1 system[637] not developed in experimental animals until birth. Dangers of simple extrapolation to man from species active at birth (e.g., guinea pig) or conducting independent embryonic existence (e.g., chick) are obvious. Greengard,[650] comparing rat and human hepatic enzyme differentiation (though with little reference to glucuronidation), introduces the AQ concept. The AQ (activity quotient) is activity of developing enzyme (units per gram of immature liver) divided by activity in adult liver.[650] Comparison of the developing liver in two species is best based on the two sets of AQ[650] with, as standard comparison, the normal adult-rat liver (sex presumably specified), and grams wet weight rather than milligrams protein. DNA is a bad basis, especially when enzyme changes are small and subject to changes in cell number (as when hematopoiesis ends); rise or fall of all enzymes is not significant, but rather their relative rise and fall per gram of hepatic mass.[650] (Hietanen[651] lists body and liver weights during rat development.)

It is probably wiser to measure fetal or neonatal glucuronidation with an unfractionated homogenate at first, rather than with microsomes. Fetal or embryonic microsomes may have different degrees of activation during the period of preparative spinning, and different sedimentation properties to those of adult liver,[4,141,142,652] due presumably, to a different type of membrane scission during homogenization. From electron microscopy, the endoplasmic reticulum of human fetal liver seems converted on homogenization partly to microsomes and partly to long, thin cisternae, losing much marker enzyme into low-speed subfractions.[141] This centrifugal difference may be[640] why the Phase 1 enzymes were at first overlooked in human fetal liver "microsomes". In rat liver, rough endoplasmic reticulum was the only form seen at and before birth,[653] the smooth form developing asynchronously at birth; human hepatocyte smooth reticulum is evident at only 6 weeks of fetal age[654] which is consistent[640] with early appearance of Phase 1 in human fetal liver. In cultured embryo liver rough membrane "flowed" into the smooth form with the enzyme still present.[143]

Changed stability of membrane and progressive change in composition may account for the age-dependent activation characteristics of UDPglucuronyltransferase,[212] complicating assay of the enzyme through development.[182,196]

II. ISOLATION OF GLUCURONIDES FROM PRE- AND POSTNATAL SOURCES

A. Isolation from Prenatal Sources

Although all evidence suggests, glucuronidation is defective in the early mammalian fetus, glucuronides, principally of estrogens, have been isolated from early fetal tissues or fluids, and must have been made or transported there. Isolation studies[655] concluded that liver was the most active site of glucuronidation in the human fetus, which was considered able to conjugate estrogen as early as 17 weeks,[656] but this conjugation included sulfation, known to be high in the fetus. Steroid glucuronides themselves have, however, been isolated in small amounts from previable human fetal liver,[657] intestine,[658] and from midterm[659,660] human fetuses perfused with labeled progesterone.

A total concentration of glucuronide of 815 μg/p 100 g wet weight of gall-bladder bile occurred in early and midterm human fetuses,[661] various C-21 steroids being glucuronidated in highest amount, and total sulfate concentration was rather less;[661] these glucuronides were considered formed in fetal tissues, probably liver, but only unconjugated aglycons could be found in liver.[661] In amniotic fluid were found estrogen glucuronides and sulfoglucuronides,[662] tetrahydrocortisone glucuronide at over twice its concentration in maternal tissue,[663,664] and sulfates and glucuronides of estriol at midpregnancy and term.[665] Meconium, the unabsorbed solid from swallowed amniotic fluid, bile, mucus, and intestinal epithelial cells, is accumulated from the first trimester, and similarly contains steroid glucuronides.[666,667]

A constant low background glucuronidation therefore may occur in the human fetus. Many reported glucuronides could be of maternal origin. Steroid glucuronides cross the placenta. The 3-glucuronides of estrone and estriol cross unhydrolyzed,[668] and midterm placentas perfused *in situ,*[669] transferred unchanged estriol 16-glucuronide. There is surprisingly little information on xenobiotic glucuronides in fetal tissues or amniotic fluid; uncharacterized conjugates of pethidine and chlorpromazine in amniotic fluid, were considered of fetal origin because their pattern of conjugation differed from the maternal.[670]

B. Isolation from Postnatal Sources

After birth, glucuronides are found in progressively increasing amounts in infant tissues and fluids. The neonatal enzyme could have been influenced by exposure to drugs at birth, *in utero,* or via milk. Several initially high levels are explicable this way, e.g., urine from newborn infants of mothers treated with diphenylhydantoin, a probable inducer of UDPglucuronyltransferase, contained the 91% conjugated drug.[671] However, the half-life of the drug in such babies is still five times that in the adult.[638] Babies treated with diphenylhydantoin from birth glucuronidated its hydroxylated derivatives quite well with glucuronic acid by the fifth day,[645] but the glucuronide of hydroxylated phenobarbital was not yet present in urine of a 2-day baby treated with phenobarbital from birth.[646]

Normal urinary levels of glucuronide from newborns and infants have been listed for several drugs[672] In the discussion to Reference 645, Levy notes that a day-old baby can excrete salicyl glucuronide. The best known example of hepatic immaturity and poor glucuronidation in infants is the "physiological" jaundice that develops in over two thirds of newborns.[673] Transferase activity to bilirubin develops slowly in man, its plasma half-life when i.v. administered being prolonged up to the second month.[674] Although glucuronidation of 4-methylumbelliferone is also "defective" up to 2 months of age in man,[675] excessive hyperbilirubinemia does not necessarily correlate with unusually defective glucuronidation of all aglycons, e.g., not with that of testosterone.[676] Renal immaturity is also a factor,[640] as is slower excretion into bile.[643] Nevertheless, individual variations are great. In 14 newborn infants given a single oral dose of salicylamide, urinary excretion varied from 45% of the dose (adult levels!) to 8%.[677] Even in strains of laboratory animals not exposed to random medication, variations within and between litters are large.[4] For bilirubin, high- and low-glucuronidating human populations may exist;[678] and consistent with genetic factors only identical twins metabolized the predominantly glucuronidated 4-hydroxyphenylhydantoin at identical rates.[679]

Increased postnatal excretion of glucuronides after treatment with specific drugs is discussed in Section IV.B.2 below.

High activity of another Phase 2 conjugation may falsely suggest deficient glucuronidation e.g., low neonatal urinary glucuronide levels of testosterone arise not necessarily from low liver UDPglucuronyltransferase, but from high testicular sulfation of the aglycon.[676]

III. FETAL AND PERINATAL DEVELOPMENT OF THE GLUCURONIDATING SYSTEM

A. Historical Aspects[4]

Relative deficiency of the neonate in glucuronidation is largely independent of transport to or from liver, as shown in 1949[680] by lower glucuronidation of 2-aminophenol in liver slices from infant than from adult rats. Many perfusions of (usually human) fetal liver (Section II.A. above) and work with slices[4] have since confirmed low perinatal hepatic glucuronidation, even of steroids.

The two first enzymic investigations, simultaneous and independent,[681,682] usefully complemented each other. UDPGlucuronyltransferase activity was virtually absent in the early fetal liver of mouse,[682] guinea pig,[681,682] and man,[682] increasing at a rate depending on species and substrate (2-aminophenol, phenolphthalein). Further work with man confirmed the deficiency in fetuses or premature infants, with substrates including bilirubin,[683,695] and suggested that defective transport inwards or outwards at the perinatal liver cell membrane[684-688] could not wholly be responsible for poor glucuronidation in vivo.[649] But possibly the low transferase activity observed is an artifact of the in vitro conditions; the enzyme might exist at adult level, but not evident because of absence of activating, or presence of inhibiting, factors. This possibility will now be considered.

B. Possible Causes of the Low UDP-Glucuronyltransferase Activity Observed In Vitro

1. Inhibition of the Transferase In Vitro

Despite search, no inhibitors peculiar to fetal liver have been proved present, nor any rate-limiting activators proved absent. Homogenates or boiled extracts of fetal or embryonic liver affected the transferase of adult liver when added to adult-liver homogenates or microsomes only by simple dilution or addition.[500,574,682] An inhibitor in egg yolk,[369] not further reported, seems to have been unspecific. Transferase activity to 2-aminophenol is depressed in liver from pregnant rats,[689] (G.J. Wishart, unpublished results), but activity in late fetal liver rises above adult male values, so any maternal inhibitor of transferase does not reach the fetus. Moreover, activity to other substrates (e.g., 1-naphthol, bilirubin, and 4-nitrophenol (G.J. Wishart, unpublished results is not obviously depressed in pregnant rats. Glucuronidation in pregnancy is further disclosed in Chapter 13, Section I.C.

2. Increased Destruction in Vitro of Added UDPGlucuronic Acid

This possibility is unlikely from the mixed-homogenate experiments mentioned above, and presence of only 10 to 20% adult UDPGlcUA pyrophosphatase activity in microsomes from perinatal rat or guinea pig liver.[310] Also, transferase activity remained absent or proportionately lower than adult when concentration of UDPGlcUA added to perinatal liver preparations was increased tenfold,[312,339,574] or when the pyrophosphatase inhibitor EDTA was added.[193] Double reciprocal plots (with V_{max} and $K_{UDPGlcUA}$) gave straight lines,[193] consistent with low embryonic pyrophosphatase activity.

3. Increased Destruction In Vitro of Formed Glucuronide

Although β-glucuronidase, fetal and adult, is largely inactive at pH values above 7 (Chapter 9), the higher activity of the enzyme in fetal tissues[594] might diminish the formed glucuronide measurable in fetal-tissue homogenates or microsomes.

However, added glucuronide was not more rapidly destroyed in fetal-liver homogenates than in adult-liver homogenates,[369] and the specific β-glucuronidase inhibitor, glucarolactone, added to fetal tissue homogenates did not increase their apparent formation of glucuronide.[369] As endogenous glucarolactone is deficient perinatally,[690] and conditions within the endoplasmic reticulum in vivo may differ from those within microsomes, fetal β-glucuronidase could still contribute to the low glucuronidation in vivo (Section III.D below).

4. Changed Activation Characteristics of Fetal UDPGlucuronyltransferase

As UDPglucuronyltransferase is strikingly activatable (Chapter 5), the fetal enzyme could be present in a fully latent form which required activation, possibly specifically, before operating in vivo. Transferase activity to 2-aminophenol in the adult Gunn rat, for example, is virtually absent unless diethylnitrosamine is added to the assay tubes, when activity appears as high as in similarly treated preparations from normal rats (Chapter 12, Section III.B). (Against this possibility is the better glucuronidation in slices than in homogenates of adult Gunn-rat liver, whereas the reverse applies in perinatal liver of normal rats). Of course, even if a fetal enzyme could be experimentally activated in vitro, it presumably exists latent in vivo from the evidence of whole animal, perfusion, and slice experiments. An "activated" fetal enzyme would be unphysiological.

On activation, the developmental pattern remains the same. The original perinatal activity was raised only in proportion, and transferase continued absent in the early fetus.[196,312,691] Diethylnitrosamine, added in case it specifically activated transferase towards 2-aminophenol as in genetically-deficient liver, had no effect,[212] or inhibited.[196] The change in activation characteristics of the enzyme towards diethylnitrosamine, UDPGlcNAc and other compounds developing progressively, e.g., in ASH/TO mice from 6 days prenatally[171] (B. Burchell, unpublished results).

Age-dependent change in activation characteristics[212] is important. Because the fetal enzyme might be "over-activated" (i.e., inhibited) by activation regimes optimal for adult enzyme, its optimal conditions must be determined separately, ideally at each developmental stage examined,[692] e.g., optimal sonication for transferase activation increases fourfold over the last 8 fetal days in mouse liver;[369] and response to other perturbations is progressive perinatally,[171,196,212] especially when based on gram of microsomal protein. Homogenization may inactivate a sensitive early-fetal transferase, and suggest deficient enzyme, but gentle manual homogenization (G.J. Wishart, unpublished work), gives no increase in observed activity. The optimal activation peak is sharp in some mouse strains and fully activated enzyme develops at a slightly different rate to unactivated enzyme.[692] In Wistar rat, the peak is fairly broad[319] but optimal activation conditions should be identified at each developmental stage for each substrate under any one set of conditions.[693] Chick-embryo-liver UDPglucuronyltransferase seems not activatable above its very low "native" levels by any process. Once induced, it can be activated by certain procedures.[193]

5. Changed Kinetic Characteristics of UDPGlucuronyltransferase during Development

The properties of UDPglucuronyltransferase are linked to its membrane environment and, as activation characteristics suggest membrane change during development, apparent K_m values, for example, might also change. No clear evidence exists, fetal or embryonic transferase activity is usually too low for accurate determination of kinetic parameters, and the artificially induced enzyme resembles the adult kinetically.[193] However, it is essential in developmental studies to assay at apparent V_{max} and to ensure linear relationship between enzyme activity and incubation time in order to minimize artifactually low transferase activity due to changed kinetics. Dissimilarities do occur; e.g., between infant and adult mouse liver in apparent K_m values with bilirubin as substrate.[694] The various kinetic and physical differences reported between perinatal and adult microsomal-bound or solubilized transferase do not prove that different fetal, infant, and adult catalytic proteins exist; this conclusion awaits complete purification of the enzyme from each source.

There is no reason to suppose that low observed level of transferase activity of fetal or neonatal tissues in vitro is due to factors other than defective UDPglucuronyltransferase activity, probably itself due to low concentration of the enzyme (as is further suggested by work on the induction of the enzyme [Section IV below]).

C. Formation and Breakdown of UDPGlucuronic Acid in Developing Tissues

UDPGlcUA is required for synthesis of uronic-acid-containing polysaccharides, and its presence at an early stage in tissues such as cartilage is not surprising.[552] It does not seem present at adult levels in tissues which are important in simple glucuronidation, however, until near birth. As noted above (Section III.B.2), there is no evidence for greater breakdown of UDPGlcUA in vitro by fetal tissues than by adult tissues, and its real deficiency in vivo is supported by its parallel development[682] with UDPglucose dehydrogenase[681] in, for example, the fetal guinea pig. Liver from chick embryos 11 days or older where (enzymically determined) UDPGlcUA is already at adult values[500] also contains high UDPglucose dehydrogenase activity,[502] at adult levels by the 15th incubation day;[502] a pronounced fall in measurable UDPGlcUA just before hatching[500] has not been further investigated.

In mouse liver, both nucleotide and dehydrogenase develop together. The dehydrogenase rose from very low fetal levels at gestation day 15 to some 10% of adult male (6% adult female) values by day 9 of infancy,[502] with a dip at day 9, followed by a steep rise on days 11 and 12 to 70 to 90% adult male values, and a sexual difference appearing thereafter.[502] UDPGlucose dehydrogenase of mouse liver would, therefore, seem to behave largely as a "weaning cluster" (see Section III.E.1 below) enzyme, the late-fetal surge being relatively minor. In man also, UDPGlcUA[682] and dehydrogenase develop together, the latter being some 25% adult liver levels at 13 weeks' gestation and still only some 50% adult levels at 1 month's infancy.[695] In fetal pig liver, chromatographic assay found only traces of UDPGlcNA at 36 days gestation, whereas UDPglucose and UDPGlcUAc were by then above adult values; it was measurable at 75 days gestation, fell together with the other two sugar nucleotides before birth. It then slowly rose to 60 days of infancy, with little increase between days 1 and 15 postnatally. At 60 days it was higher than the other sugar nucleotides.[506] Adult values of guinea pig liver UDPGlcUA were reached 10 days postnatally.[506]

Overall glucuronidation of sliced tissue rose rapidly in hatched chicken liver parallel with transferase development, but in the neonatal ASH/TO mouse liver it rose more slowly than the transferase (which was near adult levels at 18 days fetal life, whereas even a 10-day infant had only 50% of adult overall glucuronidation).[502] In chick liver, overall glucuronidation may be limited only by UDPglucuronyltransferase development, whereas in mouse liver, UDPGlcUA would seem additionally limiting.

Brodersen et al.,[696] assuming the transferase reversible for all substrates, suggested the ratio UDPGlcUA:UDP, not concentration of UDPGlcUA, as determining; onset of glycogenesis at birth would lower this ratio and diminish glucuronidation. This apparently unlikely hypothesis might throw light on the transient fall in the development of glucuronidation (and even on that of the transferase activity) noted at birth.[697] UDP inhibits the enzyme, and the role of neonatal hypoglycemia in regulating UDPGlcUA synthesis needs study. There seems no correlation between perinatal liver glycogen content and conjugating ability.[697] Livers of infant mice from large litters glucuronidate more slowly than those from smaller and presumably better-fed litters,[682] but dietary effects are complex (Chapter 11, Section III), and as the pattern persisted into transferase activity itself, it probably did not concern UDPGlcUA availability.

ATP markedly increased apparent transferase activity towards 4-nitrophenol in liver microsomes from newborn guinea pigs and, slightly, in those from newborn rats.[698] As nucleotide pyrophosphatase activity was lower perinatally, ATP was not thought to inhibit this enzyme and, thereby, spare UDPGlcUA, but rather to increase membrane permeability to UDPGlcUA,[309,310,698] a phenomenon thought[698] important in regulating glucuronidation in vivo.[698] Although work with ATP can be variously interpreted, especially using 4-nitrophenol (Chapter 5, Section II.5) and low UDPGlcUA levels, the possibility that aglycon and UDPGlcUA accessibility are hindered perinatally[698] by membrane conformation in vivo may contribute to the delayed glucuronidation in slices of perinatal mammalian liver compared with that in homogenates or microsomes.[99,502,682]

D. Breakdown of Glucuronide in Developing Tissues

The AQ (Section I above) for β-glucuronidase is higher than for transferase in fetal tissues,[594,691] whether the latter is activated or not.[691] There seems no consistent difference in developmental pattern between lysosomal and microsomal β-glucuronidase.[691,699] Rat intestinal β-glucuronidase is more active in lysosomes than in microsomes;[699] its mainly lysosomal, six-fold, increase perinatally (over two orders of magnitude above the transferase activity for bilirubin), could cause net deconjugation in the intestinal tract. β-Glucuronidase was assayed, however, at pH 3.5,[699] not at the physiological pH value, and with 4-nitrophenol, not bilirubin, as substrate.[699] There is no evidence that liver β-glucuronidase contributes other than minimally to lower the hepatic fetal and neonatal glucuronidation.

E. Recent Work on UDPGlucuronyltransferase in Developing Homoiotherm Tissues
1. Developmental Pattern in Liver

Previous sections indicate that a principal cause of low glucuronidation in developing liver is low activity of UDP-glucuronyltransferase not due to the presence of inhibitors or absence of activators (indeed, the activator UDPGlcNAc is higher in fetal pig liver than in adult pig liver[506]), to increased breakdown of glucuronide already formed, or to deficient uptake of aglycon or secretion of conjugate.

This low transferase activity has been widely examined.[4,203,339] Because many reports are difficult to interpret, assay conditions being clearly limiting or undefined, it is better not to attempt a comprehensive list here, but to employ selected examples to illustrate recent advances in our knowledge of development of the enzyme.

In all "early" fetal livers (i.e., up to the first half of gestation), transferase activity is either undetectable or below adult levels for all substrates examined, xenobiotic or endogenous. This holds even for steroids, whose low glucuronidation in early human fetuses, although at low levels, has been noted in the sections above. Wishart (unpublished results, 1977) found no significant activity for estrone or testosterone in human liver over 8 to 16 fetal weeks. Transferase activity existed towards estriol in 10-week male human fetal liver.[197] At 17 weeks it had increased tenfold, but was still only 16% of the presumed adult level (activation conditions were not shown to be optimal). Midterm human fetal liver fragments formed only traces of estrogen glucuronide.[700] Steroid glucuronide found in midterm human fetal liver, if made there, is made slowly and accumulates. Other fetal sites (Section II.E.2 below) must contribute to the net fetal glucuronide pool. In fetal mouse liver, activity to estriol was undetectable at 14 days gestation, but had risen markedly 3 days later, just before birth.[197] Fuchs et al.[231] and Lucier and McDaniel[389] found estrogen and testosterone glucuronide formation in animal liver rising post- and prenatally from presumably low early fetal values. Activated and "latent" transferase to testosterone was just detectable at 14 days in the mouse fetus, 15 days in the rat fetus, and at equivalent gestational periods in rabbit and guinea pig;[390,391] these levels had risen to 10% adult activity by day 20 and increased markedly over birth.[390,391]

There is no obvious difference between man and animals in fetal liver glucuronidation as there is with Phase 1 drug metabolism,[637] and the low fetal transferase activities in man noted above for steroids are evident also with xenobiotic substrates. Rane et al.[701] report that from 13 to 22 weeks no microsomal transferase activity towards 4-nitrophenol, 1-naphthol, or 4-methylumbelliferone could be found, even with digitonin, ATP, or UDPGlcNAc present. Over 8 to 16 weeks, Wishart (unpublished results, 1977) found no activity in activated or unactivated human fetal-liver homogenates towards 2-aminophenol, 4-nitrophenol, 1-naphthol, morphine, or the two endogenous nonsteroids, bilirubin and serotonin.

UDPGlucuronyltransferase activities towards various substrates develop at different rates.[702] The simplistic idea that transferase always reached adult levels postnatally was soon disproved,[370,378,703,704] activity for some substrates in some species reaching adult levels or over before birth. (Note that with some substrates "adult" transferase activities of pregnant or nursing mothers are below normal female values [Section III.B.1 above]). This "functional heterogeneity" of developing transferase was semiquantified some time ago,[370] but lacked provision for age-dependent latency. Recent mutually supporting work is clarifying the situation. Before discussing it, we must mention the concept of developmental "clusters" of enzymes.

Greengard[705] pointed out that although different enzymes developed at different times and at different rates their development in many fetal tissues could be assigned to "eventful periods". A surge of often-related enzymes occurred when their action would be physiologically appropriate for the fetus or infant (e.g., the glycolytic enzymes surging at birth). She listed the "late-fetal", "neonatal", and "weaning" clusters as examples of such groups of enzymes surging at these particular eventful periods in the rat.

It is hazardous to assert, and Greengard did not assert, that these are the only eventful periods and that every enzyme must fall into one or more of these clusters in every species. However, recent studies by Wishart[390,391] and Lucier's group[389] strikingly illustrate that functional heterogeneity exists among the various UDP-glucuronyltransferase activities in several mammals, and that it fits with the cluster concept. Wishart and his colleagues studied the development of activity towards twelve substrates in rat, taking account of latency, V_{max}, and other factors.[391] These activities could be arranged in two clusters: (1) one in which adult male (3 months plus) activities were achieved before birth and (2) one in which they were attained just after birth. In the first ("late-fetal") cluster were activities toward 4-aminobenzoate, 2-aminophenol, 4-methylumbelliferone, 1-naphthol, 4-nitrophenol, and serotonin; in the second, ("neonatal") towards bilirubin, chloramphenicol, estradiol, morphine, phenolphthalein, and testosterone. Both clusters include activities to endogenous and xenobiotic substrates. Within each cluster the rates of development are not necessarily identical. Activity towards bilirubin in the rat develops slower than those towards testosterone or morphine.[391] However, all three are clearly distinct from any of the first cluster activities which are at 90 to 140% adult level when they are 7, 10, and 1%, respectively. Six of the substrates (2-aminophenol, 4-nitrophenol, 1-naphthol, morphine, testosterone, and bilirubin) were studied also[391] in guinea pig, and mouse, and fell into approximately the same clusters as in rat. These clusters of transferase activity in the rat are exactly related to induction by hormones (Section IV.D below) and xenobiotics (Chapter 13, Section II). Lucier et al.[389] have independently found two developmental groups of activities. One, the "nonsteroid" group, contains activities towards 4-methylumbelliferone, 1-naphthol, and 4-nitrophenol. It develops so that adult levels are reached at day 1 of free life. The other, the "steroid" group, contains activities towards diethylstilbestrol, estradiol, estrone, phenolphthalein, and testosterone. It peaks a little later.[389] This first group falls into Wishart's first cluster, and the three activities in the second group that Wishart himself has studied fall into his second cluster. Lucier et al.[389] also find that dependency on induction by a xenobiotic may relate to a particular developmental group. Fuchs et al.[231] studied development in rat liver of activities towards estriol, pregnanediol, testosterone, and bilirubin and found all reached adult levels well after birth, consistent with their being part of Wishart's "neonatal" cluster and the "steroid" group of Lucier et al. For reasons not yet clear, the developmental rate reported by Fuchs et al.[231] is slower than that of the other investigators, e.g., activities towards estrone and bilirubin only reach adult levels on day 20 of free life, whereas in the work of Wishart,[391,707] Lucier et al.,[389] and others (see below) this occurs on days 2 to 3. Activity towards pregnanediol in the female rat is unusual in first surging at 33 days free life and then again (fourfold) at 60 days.[231] Activity to testosterone shows a second pronounced surge at 20 days.[231] As, at these late stages, stimulation by sexual maturation may overlie basic ontogeny of the enzyme, and the preparations were not deliberately activated, the results of Fuchs et al,[231] like the earlier work referred to, should not be too closely compared with those of the later investigators.

The concept of clusters must not obscure the possibility of various developmental rates of activities within each cluster, nor of a spectrum of developmental rates straddling the two clusters. At present, however, coincidence of selective induction by glucocorticoids and xenobiotics supports a functional heterogeneity displaying itself in at least two developmental patterns. Further work[145a] has extended these observations to a series of alkylated phenols with a clear break between the first and second cluster. From a study of limiting configurations of the substrates, it can be predicted what structures within this series of alkylated phenols will be readily glucuronidated by the liver transferase of infant rat on birth and which will not; implications for its extension to man and design of drugs are obvious.

Contribution of the endoplasmic reticulum membrane to this heterogeneity will appear when specificity of pure enzyme is examined ontologically.

Development of UDPglucuronyltransferase activity to bilirubin is of interest pharmacologically, for conjugation of this substrate is frequently affected, positively or negatively, by medication of pregnant mother or neonate. Also, its high plasma level is readily observed and measured. Early reports are noted by Dutton.[4] The most recent detailed observations of the digitonin-activated enzyme agree that in rat it first appears just before birth and rises to approximate adult values at the second day[389,678,706] or fourth day.[708] A slower rate noted[231] was possibly due to lack of digitonin activation. Fevery et al.[462] tabulate development of activity to bilirubin in several species and with several procedures. Early reports are especially suspect. Activity to bilirubin in liver of rhesus monkey studied by the latest techniques, including digitonin activation, showed a very similar pattern of postnatal development to that in rat liver, allowing for the longer gestational period.[649] Human 8- to 19-week fetal liver exhibited some 10% adult level or less, with one at 22 weeks some 17% adult mean.[695] The surge of transferase activity to bilirubin in rat liver only occurred on birth, whether normal, premature, or retarded.[707] A 2-day premature birth of rats resulted in precocious development of activity, but subsequent rate of development varied, apparently with nutritional state of the premature neonate.[707] Retarded birth did not prevent onset of the activity in utero.[707]

2. Developmental Patterns in Extrahepatic Tissues

a. Kidney

Fetal kidney UDPglucuronyltransferase activity is generally low but present towards xenobiotic substrates in all species, including man.[197,574,682] Overall glucuronidation in guinea pig kidney just before birth, is more active than in liver.[682] Early references are given by Dutton.[4] Activity towards steroid substrates (estrone and testosterone), absent in late fetal (or adult) rat kidney,[691] is present in low amount towards estriol in that of 12- to 17-week human male fetuses[197] and towards C-21 steroids in the premature human neonate.[709] Kidney probably never exhibits activity to certain substrates. The activity towards phenolphthalein is reported absent throughout life in rat[389,691] and swine,[710] and towards bilirubin in man (Chapter 14, Section III); where it does develop, it need not develop as in liver.[691]

Administered glucocorticoids or xenobiotics stimulated fetal kidney transferase to 2-aminophenol in the species studied.[342,390,391,711] In chick embryo kidney (meso- plus metanephros), as in liver, activity towards 2-aminophenol remained low until hatching when it rose markedly[500] and also increased spontaneously on culture.[369]

b. Alimentary Tract (see also Chapter 11, Section IV.D)

Overall glucuronidation of 2-aminophenol was over 10 times as high on a dry weight basis in strips of 5- to 7-week guinea pig fetal stomach as in liver slices, where it was often undetectable.[682] Transferase activity in this tissue was higher than observed in infant or adult-guinea-pig stomach, even with gastric mucosa as source rather than whole stomach wall.[574] The same phenomenon occurred in fetal mouse stomach (G.J. Dutton, unpublished results, 1965). Intestine of both species was less active prenatally than stomach.[574] Maximum transferase activity appeared shortly before term.[574]

Glucuronidation of 2-aminophenol was undetectable in human fetal stomach or gut at 8 to 12 weeks,[574,682] was low, but detectable, towards estriol at 10 to 17 weeks,[197,700] and could contribute to those glucuronides, largely of estriol, noted in fetal tissues and fluids (Section II.A above) and believed to be formed partly in the fetal alimentary tract.[655]

In pig gut,[710] transferase activity to phenolphthalein increased from negligible levels immediately after weaning, suggesting a dietary factor. In gut of chick embryo,[500] activity to 2-aminophenol was undetectable until hatching and then rose to adult levels within 3 days. The onset was independent of food intake or starvation, but the final level may vary with diet.[500]

Injection of phenobarbital stimulated chick-embryo duodenal transferase towards 2-aminophenol minimally, raising it markedly in liver and kidney.[342] Injection of the mother with dexamethasone stimulated development of activity in rat fetal duodenum and stomach, suggesting a hormonal cause of the natural onset.[390,391]

c. Other Tissues

An interesting new site of UDPglucuronyltransferase activity is rat uterus,[389] which prenatally glucuronidates substrates of the first group of Lucier and McDaniel[389] (Section III.E.1 above) but not those of the second group. Development perinatally in lung, another new site, has been described,[691] but not for phenolphthalein in pig lung.[710] Wishart et al.[390,391] report that dexamethasone stimulated activity in fetal rat lung towards substrates tested in their late-fetal group. Adrenal gland is suggested, on slender evidence, as a site of C-21 glucuronidation.[709]

Placenta has been a controversial site. Early workers[4] found no significant glucuronidation there, and in man, only estrone and estradiol conjugates (not specified) at over 1 μg/kg tissue at term.[712] These glucuronides were reported earlier,[713] but no UDPglucuronyltransferase activity to estriol (in 11- to 20-week tissue)[197] or to 1-naphthol,[714] 4-nitrophenol,[715] 4-methylumbelliferone,[716] or other metabolites.[648] Aitio,[326,717] employing activated placental preparations, detected only low activity towards methylumbelliferone in several species and, occasionally, in man. The highest found with this sensitive assay (measuring down to 10 pmol), was 10% fetal guinea pig liver, itself very low, and he concludes that placenta is insignificant in glucuronidation. He found no induction by cigarette smoking. However, placenta activity towards oxazepam may be induced in rabbit by phenobarbital[718] and in rat placenta towards 4-nitrophenol by 2,3,7,8-tetrachlorodibenzo-p-dioxin (TCDD).[388] UDPGlucose dehydrogenase[541,542] and UDPGlcUA occur in placenta.[4]

F. Development of Glucuronidation Enzymes in Poikilotherms

Little has recently been done on this problem. Transferase activity towards phenols is absent in larval livers of amphibians examined[4,12,719] and remains absent when the adults are wholly aquatic.[719] In livers of dogfish,[720] activity towards bilirubin appears constant during development, although UDPGlcUA breakdown is more rapid in the young. Investigators should note the low-temperature requirements of poikilotherm UDPglucuronyltransferase (Chapter 12, Sections II.D, II.E). In plants, ontogeny of the system is unknown.

IV. INDUCTION OF PERINATAL GLUCURONIDATION

A. General

Neonatal glucuronidation is often diminished further in prematurity, malnutrition, or some genetic conditions. Whereas medication with glucuronidogenic xenobiotics can be avoided in these cases, endogenous bilirubin accumulates. Precocious development of UDPglucuronyltransferase has, therefore, been sought. Two approaches are evident: (1) to assist, therapeutically, detoxication in premature and normal neonate and (2) to elucidate the natural mechanism of the system's development. Both have utilized xenobiotic and endogenous inducers.

B. Precocious Induction by Administration of Xenobiotics
1. Clinical Use

Therapeutic aspects of this procedure are expertly reviewed by Wilson,[639] particularly the ethical considerations. He lists the instances of phenobarbital treatment of neonatal or infant hyperbilirubinemia from the pioneer work of Yaffe et al. in 1966[721] up to 1970. He concludes that as postnatal treatment of full-term or low-birth-weight infants with phenobarbital produced little effect on hyperbilirubinemia until after at least 3 days,[677,722] it is not useful for acute conditions. Treatment of the pregnant mother with phenobarbital from 35 weeks, however, approximately halved the neonatal bilirubin load by the third day of life.[722] This increased clearance of bilirubin probably involved increased glucuronidation for increased excretion of salicyclamide glucuronide accompanied it.[677] Another good review is by Eriksson and Yaffe.[673]

A useful review on neonatal phenobarbital metabolism notes that its hydroxy derivative is conjugated (presumably as glucuronide) within 4 hr of birth.[723] Phenobarbital treatment appears to reduce the frequency of, and sometimes the need for, exchange transfusion, but sex and race are among factors influencing the response.[639]

As phenobarbital may have adverse psychological effects on the mother and depress respiratory function in the neonate, other drugs have been tested to stimulate bilirubin glucuronidation, but none appears satisfactory. Ethanol, if it does induce microsomal enzymes, seems little use.[639] Accelerated plasma clearance of bilirubin possibly through induction, occurs in infants of drug-addicted mothers (e.g., for heroin[724] or cannabis[640]) and chronically administered ethanol may act similarly,[462,725,726] but ethanol in acute dosage reduces transferase activity towards bilirubin.[380,381] However, as Fevery et al.[462] point out, chronic ethanol administration is possibly more likely to result in stimulation of drug-metabolizing enzymes than acute dosage, and Idéo et al.[725] administered it at a lower level for a longer period than did Hakim et al.[380,381]

2. Precocious Induction by Administration of Xenobiotics to Animals
a. Barbiturates

Remmer showed in 1964[727] that barbiturates increased glucuronide excretion by adult rats, but Catz and Yaffe[728] were among the first[4] to study the mechanism, finding increased bile flow and UDPglucuronyltransferase activity towards bilirubin in liver homogenates from infant mice and rabbits following barbiturate treatment and, in mothers treated during the last week of pregnancy, increased activity in both mother and neonates. These studies provided the basis of the clinical trials. Phenobarbital differs from other xenobiotic inducers such as 3,4-benzopyrene and 3-methylcholanthrene in increasing bile flow as well as transferase activity, and confusion of those two processes, and of hepatic uptake, has led some to deny that increased transferase activity is significant. This is discussed for adults in Chapter 13, Section II, but recent work by Gartner et al.[649] is in the neonatal context. They studied bilirubin transport and conjugation in the newborn monkey treated with phenobarbital and suggest two phases in hyperbilirubinemia. Phase I includes rapid increase of serum bilirubin by 19 hr in the newborn monkey, and its rapid decline by 48 hr to a level still 4 times that in the adult. In Phase II, this level remains until 96 hr, when it declines to the adult level. Phase I results from the six-fold greater bilirubin load presented neonatally and here deficiency of glucuronidation, not uptake or secretion, is rate-limiting.[649] However, hepatic uptake may contribute to limit Phase II, by which time the transferase has developed considerably.[649] Phase I was completely abolished by prenatal maternal and neonatal administration of phenobarbital, due to increased transferase activity. The bilirubin load, itself unaffected by the drug, was in untreated infants just too much for the transferase to handle, but phenobarbital-treated infants possessed just enough transferase to conjugate it effectively.[649] By Phase II, and during the rest of the neonatal period studied, the hepatic transferase activity towards bilirubin in treated monkeys was three-fold higher than in controls. The general pattern appeared similar in human neonates, with the time factor increased three times.[649] Ligandin, the protein assisting intracellular transport of bilirubin to the conjugating site, appears perinatally,[729] and its development is accelerated neonatally.[730] However, from the work[649] quoted above, ligandin supply appears not limiting in neonatal bilirubin glucuronidation.

Age-competence to respond to phenobarbital was investigated in chick embryo[342] by injecting fertilized eggs (the compartment injected appeared unimportant) with phenobarbital at various embryonic ages; the dose tolerated depended on age, but response depended on dose and, at higher dosage, on the period of exposure to the drug. Response occurred in liver and kidney, (meso- and metanephros), but minimally in skin and duodenum, and not in spleen, brain, lungs, or extraembryonic membranes.[342] In embryo liver, the stimulated enzyme exhibited activities up to 40 times the usual adult level. Notably, there seemed no lower age limit of response; eggs injected on day 1 yielded marked transferase activity 72 hr later in the visceral area (developing liver, esophagus, pancreas, and stomach) which in control embryos possessed activity undetectable by this assay until 6 to 7 days later.[342]

Response to phenobarbital by mammalian liver however, appears markedly age dependent. In pregnant and nursing ASH/TO mice allowed access to phenobarbital-treated drinking water after the second gestational day, transferase activity had increased after 7 days two-fold to a plateau. From at least 15 days, concentration of liver phenobarbital was similar in mothers, infants, and fetuses. Despite this high level, transferase in test fetuses increased over that in controls only on the 19th day — 1 day after the enzyme normally becomes detectable. Its stimulation was then approximately twofold and remained so into infancy. The same age-competence was reported with bilirubin, estriol, and 2-aminophenol as substrates, even though activity to 2-aminophenol reaches adult values prenatally.[377] This suggests why human neonates from phenobarbital-treated mothers still exhibit hyperbilirubinemia, though less than controls;[639] had transferase been induced earlier, one might have expected birth with adult levels of the enzyme. Certain activities of the Phase 1 system are also not stimulated in fetal guinea pigs until just before parturition, despite prolonged maternal treatment with phenobarbital; response coincided with hypertrophy of the smooth endoplasmic reticulum, similarly delayed.[731]

b. Other Xenobiotics

Chloroquine and chlorcyclizine[732] raise liver UDPglucuronyltransferase in offspring of treated rats and administration of benzo-(a)-pyrene to newborn animals[732,733] stimulates development of the enzyme.

Dietary compounds such as terpenoids (e.g., linalool) have been suggested as inducers of the gut enzyme.[734] Ethanol (Section IV.B.1), not a proven inducer of the enzyme, when infused into chick embryos failed to raise liver transferase activity to 2-aminophenol above normal negligible levels. (G.J. Dutton, unpublished results). Those polychlorinated biphenyls not accumulating significantly in maternal adipose tissue (being hydroxylated by the mother) accumulate markedly in fetal rat intestine from the 19th day of gestation, when fetal hepatic transferase activity towards them begins.[734a,734b] The hydroxylated intermediates may cross the placenta, travel to the fetal liver and become glucuronidated there, subsequently reaching the fetal intestine; there they are deconjugated, some being returned to the fetal liver and reconjugated. Accumulation did not occur after birth, when the glucuronides could be excreted.[734a] No glucuronidation occurred in fetal intestine. The structure-activity relationships necessary for accumulation were worked out.[734a] When given to the mother over days 5 to 18 of gestation, 4- and 6-chlorinated biphenyls induced transferase activity to 4-nitrophenol in the offspring at 21 days of age.[734b] Exposure to TCDD, especially via maternal milk, induced transferase activity postnatally towards "nonsteroid" substrates, but not towards steroid substrates.[389,734b]

3. Stimulation of Overall Glucuronidation by Xenobiotics

Overall glucuronidation is increased by xenobiotics, as seen with liver slices from pretreated perinatal animals[342] or in exposed cultured tissue.[735] Not only a rise in UDPglucuronyltransferase activity may contribute: UDPglucose dehydrogenase itself is stimulated. Stimulation of this enzyme by barbiturates or other xenobiotics, especially in neonates, was long debatable,[311,570,732,733,736,737] but in chick embryos and perinatal mice, it now seems reasonably certain.[502] Phenobarbital in the culture medium increased the dehydrogenase in organ-cultured embryo liver two- to three-fold (but curiously not in cultured liver cells). When injected *in ovo* at 8 days it increased the liver enzyme over five-fold, dose-dependently, 5 days later. In late-fetal, but not 5-day infant, ASH/TO mice, a two-fold stimulation of liver UDPglucose dehydrogenase was noted. The pregnant and nursing mothers were receiving the drug in drinking water.[502]

There seems no direct linkage between development of UDPGlcUA synthesis and that of transferase activity, and in adult life the increased transferase activity due to xenobiotic administration appears catered for by existing levels of UDPGlcUA (Chapter 9, Section III.F). However, the increase by xenobiotics of overall glucuronidation in neonatal mammals probably requires concurrent stimulation of both UDPglucuronyltransferase and UDPglucose dehydrogenase. Changes in the activity of β-glucuronidase and UDPGlcUA pyrophosphatase are too small and cannot contribute to this increased overall glucuronidation (J. Fyffe and G.J. Dutton, unpublished results, 1975).[171,619]

The use of glucarate excretion to monitor "microsomal enzyme induction", useful in neonates, is discussed in Chapter 13, Section II.K.

C. Precocious Induction in Cultured Tissue

A remarkable phenomenon recently demonstrated, not confined to this enzyme, is the apparently spontaneous induction of chick-embryo-liver UDPglucuronyltransferase on simple culture in chemically defined media.

Chick embryo, possessing negligible liver and kidney UDPglucuronyltransferase until just after hatching at 21 days,[500.739] is useful for studying mechanisms regulating onset of the enzyme over birth;[738] onset is sudden and embryo tissues easy to culture. Liver segments from 11-day embryos cultured immersed in serum-containing medium displayed unexpected by the appearance and rise of transferase activity, initiated within 2 days and reaching adult levels after 8 days, when transferase levels *in ovo* were still negligible. Overall glucuronidation also reached adult levels. The segments became encapsulated with spindle-shaped cells, but transferase activity was absent from the capsule and located in the "microsomal" fraction from the highly pyknotic hepatocytes of the "core".[738] On longer culture, hepatocyte degeneration progressed further, but transferase activity rose, or was maintained, up to 3 weeks. It achieved very high specific activity on a protein-nitrogen basis and remained in what spun down as the "microsomal" fraction.[738]

The rise is not from "dilution out" of other protein. Activity rose from virtually undetectable levels to several times those in the adult liver,[735.738] and later evidence (see below) supports a true induction requiring amino acid incorporation. The possibility of synthesis of transferase by cells possessing virtually unrecognizable internal structure has not been further investigated, nor the alleged inhibition by some serum samples.[738]

More "normal" cells were obtained by culture on rafts in a chemically defined medium[740] containing only glucose, amino acids, and vitamins, with no known inducers. Phenol red, useful for pH monitoring, could be omitted without effect[740] and has no inducing activity when injected into eggs.[342] Light and electron microscopy revealed healthy tissue, with considerable morphological maturation[741] for over a week, by which time the transferase had precociously increased from virtually zero to above adult levels.[740] Similar increase of enzyme and overall glucuronidation also occurred in monolayer cultures but transferase activity fell to zero when fibroblastlike cells invaded the monolayers. Increase, unaffected by X-irradiation preventing mitosis, appeared independent of tissue architecture and cell division.[740]

Initiation of liver transferase development required only isolation from the embryonic environment. In liver cultured *in ovo* on the chorioallantoic membrane, no transferase developed.[740] If these cultures were then removed from the membrane and plated as monolayers, transferase increased immediately. The membrane supported transferase activity in already developed embryo livers previously cultured in vitro, but did not permit its further development.[740] Culture of whole embryos in an open dish for the first 8 days of incubation, allowing apparently normal morphological development, provided sufficient of an "*in ovo*" environment, for the transferase was not then spontaneously induced.[755]

Phenobarbital added to the culture medium increased the rate of this spontaneous development,[735] and one series of experiments suggested that phenobarbital, more stimulatory in cultured liver from 11-day than from 5-day embryos, overrode some control of transferase synthesis breakdown which developed *in ovo* between 5 and 11 days; a half-life of 2½ days was suggested for the enzyme,[735] but with a membrane-bound enzyme of such regulatory complexity as UDPglucuronyltransferase, caution is required.

The enzyme developing on culture resembled that appearing at hatching in possessing latency (absent from embryo-liver enzyme[182]) and possessed similar "kinetic" parameters. Development in culture was inhibited by, and restored after, cycloheximide pulsing, directly with amino acid incorporation.[735] The increase, therefore, appears an induction, not activation, of the transferase, or of an enzyme whose action activates transferase: kinetics of the cycloheximide effect makes the latter interpretation less likely.[735]

Transferase is also stimulated in cultured kidney.[369] Many "explanations" can be hazarded, but the possibility of a repressor being synthesized extrahepatically and extrarenally, or extraembryonically, is no more substantiated than accumulation of inducer on culture. Cortisol metabolism lessens markedly on culture of fetal rat-liver explants,[742] which might suggest that an inducer, produced by embryonic liver and metabolized there *in ovo*, fails to be metabolized on culture and so accumulates.

The phenomenon has recently been reported with quite different enzymes, culture of chick embryo liver resulting in replacement of synthesized embryo plasma proteins by the adult forms.[743] Electrophoretic patterns[743] prove a change-over in protein synthesis and thereby suggest that induction of a transferase protein, complete with adult characteristics, is less unlikely.

"Spontaneous" rise in malic enzyme in cultured 18-day chick-embryo liver cells, requires serum in the medium appears due to a bound thyroid hormone.[744]

Sherer,[739] using the sensitive assay with harmol, detected low levels of UDPglucuronyltransferase in chick embryo liver *in ovo* after only 60 hr incubation, its spontaneous appearance there on culture, and a 1000-fold increase after 5 days culture. Culture did not increase transferase in gut or lung noticeably, as also found by Wishart and Dutton (unpublished results), and there was no trace of an embryonic inhibitor *in ovo*, nor evidence of its loss or of the appearance of an activator during culture either.[369,739]

During organ culture of chick-embryo liver, the endoplasmic reticulum hypertrophies,[740,741] and annulate lamellae rich in ribosomes accumulate.[745] The stereological work of Banjo and Nemeth[143] is described in Chapter 4, Section I.B; they suggest that embryonic reticulum in culture possesses a more constant protein framework, less exposed to hormonal changes, than in vivo.

Possible changes occurring in vivo are considered in the following section.

D. Precocious Induction by Endogenous Molecules
1. Induction by Substrate

Induction of specific UDPglucuronyltransfersase activity by accumulation of the substrate is attractive, especially with bilirubin. Fetuses suffering erythroblastosis contain some conjugated bilirubin in liver cells,[687,688] highest in the most heavily immunized babies.[745a] Bakken[746] reported that loading of pregnant rats with injected bilirubin increased neonatal transferase activity over the first 3 days, but the assay used is obscured by hemoglobin and other compounds,[747] and a reinvestigation[748] employing diazotization could not confirm this finding for either "native" or digitonin-activated enzyme in neonates of 0 to 2 and 20 to 24 hr of age.[748]

The possibility was raised again by the ingenious experiments of Thaler,[706] who mated Wistar rats with the hereditarily jaundiced Gunn strain of rat, producing two types of heterozygotes, Type 1 from jaundiced mothers and normal fathers, Type 2 from normal mothers and jaundiced fathers. Type 1 fetuses would differ only in being continuously exposed to high bilirubin concentrations crossing the placenta. Their liver transferase activity to bilirubin was about 2½-fold higher at birth than in Type 2, and its postnatal development more rapid.[706] Type 1 fetuses were born with high bilirubin levels, but cleared them rapidly, whereas Type 2 fetuses exhibited hyperbilirubinemia after birth. Thaler[706] suggested that transient hyperbilirubinemia in normal infants represented a perinatal adaptive process evoking this essential enzyme. Winsnes and Bratlid[748] question this conclusion. They ask why activity in normal Wistar neonates is as high as in Type 1 heterozygotes despite the latter being exposed to a 14-fold higher concentration of bilirubin. (However, there could be an upper limit of induced activity in heterozygotes, and Type 1 should be contrasted with Type 2 rather than with Wistar neonates). They also, more pertinently, queried the hormonal status of Type 1 fetuses compared with Type 2, for development in a jaundiced mother could expose the fetus to the possible hormonal abnormalities which enable the Gunn mother to survive under stress.

In normal mice and mice genetically deficient in erythropoiesis from 16 days *in utero*, hemolysis was provoked experimentally, and bilirubin levels judged from reduced hematocrits.[749] As development of transferase activity was similar in normal and mutant infants, the authors concluded that lack or excess of bilirubin did not affect the bilirubin enzyme. Unfortunately, they assayed activity to 2-aminophenol which develops at a different time (Section III.E.1 above) and probably under a different control (see below).

In rat intestine, transferase activity towards bilirubin develops neonatally and then declines, in parallel with β-glucuronidase, which could suggest that intestinal transferase is induced by bilirubin freed from its glucuronide by β-glucuronidase.[699]

Culture of liver explants from chick embryo or fetal rat with bilirubin, biliverdin, heme, or hemoglobin in the medium has not stimulated transferase activity to bilirubin[750] (M.T. Campbell, unpublished results), (not unexpected in chick liver, which has negligible adult activity to this substrate). One must conclude that induction by endogenous substrates of UDPglucuronyltransferase in the neonate is unproven, including induction in the gut by dietary constituents such as terpenoid derivatives (c.f., Reference 734), if these are regarded as "endogenous".

2. Induction by Hormones

Although many hormones (e.g., thyroxine, adrenaline, and several steroids) are substrates of UDPglucuronyltransferase, any induction due to such compounds reflects their status as hormones rather than as substrates.

The pattern of transferase development exhibits an "overshoot", in which adult levels are exceeded,[370,500] typical of a hormone-mediated induction.[705] Hormones were early known to affect UDPglucuronyltransferase activity in adults (Chapter 13, Section I), but little success attended attempts to induce the enzyme perinatally with these agents or to demonstrate its repression by them in the fetal environment. A report[751] that induction of liver enzymes by phenobarbital was prevented by injection into weanling rats of supernatant from fetal but not adult rat liver has not been followed up. No inhibitors of Phase 1 metabolism, such as maternal steroids, were evident in vitro when plasma from neonates or nursing mothers was examined[639] and suggestions that progesterone metabolites might be responsible[752] are capable of various explanations.

Although the work of Wilson[639] suggested a link between growth hormone and development of Phase 1 metabolism, implantation into young rats of a pituitary tumor secreting enough growth hormone to prevent the normal age-dependent rise in Phase 1 metabolism[753] did not affect development of UDPglucuronyltransferase activity (J.T. Wilson, personal communication, 1976). Other work was inconclusive, e.g., corticotropin doubled the glucuronidation in vitro of C-21 steroids,[709] but no UDPGlcUA seemed to be added to the reaction mixture.

Interpretation of work with hormones is complicated by plasma- and intracellular-binding, but recent work indicates that glucocorticoids trigger the perinatal development of certain UDPglucuronyltransferase activities. The work progressed from chick embryo to mammalian fetus, and that sequence is followed below.

a. Work with Chick Embryo

The apparent repression of the *in ovo* environment on the enzyme in cultured tissue has been noted above. Any repression was not due to embryonic pituitary gland, for its early extirpation did not allow spontaneous induction of the liver transferase *in ovo* nor prevent spontaneous induction when the liver was subsequently cultured in vitro.[369,754] Changes in oxygen or CO_2 tension during late incubation did not change the normal developmental transferase pattern.[342] Phenobarbital was then injected into the egg[342] (because a xenobiotic seemed more likely than a hormone to override regulatory mechanisms safely) and resulted in marked induction *in ovo.*[342]

A variety of injected hormones yielded no consistent results, presumably because the boundary was delicate between homeostatic destruction of hormone and death by sudden excess.[342] More physiologically, various tissues from the adult bird were grafted on to the chorioallantoic membrane so that factors, perhaps unknown, would be secreted into the embryonic circulation. Success was immediate with pituitary gland,[754] the cephalic, but not the caudal, area being effective,[754,755] confirming that these two areas differ qualitatively in chick,[756] and suggesting that luteinizing hormone, from the caudal area, and growth hormone, from both areas,[756] were not responsible for initiating the rise in UDPglucuronyltransferase. Corticotropin, thyrotropin, or follicle-stimulating hormone, derived from the cephalic area, could have been responsible. Transferase activity was stimulated from virtually zero up to adult level in both embryonic liver and kidney.[754,755] While aniline hydroxylase also rose in liver from one third to full adult value, specific activities of other microsomal constituents such as glucose-6-phosphatase, cytochrome P-450, and NADPH-cytochrome c reductase did not greatly change.[754,755] Response of transferase to the graft did not need the host embryo's pituitary gland,[754,755] indicating that host stress was not contributory, nor apart from one day for vascularization of the graft and one day for initiation of response, did it require presence of the graft itself. Response of host to factor was markedly age-dependent, unlike response to phenobarbital (see preceding sections), not occurring before day 14 of incubation.[755]

Glands from embryos younger than 17 days were ineffective and, judged on a wet weight basis by their relative effectiveness in provoking transferase development, began to secrete the factor suddenly some 24 to 48 hr before the naturally developing surge in liver transferase activity;[755] over hatching, secretion appeared higher than in adult birds. The factor was, therefore, probably responsible for the natural rise in transferase activity after hatching.[755]

A new simple continuous infusion technique[757] allowed the testing of hormones secreted by the cephalic area, and of them, only corticotropin stimulated precocious transferase development. The effect was reproducible with Synacthen, a synthetic preparation of its active peptide sequence.[711,757] Consistent with corticotropin being the active factor secreted by the gland, corticosterone and other 11 β-hydroxy corticosteroids reproduced its stimulatory effect on infusion, and the same age-dependent response of the embryo.[711] Thyroid hormones evoked precocious retraction of the yolk sac[758] but gave only small and inconsistent increases in transferase activity. They exerted a synergistic effect[711] and the infusion of corticosteroid and thyroxin together permitted a response identical in extent and rate of onset to that called out by the graft itself.[711] A wide range of other steroids and hormones had no effect, and in the chick, development of UDPglucuronyltransferase activity towards 2-aminophenol and 4-nitrophenol was concluded to be triggered by corticotropin and assisted by thyrotropin. Information on hormone levels in vivo over hatching, incomplete or unsatisfactory,[711] suggests a rise in embryo plasma corticosteroid just before hatching. It is consistent with such a role for corticotropin. Competence experiments indicate that competence is not primarily through supply of corticosteroid *via* the pituitary-adrenal axis at that age (days 13 to 14), but in acceptability of corticosteroid, probably at the liver cell. Corticosteroids act at the liver cell, from their stimulation of transferase in cultured embryo liver.[711]

A surge in Phase 1 enzymes reproducing precociously the pattern seen on hatching was also noted,[759] the lag being 24 hr less than with the transferase, consistent with their relative half-lives.[759]

b. Work with Mammalian Fetus

Although there is no *a priori* reason for hormonal analogies between the independent avian embryo and the maternally governed mammalian fetus, stimulation of embryo-liver transferase by mammalian pituitaries grafted on the chorioallantoic membrane[755] suggested that possibility.

The disadvantage of studying a maternally dependent fetus is offset by absence of spontaneous induction of transferase in cultured fetal tissue such as hampered experiments with corticosteroids on cultured-embryo tissue.[711]

Liver explants from 14-day rat fetuses cultured 3 days in a chemically defined medium containing the synthetic corticosteroid dexamethasone (5-fluoro-11β,17,21-tri-hydroxy-16α-methylpregna-1,4-diene-3,20-dione) had by then developed transferase activities towards 2-aminophenol from negligible to within adult-male values.[319,760] Transferase activity attained *in utero* by 17 days or in explants cultured without dexamethasone was still negligible. Compounds giving this stimulation at 2μM all possessed a pregn-4'-ene or pregna-1,4-diene structure and a hydroxy or an oxo group at C-11. Other steroids were inactive even at 20 μM. Transferase activity was assayed with "native" enzyme and with that activated by digitonin concentrations up to and beyond optimal.[319,760]

In case glucocorticoids merely maintained viability of the organ fragment and so allowed it to exhibit culture-stimulated development of transferase, the explants were maintained in a dexamethasone-free medium for 2 days and then transferred to one containing the glucocorticoid. Transferase activity, previously negligible, developed over the next 2 days at similar or greater rates than if cultured for 2 days from excision in dexamethasone.[319] Protein loss was similar in both cases and not above 40% after 4 days culture.[319]

On pulsing with cycloheximide, incorporation of ^{14}C from labeled leucine into protein fell by two thirds and the rate of increase of transferase, activity had also fallen by two thirds when measured 24 hr later;[319] within a further 24 hr, both the incorporation and rate of increase of transferase activity returned to prepulse level.[319] This confirmed what had seemed likely from observation of the precocious stimulation in preparations optimally activated by digitonin — that induction rather than activation occured on culture with glucocorticoids.[319,760] Some degree of transferase activation, however, may occur in cultured fetal mouse liver.[182,692]

In case the phenomenon arose only in culture, pregnant rats were injected on each of days 14, 15, and 16 of gestation, with glucocorticoids (which cross the placenta)[761,762] and the transferase activity was assayed in fetal liver homogenates,[693,763,764] with the same precautions regarding optimal activations and V_{max} as observed with cultured liver. Fetal-liver transferase activity rose progressively on injection of dexamethasone, reaching adult-male values by day 17 when values *in utero* and in controls injected with carrier (arachis oil) were still negligible. Rate of this precocious development was similar to that appearing naturally over the last few days *in utero*. Notably, transferase at extrahepatic fetal sites, e.g., kidney, lung, and upper alimentary tract, was also stimulated up to some four- to fivefold by dexamethasone.[693] This, together with rise to adult levels of enzyme activity from virtually zero, rules out premature loss of hepatopoietic cells (reported to follow cortisol injection)[765] as an explanation of the observed rise. Natural corticosteroids, equally effective as dexamethasone in culture, were less so on injection,[693] presumably because of their shorter biological half-life;[766] cortisol clearly raised activity, but corticosterone and progesterone were only marginally effective.[693] Progesterone is inactive in culture[319] and in vivo is probably metabolized to glucocorticoids.

Relevance of these findings to the regulation of perinatal development of UDPglucuronyltransferase may now be discussed.

Rat-fetal adrenocortical activity increases between days 17 and 20 of gestation,[761,767,768] being stimulated by corticotropin from the fetal hypophysis[769] and raises the corticoid level of fetal plasma above that of maternal plasma.[770] Over this period, liver UDPglucuronyltransferase activity in the rat colony employed rose from negligible to adult values.[319] Ablation of fetal hypophysis and previous maternal adrenalectomy prevented this normal rise.[693]

Wishart and Dutton[698] conclude that, for 2-aminophenol in rat, the prenatal surge of UDPglucuronyltransferase activity is brought about initially by fetal corticotropin stimulating production of fetal glucocorticoids which act at the liver cell to induce the enzyme. The mechanism therefore resembles that proposed for initiating the enzyme in chick-embryo liver and closely parallels the initiation of glycogen synthase activity in fetal-rat liver.[771]

Factors triggering perinatal onset of an enzyme need not be required for its subsequent maintenance,[705] and dexamethasone did not affect the maternal transferase;[693] adrenalectomy of adult males did not, within 10 days, lower transferase activity in their livers (M.T. Campbell, A.M. Donald and G.J. Wishart, unpublished results, 1977) as it similarly did not lower glycogen synthase in adult-rat liver although preventing its rise perinatally.[772]

Overall glucuronidation in rat-fetal liver under these conditions also rose precociously.[319,693,763,764] Stimulation by glucocorticoids of the other rate-limiting enzyme, UDPglucose dehydrogenase, has been noted (C. Petrou and G.J. Wishart, unpublished results, 1977) and may be involved.

Human fetal blood glucocorticoid levels (cortisol and cortisone), analyzed by competitive binding, rise threefold in late pregnancy and abruptly again during labor,[773] but the complexities of binding and cellular uptake demand direct studies before dependable extrapolation of the situation in rat to that in the human can be made.

Importantly, prenatal induction by glucocorticoids extended only to transferase activities glucuronidating substrates in the first, late-fetal group of Wishart[390,764] (Section III.E.1 above), comprising 2-aminophenol, 4-nitrophenol, 1-naphthol, 4-methylumbelliferone, 2-aminobenzoic acid, and serotonin. In contrast, activities towards chloramphenicol, phenolphthalein, morphine, testosterone, estriol, and bilirubin are not inducible in fetal rat by glucocorticoids, and though enhanced in neonatal rat by these compounds, they need additional, so far unknown, birth-dependent factors for their initiation.[391,707]

Of a series of substituted phenols, those below a certain limiting structure fall into the late fetal cluster of transferase substrates, and their glucuronidation is not induced by glucocorticoids; those bulkier than this structural limit fall into the neonatal cluster and their transferase activities are inducible by glucocorticoids.[143a]

As corticoids rise proportionally in nucleus as well as in plasma by the time of late fetal life,[774] the frequent assumption (e.g., Reference 775) of correlation between neonatal surge of an enzyme activity and development at that time of cytosol receptors for corticoids appears oversimplified.

V. OTHER TIME-DEPENDENT FACTORS IN GLUCURONIDATION

A. Cell Morphology

The cell population of mammalian fetal liver changes through development, but does not significantly contribute to onset of enzyme development,[705] the changes being too small to explain, for example, rise of UDPglucuronyltransferase activity from zero to adult levels.

The relationship of developing hepatocyte morphology to the transferase has been discussed[143] (Chapter 4, Section I.B), and to drug metabolism, generally.[731,776] The surface area of smooth endoplasmic reticulum of Fischer rats doubles between 1 and 20 months, and falls between 20 and 30 months. Surface area of the rough membrane at the Golgi apparatus remains unchanged.[777]

B. Circadian Rhythms

Little has been done on glucuronidation. Lagoguey et al.[778] found circadian and circannual changes of the urinary content of two steroid glucuronides.

C. Aging

Again, little has been done specifically on glucuronidation in senescent animals. From a good review,[779] glucuronidation appears, like most drug-metabolism reactions, to fall in the elderly.

D. Malnutrition During Development

Apart from limitation of carbohydrate for UDPGlcUA synthesis and of protein for synthesis of the transferase, malnutrition may affect the endoplasmic reticulum microenvironment to cause apparent gain or loss in UDPglucuronyltransferase activity. This phenomenon is best considered with diet in the following chapter.

VI. DEVELOPMENT OF OTHER DRUG METABOLIZING ENZYMES

Phase 1 drug-metabolizing enzymes develop in animals in a manner generally similar to the development of the glucuronidating enzymes, but in man appear present at quite a high level in mid- to late-fetal life. Evidence for this and the implications, have been well reviewed.[637,637a,637b]

Most conjugating enzymes follow the developmental pattern of UDPglucuronyltransferase, but very little is known either of the detailed pattern or of any mechanism of their initiation.[636] Development of glycosylation and sulfation is mentioned in Chapter 17.

VII. ADDITIONAL NOTES

Section III: This section is effectively updated to early 1980 by a fairly detailed review.[636a] Important points include the following. A uniquely comprehensive survey of hepatic transferase activity in human infants and fetuses towards 2-aminophenol and bilirubin[636b] finds both activities in the neonatal cluster, indicating that the cluster components in rat (Section III.E.1.) do not hold for man. A fuller report[636c] of Reference 707 details the birth-dependent triggering of neonatal cluster activities in rat, and another[636d] notes that activities in both clusters in the rat respond to glucocorticoids postnatally. Bilirubin injected into rat fetuses did not stimulate transferase activity to bilirubin,[636c] consistent with conculsions from other approaches. Study of microsomal membrane change developmentally in relation to transferase activity has begun.[636e] The possible perinatal triggers of drug-metabolizing enzymes are discussed in some detail in Reference 636a, with particular attention to the transferase.

Chapter 11

THE INFLUENCE OF DIET ON GLUCURONIDATION

I. DIETARY CHANGE

A. General

Dietary restriction affects UDPglucuronyltransferase and glucuronidation. Caloric restriction may reduce the availability of UDPGlcUA, and protein restrictions change the membrane composition,[780] and hence the constraint, compartmentational and/or conformational, of UDPglucuronyltransferase. As there is decrease in total liver protein, in the large areas of liver glycogen, and in the "concentration" of liver tissue during fasting, transferase activity expressed on a protein or wet-weight basis tends to be falsely high.[781] Many "activations" recorded result from neglect of calculating activity also on the basis of the whole liver during these conditions.

Early work on defective overall glucuronidation after starvation (i.e., after caloric and protein restriction) generally agreed on a fall with a rise again on administration of glucose.[4] Those reports and some recent ones did not allow for the membrane change accompanying dietary restriction, which could, depending on extent and probably also on substrate assayed, either stimulate or inhibit observed UDPglucuronyltransferase activity; contradiction is therefore evident.

Starvation depletes liver glycogen, but the transferase need not fall from this depletion. Artificial depletion of liver glycogen in mice following cysteamide injection did not decrease transferase activity there (W.S. Myles, private communication, 1964), although possible, though unlikely, induction of the enzyme by cysteamide was not controlled.

B. Influence of Diet on UDPGlucuronyltransferase Activity

1. General Substrates

Early work reported that starvation stimulated UDPglucuronyltransferase activity,[320] and recently Graham and Wood's group has helped to explain such observations.

Young female rats fed a protein-free diet for 7 days, markedly increased their liver transferase activity towards 2-aminophenol and 4-nitrophenol.[782,783] To distinguish between protein deficiency and calorific deficiency due to lack of appetite in protein deficiency, one group of rats was force-fed a nonprotein diet so that its total calorific intake remained as in a control group on a normal diet. Dietary stress in force-fed animals is greater because liver shows histologic abnormalities after such treatment.[783] Notably, transferase activities to 4-nitrophenol were even higher in these force-fed animals than in the simply protein-deficient caloric deficient rats.[783] More UDPGlcUA added in vitro, to allow for its possibly increased breakdown in homogenates from treated rate (from stimulation of UDPGlcUA pyrophosphatase), did not abolish these differences in transferase activity. Added glucarolactone indicated that increased β-glucuronidase activity was not responsible.[783] Since sulfotransferase was unaffected and as similar treatment lowers markedly Phase 1 enzyme activity (for references see Reference 783), the response of UDPglucuronyltransferase is unusual. The reason appears to be its activation by changes in endoplasmic reticulum.

To demonstrate this, male rats were forcibly fed a protein-free diet for 5 to 7 days or a low-protein (5% w/w, casein) diet for 60 days.[780] In both cases hepatic transferase activities towards 2-aminophenol and 4-nitrophenol increased some 2.4-fold and microsomal protein and phospholipid decreased together. Phospholipid in liver microsomal membranes was altered, control rats possessing very little or no lysophosphatides there, but protein-deficient rats possessed significant amounts of lysophosphatidylcholine and lysophosphatidylethanolamine. If liver microsomes from normal rats were pretreated with lysophosphatidylcholine, their transferase activity increased,[252,780] and it also increased if they were pretreated with phospholipase A, which liberates lysophosphatides.[780] Increased transferase activity in microsomes from protein-deficient rats therefore probably arose from its activation by lysophosphatides.[780] Supporting this activation in vivo, preliminary work[780] indicated increased glucuronidation of 4-nitrophenol[780] and enhanced bilirubin excretion[784] by livers of protein-deficient rats. Although specific and total transferase activity towards chloramphenicol fell in microsomes of protein-deprived guinea pigs,[785] transferase activity in guinea pig liver is more susceptible to inactivation by membrane perturbation than in rat liver (Chapters 5 and 12), and any lysophosphatide release should cause "over-activation", i.e., inhibition, of liver transferase activity in this animal. Two other reports[786,787] confirm a slight rise in transferase activity (to 4-methylumbelliferone) in protein-deficient rats. Liver transferase activity fell in rats fasting 72 hr,[788] but the explanation could be (a) calorific insufficiency, or (b) the preliminary storage of the liver at $-30°$ for 2 days partially activating the enzyme so that further perturbation from fasting resulted in inactivation; (a) is consistent with the protective action of UDPglucose injected intraperitonally in milligram amounts beforehand,[788] but specific effect of UDPglucose rather than glucose[788] is not clear. Simple starvation (3 days) did not lower specific transferase activity towards 4-nitrophenol[789] in rat liver or intestine, though total and organ weight fell; surprisingly, stimulation of transferase activity by phenobarbital pretreatment was greater in the starved animals.[789] Transferase activity in rat liver remained constant despite various pelleted diets, and in gut changed slightly,[790] but never as much as benzo(a)pyrene hydroxylase;[790] dietary soya protein tended to increase transferase activity.[791]

"Cholesterol fatty liver" did not change activity of transferase to 2-aminophenol;[99] more recently, influence of dietary lipids on UDPglucuronyltransferase activity has been studied by Laitinen[792] and colleagues. Here again transferase behaved differently to Phase 1 enzymes. A 4-week diet containing cholesterol, cacao butter, or olive oil (or in combination) at 24 to 34% of the total dietary weight modified composition of liver endoplasmic reticulum and, to a lesser extent, that of the postmitochondrial supernatant of a gastro-duodenal mucosa homogenate.[793,794] Cholesterol and olive oil alone, but not together, increased cholesterol in microsomes. Transferase activity to 4-nitrophenol in "native" liver microsomes increased somewhat on a cholesterol-rich diet, but in trypsinized microsomes it decreased: e.g., a tenfold activation in controls, but a 6-, 4-, and 3-fold activation in those fed cholesterol, olive oil, and cacao butter, respectively. In duodenal mucosa, a decrease also occurred.[794] Alteration in membrane structure had presumably decreased transferase latency.[794] A fat-free diet for 4 weeks left liver microsomal protein, cholesterol and phospholipid unchanged, but fatty acid composition of the phospholipids changed markedly, palmitoleic and eicosatrienoic acids increasing from virtually zero to 6.9 and 9.5% respectively, oleic acid doubling, and linoleic and arachidonic acids notably decreasing;[795] whereas Phase 1 enzyme activity remained unaffected, transferase activity to 4-nitrophenol in the native microsomes fell somewhat, and was almost halved after trypsinization.[795] Again, membrane changes appear to have reduced transferase latency. To study the influence of cholesterol itself, rats were fed a fat-free diet and a fat-free diet supplemented with cholesterol. The nutritional status of the two groups was similar, but in cholesterol-fed rats microsomal palmitoleic acid was twice, and arachidonic acid half, that in the controls.[796] Transferase activity to 4-nitrophenol was higher on the cholesterol diet in both native and trypsin-activated hepatic microsomes[796] and in preparations from duodenal mucosa.[797] Cholesterol in vitro had no effect.[796,797] Changed membrane environment following a cholesterol-rich diet may also lower the enhancement of transferase activity by polychlorinated biphenyl pretreatment in liver[798] and duodenal mucosa.[797] Changes in microsomal structure following cholesterol feeding have been detected with fluorescent probes.[799]

2. Bilirubin Glucuronidation

Dietary influence on the glucuronidation of bilirubin has been particularly studied. Fasting has long been known to evoke hyperbilirubinemia in animals[800,801] or man, especially when chronic nonhemolytic jaundice[802,803] or hemolytic anemia[804,805] is present. Evidence suggests increased bilirubin production through increased heme oxygenase[806] or decreased hepatic bilirubin clearance.[801,807] Suggested changes in liver UDP-glucuronyltransferase activity towards bilirubin[804] were not seen in fasted adult rats or in adult fasted patients with Gilbert's disease,[807,808] and as Gunn rats (which congenitally lack this enzyme activity) developed hyperbilirubinemia on starving,[809] a role for the enzyme appears excluded. Diets predominantly of carbohydrate or protein also increased hyperbilirubinemia in Gunn rats[809] and in man;[810] carbohydrate supplementation did not lessen fasting hyperbilirubinemia in Crigler-Najjar patients (who also congenitally lack transferase activity to bilirubin).[811] In Gilbert's syndrome, dietary lipid did reverse fasting hyperbilirubinemia;[800] neonates, presumably, should not risk lipid deprivation.

Transferase activity to bilirubin based on specific activity, but not total liver weight, rose 33% after complete starvation for 3 days; as this rise was additive to digitonin activation it might result from actual induction of the enzyme,[812] as with heme oxygenase. The fall on the total liver basis may be linked with fasting hyperbilirubinemia.[812]

II. DIET AND OVERALL GLUCURONIDATION

Some aspects of this subject have been treated in Section I.A above. Specific reports on the effect of diet on UDPglucose dehydrogenase are few. Some find the enzyme lowered on starvation,[320,813] others that it slightly (0.5-fold) increased,[789] although as liver weight decreased, the total enzyme fell. The discrepancy is attributed[789] to different techniques. Total β-glucuronidase activity increased in liver and duodenum during starvation,[320,789] possibly from increased lysosomal fragility, as microsomal β-glucuronidase was unchanged.[789] UDPGlcUA pyrophosphatase decreased during fasting,[320] so increased UDPGlcUA breakdown probably plays no part in the decreased overall glucuronidation during starvation noted in Section I.A above. No changes in rat liver UDPglucose level were observed during dietary stress.[814]

III. DIET AND DEVELOPMENT OF GLUCURONIDATION

This subject is also problematical; it involves progressive synthesis of both membrane and enzyme as well as dietary change.

Neonatal development of transferase activity was retarded towards 2-aminophenol in mice[682] and bilirubin in rabbits[815] following undernourishment, but supplementing an already adequate diet did not accelerate the latter activity.[815] Specific hepatic transferase activity towards 4-methylumbelliferone was similar in rats from large and small litters[816] but towards 4-nitrophenol it rose with increasing litter size;[817] substrate specificity is possible, because on experimental protein deprivation activity towards 4-nitrophenol fell in infant rats, but rose towards 2-aminophenol.[817] Activity towards bilirubin in newborn rats (0 to 2 days old) was consistently retarded on starvation or restricted milk intake, not usually developing further than the negligible levels of the first few hours (M.T. Campbell, unpublished results).

When jaundice was present in human infants up to 1 month old suffering caloric restriction through pyloric stenosis, an abnormally low transferase activity to bilirubin was confirmed by liver biopsy.[818]

IV. THE BREAST MILK FACTOR

Neonatal hyperbilirubinemia can be caused by breast milk (for references see Reference 819). Pregnane-3α,20β-diol was found in such milk; on administration to neonates it caused hyperbilirubinemia[820] and also appeared to inhibit UDPglucuronyltransferase activity towards 2-aminophenol in vitro.[821,822] The in vitro work was not confirmed.[819] Pregnane-3α,20β-diol did not inhibit human liver transferase activity towards bilirubin,[819] but it did inhibit slightly bilirubin glucuronidation in liver slices from certain species,[819] probably by inhibiting secretion of bilirubin glucuronide from the slice without affecting the transferase itself. An inhibitory "factor", thought to be a fatty acid,[267] for transferase itself, did however exist in the milk and free fatty acids, added alone or in milk,[823] inhibited bilirubin glucuronidation by liver slices; added stored milk, where hydrolysis increasingly liberates fatty acids, was particularly inhibitory. Unsaturated fatty acids of length C_8 to C_{20} (especially C_{12}), but not the saturated acids, decreased bilirubin glucuronidation by both slices and microsomes.[268,270] Milk from the affected mothers contained these fatty acids, and released them more readily on storage than normal breast milk.[270] Fatty acids can activate transferase activity towards other substrates (e.g., 4-nitrophenol, Chapter 5, Section II.M) so studies on the "factor" must always employ bilirubin as substrate;[462] in fact, bilirubin solubility or binding may be the critical factor and not the transferase itself.

The possible occurrence of glucuronides of fatty acids in fresh cow's milk[824,825] seems not to have been considered; these conjugates, or products of their hydrolysis, could contribute to transferase inhibition by milk.

Chapter 12

THE INFLUENCE OF SEX, SPECIES, AND STRAIN ON GLUCURONIDATION

I. SEX DIFFERENCES IN GLUCURONIDATION

The influence of sex on glucuronidation is an example of the influence of hormones, and so is a special aspect of Chapter 13, Section I.A.

In some species, sex differences in glucuronidation are considerable and in all cases where hormonal aspects have been studied appear due to the sex hormones themselves. Insofar as these effects concern the microsomal lipid environment, apparently contradictory reports are predictable. Activities observed will reflect substrate, method of tissue preparation, and diet as well as the animals' hormonal status; degree of transferase latency could also differ with sex. As kinetic parameters of the crude enzyme probably differ in males and females,[173] "sex differences" in activity may arise artifactually by assay conditions being optimal for one sex and not the other; e.g., transferase activity to testosterone showed a sex difference in the rat liver only at subsaturating substrate concentrations.[138]

Early reports[4] suggested that where a difference exists the female usually possesses the lower level of liver transferase activity. There are exceptions, e.g., female hamster forms more glucuronide of progesterone metabolites than male;[826] but female hamsters need not glucuronidate all substrates better than males. Sex differences can be substrate specific. Male rats possessed higher transferase levels for most substrates examined, but that towards 4-hydroxyamphetamine is higher in females,[408] which excrete the drug as 70% glucuronide, whereas males excrete it largely unconjugated;[827] female rats also possess a higher specific liver activity towards estrone and estradiol-17β.[138] Steroids are substrates for which sex differences might be particularly evident; slices of male-rat liver glucuronidate both 5(α)- and 5(β)-testosterone, female liver slices only the latter.[828]

Sex differences extend to relative activities in different tissues. Transferase activity to 4-nitrophenol was higher in male-rat liver than in female, but lower in male kidney than in female;[829] gut and lung displayed no sexual difference in activity. Against this, on a fresh weight basis, transferase activity towards 4-nitrophenol appeared rather higher in female rat liver than in male, but much lower in female than male intestine.[830] Castration raised the male-liver enzyme activity, but not that in the gut.[830] Whereas one report[829] found no sex-dependent latency in liver and kidney another[212] using fixed concentrations of activators, claimed greater activation in males for some substrates including 4-nitrophenol, but not for others, including bilirubin. Sex differences in the transferase activity persisted into overall glucuronidation of 2-aminophenol in rat liver slices, possibly because of differing apparent $K_{UDPGlcUA}$.[173] Higher transferase activity of male-rat liver was thought related to higher UDPGlcUA levels found there,[512,831] but this is difficult to reconcile with the occasional lower activity in males towards certain substrates; moreover, ASH/TO mice possess preponderant transferase activity to 2-aminophenol in males, but a much higher activity of liver UDPglucose dehydrogenase in postpubertal females.[502] However, this is not inconsistent with higher turnover of UDPGlcUA in females.

Excretion ratios of various neutral steroid glucuronides correlate with change from follicular to luteal phase of the menstrual cycle.[832]

The unsatisfactory evidence for sex differences in bilirubin glucuronidation in man has been reviewed,[639] and also the broader aspects of sex-related differences in drug metabolism, primarily Phase 1.[833]

Following Inscoe and Axelrod's observations[733] on administered estradiol and testosterone respectively feminizing and masculinizing transferase activity in male and female rats, Lucier[734b] "imprinted" transferase sex differences in adult male rats (4-nitrophenol as substrate) by exposing them on days 2, 4, and 6 of life to testosterone. As hypophysectomy prevented this, the pituitary gland is presumably responsible, as reported[834,835] with the monoxygenase system for both endogenous and xenobiotic compounds. Neonatal exposure not only to testosterone but also to diethylstilbestrol or 3,4-3′,4′-tetrachlorobiphenyl "feminized" transferase activity subsequently in male rats.[734b] Female rats were resistant to such imprinting.[734b]

II. SPECIES DIFFERENCES IN GLUCURONIDATION

A. Glucuronidation in Prokaryotes

Virtually nothing is known of glucuronidation in simple organisms. A UDPglucuronyltransferase exists in *Pseudomonas diminuta* ATCC 11568[108] (Chapter 7, Section II.D.2) accepting exogenous diacylglycerols containing ester-linked fatty acids; it is notably heat labile, as usual in poikiotherms, and does not conjugate 4-nitrophenol (B. Burchell, unpublished results).

UDPGlcUA participates in biosynthesis of several bacterial capsules (see Reference 597 for early references), but bacterial glucuronidation of small molecules has not yet been found (G. J. Dutton, unpublished work). Simple glucoside formation via UDPglucose occurs in bacteria.[836] Microbial drug metabolism has been recently reviewed.[837]

B. Glucuronidation in Plants

Plants readily form glycosides with xenobiotics through an *O*-, *S*-, or *N*-atom (see Reference 838 for a recent review); but although UDPGlcUA is abundant in plants,[4] glucuronides seem not among these conjugates.

Plants do form simple glucuronides,[602,839] but aglycons are endogenous; e.g., soluble preparations from *Phaseolus vulgaris* leaves transferred radioactivity from labeled UDPGlcUA to a compound travelling like their authentic endogenous quercetin-β-D-glucuronide on chromatography or electrophoresis.[546] When foreign compounds (e.g., phenol, *l*-menthol, benzoic acid) were offered, no glucuronide formation was detected.[546] Similarly, *Scutellaria galericulata* forms an endogenous glucuronide, but homogenates of its leaves conjugated 2-aminophenol with glucose, not glucuronic acid, when the relevant uridine nucleotides were offered.[840] Glucuronic acid was not transferred from UDPGlcUA to hydroquinone or phenyl β-glucoside by a wheat-germ extract capable of synthesizing the glucosides.[841,842]

Plants, therefore, possess a UDPglucuronyltransferase of great specificity catalyzing glucuronyl transfer to aglycons endogenous in that species. More work on specificity is required.

C. Glucuronidation in Invertebrates

Glucuronidation is rare in invertebrates. A good recent review discusses the evolutionary aspects.[3]

The parasitic worms *Ascaris lumbricoides* and *Moniezia expansa* produced no glucuronides with 4-nitrophenol, 2-aminophenol, or 4-methylumbelliferone,[843] but the protochordate *Balnaglossus minutus* and the sea urchin *Erechinus chloroticus* formed traces of ether glucuronides with substituted phenols.[844] Jordan and Smith (unpublished work quoted by Smith[3]) found that UDPGlcUA, but not UDPglucose, incubated with 2-aminophenol in homogenates of sea urchin or holotherian gut, formed a conjugate, but only during sexual maturation in summer.

Molluscs form sulfates rather than glycosides,[34] but glucosides, both *O*- and *S*-linked, are synthesized in slug *(Arion ater)* preparations by a UDPglucosyltransferase,[395,845] phenobarbital pretreatment enhancing glucosylation, but not evoking any detectable UDPglucuronyltransferase activity.[846]

In crustacea findings are negative[3,4] but for the glucuronide of 4-nitrophenol being formed from added UDPGlcUA in homogenates of digestive gland of the blue crab, *Callinectes sapidus*, known to contain UDPGlcUA.[847]

In insects, intensively studied because of their economic importance, no UDPglucuronyltransferase activity has been detected, even with endogenous β-glucuronidase inhibited. Instead, a UDPglucosyltransferase formed glucosides of various xenobiotics in cockroach[848] and locust.[4,849] *S*-glucosides are also formed in insects by this pathway.[850] A report that houseflies fed 1-naphthol excreted glucuronides is not substantiated.[3] Only UDPglucosyltransferase activity was detected towards this and other substrates,[851] the original conjugate being a phosphate.[849] A glucoside phosphate[852] is formed in flies dosed with 4-nitrophenol or 1-naphthol and is resistant to chemical and enzymic hydrolysis. No evidence of a glucuronide phosphate yet exists.

Insects may, like plants, form glucuronides of endogenous aglycons. Silkworm pupae reportedly contain the glucuronide of 3-hydroxykynurenin,[4] and an ecdysone metabolite may be glucuronidated by locust.[853] The question should be readily answered.

D. Glucuronidation in Fish

Increasing pollution of water resources has focussed attention on metabolism of xenobiotics by fish. A very useful review is by Sieber and Adamson.[854]

Brodie and Maickel[12] suggested that fish, able to dialyze against an infinite volume of water, would not need glucuronidation as much as land animals. As saltwater fish have greater difficulties in water conservation than freshwater fish, their glucuronidation might be more evident,[2] but gills appear relatively impermeable to foreign compounds,[855] and glucuronidation is common in both freshwater and marine fish.

Earlier work[4] concluded that low liver UDPglucuronyltransferase activity existed in most fish examined, whatever their environment. The aglycons were phenols and bilirubin. When some workers could not detect glucuronide formation in vivo, the reason proposed was demonstrable lack of UDPglucose dehydrogenase. Nevertheless,[4] UDPGlcUA is formed in fish, though not necessarily *via* UDPglucose dehydrogenase, and the anomaly probably arose through the difficulty of assaying the small amount of glucuronide released into the water; in *Salma trutta*, for example, possessing liver transferase activity to 2-aminophenol in liver homogenates, liver slices glucuronidated that substrate at much the same low efficiency as the homogenates.[856] Moreover, UDPglucose dehydrogenase is demonstrably active in several freshwater fish including rainbow trout.[857] Transferase activity or glucuronidation in sharks and dogfish, is reported for substrates including bilirubin.[858,859] Dogfish (*Squalus acanthias*) forms the glucuronide of phenol red,[855] 23% of the dose excreted in the 48-hr bile and urine being glucuronide.[860] Up to 73% of biliary 1-naphthyl-*N*-methylcarbamate (carbaryl) is glucuronidated by the rainbow trout,[861] but considerable variation in percentage glucuronidation among substrates and species has inevitably been recorded in fish. Some 4, 48, and 95% of the 2-, 3-, and 4-isomers, respectively, of aminobenzoic acid were found in dogfish-shark urine as glucuronides, whereas in flounder, the figures were respectively, 27, 2, and 43%, and in goosefish under 2% appeared as glucuronide.[862] Goldfish conjugate phenols preferentially with sulfate.[863]

Dewaide[864] studied glucuronidation of 4-nitrophenol in 17 species of fish. Endogenous glucuronides have also been reported of testosterone in salmon,[865] and of iodothyronine in bile of the brook trout, *Salvelinus fontinalis*, and other species.[866]

Selective toxicity dependent on glucuronide formation is suggested by the work of

Lech and colleagues, who found trout forms glucuronides of the lampricide 3-tri-fluo-romethyl-4-nitrophenol (TFM) and of its synergist 2′-5-dichloronitro-salicylanilide.[867,868] Lamprey, more susceptible to TFM than trout, possesses less liver UDPglucuronyltransferase activity to TFM.[869] Both novobiocin[870] and salicylamide,[871] known inhibitors of mammalian glucuronidation, increased TFM tox-icity in trout and inhibited its glucuronidation there.

Fish possess about a tenth of the mammalian capacity to metabolize xenobiotics; as the temperature optimum of many of these reactions is lower than in mammals,[854] they can easily be overlooked experimentally. Virtually no UDPglucuronyltransferase activ-ity exists towards 2-aminophenol in trout-liver homogenates at 37°C, but is clearly demonstrable there at 18 to 24°C.[856] The enzyme was very thermolabile and, pre-sumably, contributed to the similar temperature-dependency of overall glucuronida-tion in trout liver slices.[856]

As xenobiotics inducing Phase 1 reactions in mammals also do so in fish,[854] glucu-ronidation should likewise respond to them in fish, but reports have not been found.

The idea of Smith[3] that marine fish may need glucuronidation more than freshwater fish was examined by acclimatizing two species of *Tilapia* and *Salmo gairdnerii* from fresh to salt water;[857] liver UDPglucuronyltransferase activity to 4-nitrophenol re-mained unchanged, and UDPglucose dehydrogenase decreased.

E. Glucuronidation in Amphibia and Reptiles

Amphibia form glucuronides and possess liver UDPglucuronyltransferase, provided the adult lives partly on land, e.g., *Rana pipiens, R. catesbiana*, and *Bufo marinus*. Amphibia living wholly in water appear not to possess the enzyme or form glucuron-ides, e.g., *Xenopus laevis* and *Necturus maculosis*.[4] Xenobiotic phenols were the only substrates offered, although one report[872] notes glucuronidation of administered estro-gens by UDPglucuronyltransferase in liver microsomes of the newt *Pleurodeles waltlii*. Larval forms of all species investigated lacked glucuronidation by the criteria em-ployed.[4] Full references to the earlier work exist.[4] A thermolabile form of the enzyme may exist in aquatic amphibian species and in larval forms, as in fish, and consequently could have been overlooked.[4]

Liver microsomes of alligator and snake possess transferase at mammalian levels.[12]

F. Glucuronidation in Birds

Distribution of UDPglucuronyltransferase in adult birds is similar to that in mam-mals; liver, kidney, and gut being major sites (for early references see Reference 4). Major development of the enzyme in chicken appears just after hatching (Chapter 10, Section IV).

Interest in agriculture and in wildlife conservation has prompted study of metabo-lism of pesticides and other xenobiotics in birds. Some facts relevant to glucuronida-tion have emerged, e.g., formation of glucuronide as a major metabolite of isopropyl carbanilate (Propham®) in chicken,[873] and the apparently negligible influence of ex-posure to DDT on the transferase in liver microsomes of pheasant.[874] However, avian tissues may not possess the wide mammalian range of transferase activities to xeno-biotic and endogenous compounds. In chicken liver, UDPglucuronyltransferase activ-ity towards 4-methylumbelliferone,[369] chloramphenicol (B. Burchell, private commu-nication), and serotonin[426a] is absent. The status of bilirubin glucuronidation in chicken is uncertain, transferase activity towards bilirubin being either undetectable[369] or very low.[462,475] The bilirubin glucuronide of chicken bile[475,496,875] and the mono- and diglucuronides of goose serum[876] are possibly made by a very low transferase activity of liver, gut, or kidney. As biliverdin reductase is absent from chicken liver,[877] bilirubin would not be expected in any amount in chicken bile. Biliverdin is the major excretory product and is present to some extent as glucuronide,[495] which could be oxidized to bilirubin glucuronide artifactually.

Steroids are conjugated by chicken as glucuronides, not only as sulfates as once believed,[436] the urine containing esterone and estradiol glucuronides after injection of estrone,[878] and transferase activity towards estriol being noted in chicken liver and kidney;[197] chicken adrenal gland, also, forms steroid glucuronides.[879]

G. Glucuronidation in Mammals

1. General

Mammalian species vary in their relative UDPglucuronyltransferase activities to any one substrate, as previous chapters have demonstrated together with probable reasons for these differences. Guinea pig, for example (see Reference 4 for early references),[253,406,407,880] may possess higher native liver transferase activity towards substrates tested (with the possible exception of bilirubin[881]) than other laboratory animals. Guinea pig-liver enzyme has been claimed to possess a lower $K_{2-aminophenol}$ than that from rat.[285,366] It seems more readily inhibited by proteases and perturbants, behaving as if less latent than in other species. Constraint of transferase may be less in guinea pig liver,[253] and/or the enzyme may be more superficial than in rat. However, similar activation characteristics have been shown for the liver enzyme from both guinea pig and rat, except that the guinea pig enzyme is readily activated by washing the microsomes in KCl;[162] the other "differences" in activation between the species may arise from use of microsomes prewashed or prepared in KCl, and so in the case of guinea pig, already almost wholly activated.[162]

Species differences in transferase activation towards various substrates[229,475] and in its induction after xenobiotic pretreatment have been noted[336,383] (Chapter 13, Section II. H).

Species differences in liver UDPglucuronyltransferase activity and in overall glucuronidation, (measured by biliary excretion) have been correlated for ten mammalian species with bilirubin as aglycon.[475] Species differences exist in mono- and diglucuronidation of bilirubin and in extrahepatic distribution of the enzyme.[475] The kidney transferase was significant only in rat and dog.[475]

Some species differences in "glucuronidation" arise other than from the transferase, e.g., the N-glucuronide of Dapsone® (4,4'-diaminodiphenylsulphone) is unstable in acid urine and, therefore more evident in alkaline urine of rabbit than in acid urine of rat.[882]

Living organisms have developed enzyme systems to suit their ecological niche (Chapter 1), and an obvious absence of glucuronidating capacity in a successful species is due either to absence of certain glucuronidogenic substrates in its environment or to use of analogous conjugations. Although of some 242 living species of carnivora only seven have been investigated for xenobiotic conjugation,[883] this idea appears in a recent useful review[883] applied to the cat, or more generally the *Feloidae*, which are deficient in some glucuronidation reactions performed by other species. An earlier review on interspecies glucuronidation also discusses cat.[884]

2. The Cat

Low glucuronide synthesis in cat, first demonstrated[885] using slices of cat liver, kidney, and other organs and 2-aminophenol as substrate, was shown[886] to be due to lack of UDPglucuronyltransferase activity to that phenol. Only negligible amounts of urinary glucuronic acid appeared in cats dosed with several other normally glucuronidogenic xenobiotics (i.e., phenol, borneol, 4-methylumbelliferone, 2-amino-4-nitrophenol, 8-hydroxyquinoline, 2-methoxybenzoic acid),[887] such compounds being excreted as urinary sulfates by cat, (e.g., phenol, 1-naphthol, 2-naphthol, morphine,

4-acetaminophenol),[373,883,888] or in one case as a phosphate;[889] only 1 to 2% glucuronide of some of these compounds could be found and only traces in cat bile or urine of morphine 3-glucuronide,[22,890] of quinol or of 2,6-dimethoxyphenol glucuronides,[891] or of glucuronides of hydroxylated metabolites of 2-acetamidofluorene.[892] Ester glucuronides of benzoic acid[893] and of several phenylacetic acids[883] appeared absent from urine of cats dosed with the aglycons.

For several substrates, the cat is not alone in its deficiency: with the mouse, it fails to glucuronidate 4-hydroxy-3-methoxyphenylethanol,[894] and with the hyena, phenol.[895] Moreover, the cat can form certain glucuronides. Even a phenol, phenolphthalein, is extensively glucuronidated.[373] The conjugate appears in urine and bile. Other xenobiotic aglycons glucuronidated include hydrotropic acid,[883] diphenylacetic acid,[371] various radioopaque compounds,[896] and lorazepam metabolites.[897] Low activity to 4-nitrophenol also exists, apparently not stimulated by phenobarbital pretreatment,[383] though activation was not optimal. Endogenous compounds such as bilirubin,[683] thyroxine,[898] progesterone,[899] testosterone,[900] estradiol metabolites,[901] estradiol, and estrone[112] are all actively glucuronidated by the cat, and for some (e.g., bilirubin[383,475,902]), transferase activity has been demonstrated; it is absent for serotonin.[426a]

The cat, therefore, is not unique but forms glucuronides of fewer xenobiotics than most animals. This could be due (a) to absence of the relevant UDPglucuronyltransferase protein(s), the "trace" glucuronidations observed arising from nonspecific conjugation by other UDPglucuronyltransferase activities, or (b) the specific transferase present in low activity. Either (a) or (b) is compatible with the observed higher apparent K_m value towards 4-nitrophenol in the cat than in the rat.[383] Hirom and colleagues (unpublished, quoted by Hirom et al.[883]) find that sulfation tends to replace glucuronidation in cat, and whereas apparent K_m value for 1-naphthol glucuronidation in rat liver is 12 μM and in cat liver is 3000 μM, corresponding values for 1-naphthol sulfation are 20 μM and <1 μM. As sulfation appears more readily saturable, phenols are predictably toxic to cats; the lethal dosage is five to ten times less than for rats and herbivores.[903]

Interest in wildlife conservation has enabled the greater cats to be investigated, and it appears that all Feloidae tested, and civet and genet of the related Viveridae, glucuronidate little or no phenol, benzoic acid, 1-naphthylacetic acid, or sulfadimethoxine.[373,883]

3. Influence of Diet on Species Variation in Glucuronidation

True carnivores, such as the Feloidae, are not exposed in their ecological niche to the abundance of dietary plant phenols encountered by omnivores or herbivores, such as the common laboratory animals, and so may not have needed to develop the range of UDPglucuronyltransferase activities the latter already possess.[883] The Canoidae seem better equipped for glucuronidation, but less so than herbivores such as rabbit or guinea pig, and other carnivorous superfamilies require study before the Feloidae can be considered exceptional.

Diet might influence especially the gut, and transferase activity towards 4-nitrophenol in rat small intestine apparently rose to that in the guinea pig organ on feeding rats a guinea pig's diet.[904] Many extrahepatic species differences may be due similarly to activators or inducers[734] in the diet. Genetically defective levels as in cat,[237] cannot be "normalized" by any inducers or activators tried.

Marsupials were thought[4] to excrete free urinary glucuronate, now known to be from hydrolysis by high urinary β-glucuronidase[905] which, being only slightly inhibited by glucarolactone, is probably of microbial origin. Eucalyptus leaves in the diet increased excretion of urinary glucuronide and inhibited hydrolysis,[905] and urinary glucuronide output from ingestion of eucalyptoid terpenes may be large.[883]

4. Primates and Man

In a good recent review of drug metabolism in nonhuman primates,[392a] most of the information concerns rhesus monkey, which, like man, forms the major types of glucuronide. Many reports link man with other primates in drug conjugation; e.g., whereas rat, guinea pig, and pig conjugate 4-hydroxyamphetamine with glucuronic acid, man and four species of monkeys conjugate it with sulfate.[408,827]

Although rates of glucuronidation in man and experimental primates may be quite different, especially for persistent drugs, they are often similar enough to be useful; as with amphetamine, norephedrine, and phenmetrazine.[392a,906]

Capacity to form an N'-glucuronide with sulfadimethoxine appears only in primates and by this taxonomic criterion ("pharmacozootaxonomy",[907]) the controversial tree shrew *Tupaia* is a primate. The pyrazole-linked *N*-glucuronide of the drug EMD 16923 (Chapter 7, Section II.B.) was only formed in man and to a lesser extent in rhesus monkey among species tested. Man and old-world monkeys conjugate phenol largely with sulfate rather than with glucuronic acid, while in the two new-world monkeys examined,[908] the converse is true. A similar divergency exists with 1-naphthol (P. C. Hirom and R. Mehta, unpublished results quoted in Reference 392a).

Davies[906] has compared drug metabolism in man with that in laboratory animals.

III. STRAIN DIFFERENCES IN GLUCURONIDATION

A. General

"Strain differences" in enzyme activity are perennially invoked to account for otherwise inexplicable observations (or artifacts), but do exist in glucuronidation. For example, at 17 days fetal age ASH/TO mice possess 100% adult transferase activity towards 2-aminophenol, ICR/IAR strain only 20% of this activity (J. E. A. Leakey, 1976, unpublished results). Differences in relative transferase activities between Wistar and Sprague-Dawley rats are informally acknowledged but not published; Burchell (personal communication) finds specific transferase activity of activated liver microsomes toward 2-amino- and 4-nitrophenol is 5 to 6 times higher with Wistar than with Sprague-Dawley rats.

In one intensive genetic study,[909,910] liver transferase activity towards 4-methylumbelliferone was induced by 3-methylcholanthrene or β-naphthoflavone in mouse strains C57BL/6N, A/J, PL/J, C3HeB/FeJ, and BALB/cJ, but not in DBA/2N, AU/SsJ, AKR/J, or RF/J. In offspring from appropriate backcrosses and intercross between C57BL/6N and DBA/2N parents, genetic expression of 3-methylcholanthrene-inducible transferase activity was inherited as an additive (codominant) trait.[910] Differential rates of induction and breakdown and pH and heat stability profiles suggest the "transferases" in the two strains may differ,[910] possibly in environmental constraint rather than primary structure. As C 57 BL/6N mice conjugate 9-hydroxybenzo(a) pyrene directly, but DBA/2N mice only if benzo(a)pyrene itself is source of the 9-hydroxy derivative, accessibility of the transferase active site for the derivative may differ in the two strains.[910a]

Within any species, such slight hereditary variations in UDPglucuronyltransferase activity must occur unnoticed. The condition is recognized only when grossly defective transferase activity towards endogenous substrates disturbs metabolism, and the mutants survive until birth. In rat and man, certain strains grossly defective in activity towards bilirubin have been identified; they are discussed in Sections III.B, III.C, and III.D below.

B. The Gunn Rat

This rat strain, first described by Gunn,[911] exhibits nonhemolytic unconjugated hyperbilirubinemia. Excretion of injected bilirubin glucuronide, retention of free pigment, levels of UDPglucose dehydrogenase, of UDPGlcUA and of associated enzymes exist as in the normal Wistar strain from which it is derived.[912-915] The jaundice is probably due to the undetectable UDPglucuronyltransferase activity towards bilirubin in liver,[4,476,912,916,917] kidney, and gut.[918,919] This deficiency is autosomally recessive, heterozygotes possessing subnormal enzyme activities to bilirubin.[916]

As the Gunn rat reproduces almost exactly the human patient with the lethal Crigler-Najjar syndrome (Section III.C below), it has been extensively studied.[920,921] However, the Gunn "strain" is of doubtful genetic purity (see discussion in Reference 207), and its relative transferase activities to substrates other than bilirubin have long caused dispute.[4] Activities very low, but not absent, are to 2-aminophenol, menthol, 2-aminobenzoate,[912] 4-methylumbelliferone,[919] paracetamol,[234] phenolphthalein,[922] and to some extent, thyroxine;[918,923] glucuronidation in homogenates may be lower than in the intact cell, as for 2-aminophenol.[173] Among activities reported at normal Wistar-rat level (for early work, see Reference 339) are notably those towards 2-aminothiophenol (the substitution of S for O being critical),[206] tetrahydrocortisone,[362] diphenylacetate,[386] substrates forming N-glucuronides,[350] morphine,[372,924] chloramphenicol,[924] and serotonin.[426a] Activity towards 4-nitrophenol is controversial, and possibly two Gunn rat populations exist, one with low[207,209,234,362,915,927] and one with normal "high"[235,366,378,383,966] activities, the latter possibly slow on developing.[927] In Gilbert's disease (see Section III.C below), a condition often resembling that in heterozygous Gunn rats, a similar situation may exist: of two groups of patients both exhibiting some 25% normal activity, one group possessed down to 50% activity to 4-nitrophenol and the other 100% activity.[928] However, many reported differences in Gunn rat may arise from different laboratories using microsomes at different stages of activation toward 4-nitrophenol.

Wistar and Gunn rats seem to differ in constraint of transferase activity towards 4-nitrophenol and, possibly, other substrates, including steroids.[213] Studies of binding energy[236] attributed the low activity to 4-nitrophenol in one colony (but see Reference 235), where it is "high" in the same colony) to abnormal interactions between enzyme and UDP moiety of UDPGlcUA, although other kinetic parameters and activation by UDPGlcNAc[236] were similar in the two strains.

Phospholipase A and Triton® X-100, activators of transferase towards 2-aminophenol and 2-aminobenzoate in Wistar rats, have no effect on the low activity present in Gunn rat-liver microsomes;[235] activity to 1-naphthol was similar in Gunn and Wistar rats when native microsomes or those treated with UDPGlcNAc or diethylnitrosamine were used, but was preferentially activated in Wistar rat by the detergent Brij 58®[924]. In perfused liver, no difference between strains was seen, confirming that transferase in vivo is not in the detergent-activated form (Chapter 5, Section I).[924] Although activity to 4-nitrophenol was partially resistant to perturbant activation in Gunn, but not Wistar, rat,[915] other workers found that deficient activity towards 4-methylumbelliferone[929] or 4-nitrophenol[209] in Gunn rats was activated by phospholipase C, digitonin, or trypsin proportionately as in Wistar rats, suggesting a similar microenvironment for some transferase activities in the two strains.[209]

A more informative approach employed diethylnitrosamine. When this compound was added to Gunn rat-liver microsomes or homogenates, their genetic deficiency in activity towards 2-aminophenol and paracetamol disappeared and transferase activity was activated to equally high levels in preparations from both Gunn and Wistar rats.[234] These observations have been often confirmed and recently extended to 2-aminobenzoate.[235] Modification of protein-lipid interactions may be involved. Prior treatment of Gunn rat-liver microsomes with diethylnitrosamine produced a form of transferase activity towards 4-nitrophenol which, unlike the untreated form, was wholly activated by phospholipase A or Triton® X-100 just like Wistar rat transferase, and the defective affinity of Gunn rat enzyme for the UDP moiety of UDPGlcUA was increased.[236] This suggests a difference of transferase constraint could exist between Gunn and Wistar rat for all substrates, but no evidence yet points to membrane composition being the cause. Delipidated Wistar-rat transferase to bilirubin is reactivated by lipids or membranes from Gunn rats,[257] but the reactivation is fairly nonspecific, even lecithin serving. Equally concave Lineweaver-Burk plots for Wistar and Gunn rat-liver transferase[929] is unconvincing evidence for a similar microenvironment. Use of purified enzyme from each strain is necessary and recent work demonstrating that activation by diethylnitrosamine persists into an apparently homogeneous enzyme preparation[240] containing minimal phospholipid,[242] suggests its effect is on the enzyme and not on the membrane. The protein moiety of pure UDPglucuronyltransferase seems largely alike in Wistar and Gunn rats, for their antibodies cross-react and molecular size and charge are similar.[323] Transferase activities to 2-aminophenol and 4-nitrophenol remain lower through all stages of purification of the Gunn-rat enzyme than during purification of the Wistar-rat enzyme.[323] They disappear after the final stage, but reappear on addition of diethylnitrosamine, to much higher specific activity than in previous stages.[323] Purification appears to increase the diethylnitrosamine-sensitive latency of transferase activity to 4-nitrophenol in Gunn rat, much as the genetic defect increases its diethylnitrosamine-sensitive latency to 2-aminophenol, suggesting that in both cases delipidation changes constraint on the protein. Activity to 1-naphthol is lower in Gunn than in Wistar rat enzyme throughout purification and remains unenhanced by diethylnitrosamine.[323]

Weatherill and Burchell[323] consider that diethylnitrosamine interferes with either protein-phospholipid, or (as phospholipid is extremely low in their purified preparations), with Lubrol-protein interaction or even directly with the transferase protein. An intramolecular conformational change[238,323] could then facilitate access of the two phenols to the reactive site. This effect of diethylnitrosamine can be reproduced with simple alkyl ketones.[222a]

Although UDPglucuronyl-,[465] UDPglucosyl- and UDPxylosyltransferase[479] activity to bilirubin as IXα- or non-IXα isomer is undetectable in homozygous Gunn rats even with digitonin present,[479] traces of β-D-glucosidic conjugates, including glucuronide, of bilirubin IXα are found in bile from Gunn rat[930] and from children with Crigler-Najjar disease,[931] no transferase being detectable. Blanckaert et al.[930] offer two explanations, (a) transferase activity for bilirubin exists in the Gunn rat, but is wholly latent, (b) that a small amount of bilirubin glucuronide is formed by transferase activities specific for other substrates.

Activity in Gunn-rat liver, kidney, or gut to bilirubin has never been evoked by activators or inducers. Diethylnitrosamine has no effect in vivo.[237] Phenobarbital has been administered without success,[925,932,933,935,936] as has glutethimide,[937] but activities already present in Gunn rats are usually induced (e.g., activity to steroids[213] and 4-nitrophenol[383,925]) by phenobarbital. Some differences in inducibility of other transferase activities between Gunn and Wistar rats have been noted (e.g., methylcholanthrene in Gunns not inducing activity towards 4-nitrophenol[209] or 1-naphthol).[924]

Transferase activity to bilirubin being noninducible clinically in Crigler-Najjar cases,[639] grafting techniques in animals have been studied for possible therapeutic applications.

A minimal-deviation hepatoma possessing transferase activity for bilirubin, grafted into Gunn rats, increased the bilirubin conjugation.[938] Wistar rat kidney grafted into nephrectomized Gunn rats[939] permitted conjugation of 3/5 their total bilirubin pool within 48 hr, conjugates being taken up hepatically and excreted into bile, and only appearing in plasma or urine on bile duct ligation;[939] evidence that the hyperbilirubinemia of the Gunn rat stimulated the transferase of the grafted kidney (which doubled its activity in the 48 hr)[939] suffered from unsatisfactory controls. As human kidney[475] does not possess UDPglucuronyltransferase activity towards bilirubin, renal grafting would not seem helpful in Crigler-Najjar cases.

A report[940] that Wistar-rat liver cells grafted into the right liver lobe of Gunn rats specifically lowered plasma bilirubin after 12 weeks and evoked transferase activity to bilirubin in both liver lobes, was not confirmed.[941] Hyperbilirubinemia decreased no more than in sham-operated animals, and there was no bilirubin glycoside in bile nor UDPglycosyltransferase activities in liver.[941]

C. The Crigler-Najjar Syndrome

This hereditary hyperbilirubinemia first described in 1952,[942] is analogous to that of the Gunn rat (Section III.B above). The classical "Type 1" syndrome is autosomal and recessive, exhibits serum bilirubin above 342 μM (20 mg/100 mℓ) and possesses no biliary bilirubin and no detectable liver UDPglucuronyltransferase activity towards bilirubin.[382,916,944,947] In "Type 2", serum bilirubin is usually, but not necessarily, below 342 μM; bilirubin is detectable in bile and some transferase activity is measurable.[916,943]

Transmission might be autosomal and dominant[944] or due to transmission of two abnormal genes, one from each parent,[945] when each parent could exhibit a different form of Gilbert's syndrome (Section III.D below); however, although Crigler-Najjar and Gilbert's syndrome patients all possessed a higher proportion of monoconjugated bilirubin than normals, Type 1 and Type 2 Crigler-Najjar conditions could not be distinguished by this means.[482] Hyperbilirubinemia of Type 2 is usually lessened by treatment with phenobarbital or other xenobiotics,[721,916,946] but Type 1, like that of homozygous Gunn rat, is resitant to phenobarbital treatment.[916]

Kernicterus develops frequently, and most patients with Type I syndrome die in early childhood. Glucuronidation of other substrates is, as in Gunn rat, abnormal. Glucuronidation is normal for salicylamide[948] and estriol,[184] but low for menthol and N-acetyl-4-aminophenol[4] (for other substrates, see Reference 916), and absent for sulfadimethoxine[949] although the last is excreted normally as an N-glucuronide, a type sometimes formed by the analogous Gunn rat.[350] Transferase activity itself was low for 4-methylumbelliferone and 2-aminophenol,[944] but almost normal for 4-nitrophenol,[946] except in Type 2 where it is reduced.[944] Gradation between Type 1 and 2 is indistinct, as also probably between Type 2 and Gilbert's Syndrome, which will now be discussed.

D. Gilbert's Syndrome

Gilbert's syndrome, first described early in the century[950] is a benign, chronic, low-grade, unconjugated hyperbilirubinemia in man with absence of structural liver disease or overt hemolysis. Plasma concentration of conjugated bilirubin is normal. The genetics are reviewed by Schmid and McDonagh.[916]

It seems due to an autosomal dominant trait with incomplete penetrance.[951] A recent study[952] of 197 "normal" males and 102 "normal" females revealed bimodal distribution of serum bilirubin concentration. Men possessed an antimode at 24 μM, women, at 12 μM. If subjects possessing higher levels are considered[952] to exhibit Gilbert's syndrome, incidence for both sexes is 6% total population.

The unconjugated hyperbilirubinemia arises from decreased hepatic bilirubin clearance,[948] which can itself arise from defective bilirubin uptake into the cell;[948,953] from defective bilirubin transport there by an organic anion-carrying protein,[954] and/or from defective bilirubin conjugation by UDPglucuronyltransferase.[917,955,956] Not only bilirubin is affected: there is slower clearance of tolbutamide,[957] bromosulfophthalein[958] (neither glucuronidated) and of estrogen.[959] Overproduction of bilirubin might also occur.[928]

Relationship between Gilbert's syndrome and the "Type 2" Crigler-Najjar syndrome is obscure, both conditions being poorly characterised biochemically. Auclair et al.[928] investigated UDPglucuronyltransferase activity towards bilirubin and 4-nitrophenol in patients with either syndrome. They distinguished three groups (*A, B, C*) with Gilbert's syndrome; *A* possessing normal activity to 4-nitrophenol and 25% of the normal activity to bilirubin, *B* only 50% of the first activity and 25% of the second activity, and *C* normal activity to 4-nitrophenol and 50% of activity to bilirubin.[928] Group *C* was thought heterozygous to the homozygous condition of group *A*. Group *B* was supposed heterozygous to a homozygous condition found in Type 2 Crigler-Najjar syndrome where subjects possessed only 10% normal control activity to 4-nitrophenol and 25% activity to bilirubin.[928] Only 27 patients and 30 controls were studied, and existence of these groups conflicts with some other findings.[808,917] However, Gilbert's syndrome can be divided into at least three groups on the basis of estrogen glucuronidation and clearance.[959] The hypothesis of homo- and heterozygous groups is warrantable because of the frequency of Gilbert's syndrome and the difficulty of accounting for a 75% decrease in activity as due to a heterozygous condition, but a full study of this benign condition is scarcely justifiable clinically. The greatest danger to a patient with Gilbert's syndrome would seem to be cholecystectomy from mistaken diagnosis.

Patients with Crigler-Najjar or Gilbert's syndrome possess a strikingly increased proportion of monoglycosides of bilirubin in bile.[482] In normal human bile, the monoglycosides (predominantly glucuronide) were some 27%, in Gilbert's disease some 49%.[482] In bile from Crigler-Najjar children, the little conjugated bilirubin found was predominantly monoglucuronide.[482] Unconjugated hyperbilirubinemia from hemolysis gave rise to the normal proportion of monoconjugate.[482] Fevery et al.[482] suggest that Gilbert's disease is a milder expression of the biochemical defect found in the Crigler-Najjar syndrome. If the defect is due to less transferase activity, one explanation would be greater affinity of the enzyme for free bilirubin than for monoglucuronide, which is preferentially excreted. If the two enzymes of Jansen et al.[125] (Chapter 7, Section III) exist, then a deficiency (not seen by Jansen et al.[125]) or inhibition of the second enzyme should occur,[482] or perhaps its reversal.

The effect of diet on this condition (Chapter 11, Section I.B.2) is probably responsible for the marked and unpredictable fluctuations seen in the serum bilirubin of patients. Stress such as caloric deprivation in achalasia,[960] also evokes symptoms of Gilbert's disease in subjects otherwise normal.

Characteristic of Gilbert's syndrome is enhanced clearance of bilirubin following phenobarbital treatment,[961] due probably to increased transferase activity to bilirubin[639] or to increased aglycon uptake by intracellular anion transporters.[957,962] Gunn rats and patients with Crigler-Najjar or Gilbert's syndrome excrete the unconjugated nonnatural bilirubin IX isomers as readily as normal rats or man, only the IXα isomer requiring conjugation before excretion by mammals.[963]

E. Other Strains with Lower Excretion of Bilirubin Conjugates

The Corriedale sheep strain[964,965] exhibits reduced ability to excrete bilirubin, and defective metanephrine glucuronide excretion;[966] Southdown sheep remove bilirubin slowly from plasma.[967,968] No decrease, but a slight increase, in UDPglucuronyltransferase activity towards bilirubin occurs in these sheep strains;[969] as activity of β-glucuronidase, is normal,[969] their hyperbilirubinemia seems unconnected with glucuronidation defects.

In the human Dubin-Johnson syndrome, defective excretion of bilirubin glucuronide and estriol 3-glucuronide into bile[969] appears also unconnected with glucuronidation itself.

IV. ADDITIONAL NOTES

Section II.B: The fungus *Cunninghamella elegans* formed conjugates of hydroxybiphenyl which were hydrolysable by a β-glucuronidase preparation.[969a] As the specific inhibitor glucarolactone was not employed, this apparent first report of a fungal xenobiotic-glucuronide remains uncertain.

Section III.C: The Gunn rat is not a good animal model for the Crigler-Najjar infant, for liver transferase activity to 2-aminophenyl glucuronide, virtually absent in the Gunn rat, is normal in the Crigler-Najjar infant.[636b]

Chapter 13

THE INFLUENCE OF HORMONES AND XENOBIOTICS ON GLUCURONIDATION

I. THE INFLUENCE OF HORMONES ON GLUCURONIDATION

A. Sex Hormones

Influence of sex hormones on glucuronidation is implicit in the sex differences described in Chapter 12, Section I. Evidence concerned with sex hormones themselves or tissues producing them, will now be discussed.

Early observations[733,831] showed that adult male rats possessed UDPglucuronyltransferase activity towards 2-aminophenol some four times greater than that of females and that chronic administration of estradiol halved male activity whereas chronic male administration of testosterone more than doubled female activity.[733] No effects being noted on addition of hormones in vitro, transferase synthesis and/or breakdown was considered responsible.[733] However, no sex difference towards this substrate was found in slices of various tissues from another rat colony,[970] but sexually immature animals glucuronidated it faster than adults; on treatment with estradiol or testosterone, respectively, the immature 4-week female and male animals glucuronidated at the lower adult rate. Castration at 4 weeks did not lower glucuronidation rate.[970] In another colony[831] castration of male rats lowered their glucuronidation rate. Because the sex difference with rat-liver microsomal transferase disappeared after dialysis, removal of an inhibitor was suggested,[312] but no such inhibitor has been demonstrated by "mixing" experiments and from the work[733,831] quoted above, is likely to be estradiol or testosterone itself. The antiandrogen, Cryproteron, did not reverse enhancement by testosterone, but itself caused stimulation.[831] From assay of UDPGlcUA levels in liver, sex differences in vivo have been considered due to higher UDPGlcUA formation by males,[831] but this cannot explain persistence of sex difference into microsomes.

Sex hormones influence their own glucuronidation. Specific activity of the UDP-glucuronyltransferase glucuronidating estrone and estradiol-17β was four- to tenfold higher in livers from female rats than from males.[138] Ovariectomy decreased activity to estrone 56%; this decrease, evident 9 days later, still existed after 30 days.[138] If estradiol-17β was injected intramuscularly each day, transferase activity to estrone increased after 6 days, rising over the next 6 days or so to reach a value 2.6-fold above the original low activity, and a little above the activity in normal female liver.[138] Further injections produced no further rise, and activity in nonovariectomized controls never responded to estradiol injection.[138] The rise could be largely blocked by cycloheximide and actinomycin D. Estradiol-17β may, therefore, induce its own glucuronidating enzyme after a lag period of 6 days.[138] UDPGlucuronyltransferase activity to testosterone was not changed by repeated injection of estradiol-17β[138] nor in male liver lowered by castration.[138]

UDPGlucuronyltransferase activity towards estrone did not vary over the estrus cycle in rats[138] but some change in the neutral steroid content of human urine has been noted during the menstrual cycle.[832] "Imprinting" has been mentioned in Chapter 12, Section I.

B. Other Hormones

Earlier work was largely inconclusive.[4,339] Injection of hydrocortisone into rats raised liver UDPGlcUA,[971] but adrenalectomy of adult male rats either tended to lower liver UDPglucuronyltransferase activity or have no effect.[693,972] Perinatal effects are described in Chapter 10, Section IV.D.2.

Injection of growth hormone into mouse[973] or rat[639,753] retards development of Phase 1 enzymes, but does not affect transferase activity towards bilirubin,[973] and may even raise it towards 4-nitrophenol (G. J. Wishart unpublished results, 1977). Hormonal effects vary between substrates and may involve the membrane. Thyroidectomy or hypophysectomy lowered rat liver UDPglucuronyltransferase activity towards 2-aminophenol, UDPGlcUA concentration not being rate limiting.[237,376] After hypophysectomy, activity rose for 4 days before falling to some 30% normal, and after thyroidectomy a decline began within 8 to 10 days. The lowest activity, reached after some 3 months, was similar following either operation.[376] Combined ablation gave the initial rise, followed by a more rapid decline.[376] Administration of *l*-thyroxine restored activity to normal more promptly with the thyroidectomized rats, where activity rose above normal after 15 to 22 days.[376] Thyroid hormones appeared to be critical here: addition of corticotropin, growth hormone, oxytocin, pitressin, or diethylstilbestrol had no effect.[376]

Two notable observations were made.[237] First, activity to bilirubin was not decreased by hypophysectomy and/or thyroidectomy; it even increased if suboptimal UDPGlcUA concentrations were used in the assay, this effect disappearing after thyroxine treatment. With optimal activation by digitonin, activity towards bilirubin doubled in rat liver microsomes on hypophysectomy.[237]

Second,[237] activity to 2-aminophenol lost by hypophysectomy and/or thyroidectomy was restored by addition in vitro of diethylnitrosamine, exactly as for the genetically deficient Gunn rat (Chapter 12, Section III.B; Chapter 5, Section II.J).

This work need not indicate two separate transferase activities for bilirubin and 2-aminophenol, but that thyroidectomy reversibly changes enzyme constraint so that activity to bilirubin is increased, but that to 2-aminophenol is decreased.[376] Diethylnitrosamine would repair the latter decrease (with Gunn rat, diethylnitrosamine affects the protein, not the membrane[323]). The fall in activity specific to 2-aminophenol on thyroidectomy is unlikely to reflect general UDPGlcUA deficiency, especially as UDPglucose dehydrogenase activity is reduced by thyroxine administration.[813] Male rats, hypothyroid after a month's thiouracil treatment, possessed rather higher specific activities of UDPglucuronyltransferase (towards 4-nitrophenol) and UDPglucose dehydrogenase than normal rats;[975] this could have a bearing on the reported increase in glucuronide conjugation during hypothyroidism,[976] but thiouracil itself may induce the transferase (Chapter 13, Section D) even though phenobarbital induction is still seen in thiouracil-treated rats.[975] Injected glucagon or dibutryl cyclic AMP raised rat-liver activity to bilirubin 60% within 2 hr unless Actinomycin D was present.[976a]

C. Hormones and Glucuronidation in Pregnancy

Glucuronidation in pregnancy essentially comes under hormonal regulation.

Specific liver UDPglucuronyltransferase activity is sometimes lower in pregnant than in normal females,[312] especially with 2-aminophenol as substrate,[689] (G. J. Wishart and G. J. Dutton, unpublished work). Serum from pregnant women, and several steroids, inhibited glucuronidation of bilirubin in rat-liver slices or the excretion of the glucuronide from the slices;[689] $3\alpha,20\alpha$-pregnanediol, as its glucuronide, inhibited excretion even at 1 μM.[689] Inhibition by estrogenic and progestational steroids has often been invoked (e.g., Reference 1289) to account for the fall in specific transferase activity during pregnancy, but when specific liver transferase activity to 4-methylumbelliferone in rats 19 to 20 days pregnant had fallen,[258] the liver weight had increased by 40%, and there was no change in total liver transferase activity.[1289] In full-term rabbit, liver transferase was similarly lower by 20% in specific activity.[1289]

Recurrent jaundice of pregnancy (RJP) and its milder form, *pruritus gravidarum*, appear due to exaggeration of changes in liver function normally observed during pregnancy.[977] Because symptoms develop readily in these subjects after administration of oral contraceptives or synthetic estrogens,[977,978] estrogens are implicated. The primary cause is cholestasis,[977] which impairs the enterohepatic circulation of estrogens so that estrogen levels are elevated and evoke the abnormal reaction in susceptible women, seen as RJP or pruritus gravidarum. In the normal human subject, estriol is conjugated as estriol 3-sulfate,16-glucuronide in liver, excreted as this conjugate in the bile, and liberated in the intestine by bacterial action. In the human intestinal mucosal cells, this freed estriol is conjugated again, this time as estriol-3-glucuronide (Chapter 14, Section IV.F.2; references in Reference 977), and then excreted in the urine. Excretion of urinary estriol 3-glucuronide is much decreased in RJP and rather less so in pruritus gravidarum.[977] This is consistent with the above sequence. The decrease is not quite down to the 3% normal seen in the actual rate of biliary secretion of estriol. Tikkanen and Adlercreutz[977] discuss possible reasons for the discrepancy.

II. THE INFLUENCE OF XENOBIOTICS ON GLUCURONIDATION

A. General

Influence of xenobiotics on glucuronidation is so extensive that it must be described in two divisions. The first concerns the influence on the *activity* of the existing glucuronidating enzymes their inhibition or their activation. The second concerns the influence on the *amount* of glucuronidating enzymes available their decreased synthesis and increased breakdown or their increased synthesis and decreased breakdown.

It is important to distinguish between activation (in the former division) and induction (in the second) To demonstrate any increased transferase activity is due to induction, the increase must be arrested with a protein synthesis inhibitor such as cycloheximide and must resume at the same rate on its removal, incorporation of amino acid into protein similarly halting and resuming in the same preparation at the same time. Culture techniques, with their ease of pulsing and localization of the inhibitor's effect at the desired cells, are obviously the most suitable. Additionally, or at second best, the increased activity must be shown activatable by a perturbant to the same extent as in a control preparation not previously exposed to the suspected inducer. Identical additional activation would not occur if the suspected inducer had itself already acted as an activator. The question has been discussed[130,171,182] and one study[979] neatly uses activators to illustrate that methylcholanthrene probably induces the transferase rather than activates it.

Where, a compound, e.g , piperonyl butoxide, both inhibits and induces the transferase, its different actions will be treated separately.

Two points need emphasis. First, induction of transferase activity to one substrate need not imply induction of activity to all substrates, e.g., pretreatment with morphine increased specific activity towards estrogens, but not towards 4-nitrophenol[112] (Chapter 6). Second, as a consequence of the first, drug tolerance cannot be explained simply as induction of enzymes conjugating the drug, e.g., pretreatment with amphetamine increased specific activity of rat-liver UDPglucuronyltransferase towards estrone and estradiol 17β,[112] but, even when the rats became amphetamine tolerant, not towards 4-hydroxyamphetamine itself.[408]

B. Inhibitors and Activators of Glucuronidation

Effects of many of this first group have already been encountered (Chapter 5), e.g., activators such as diethylnitrosamine and certain substrates, and inhibitors such as competing substrates, certain thiol reagents, and UDP. These, and the effects of ions, will not be discussed further.

Competitive inhibition of the transferase is scarcely demonstrable, interpretation of observed kinetics being equivocal with the microsomal or crudely solubilized preparations employed; e.g., oxazepam,[980] a known substrate, is reported to competitively inhibit glucuronidation of acetaminophen,[980] 4-hydroxyphenylhydantoin, another substrate, to inhibit transferase activity competitively towards two other xenobiotics.[374]

Imipramine and related drugs have been the subject of several reports. Chlorpromazine,[981] imipramine, promazine, and thioridazine[982] inhibited glucuronidation of, and transferase activity towards, 4-aminophenol in hepatoma cells; imipramine, desmethyl-imipramine, and some other monoamine oxidase inhibitors inactivated the transferase after preliminary activation,[983] and potentiation of the pharmacological activity of morphine by monoamine oxidase inhibitors was suggested due to their inhibition of UDPglucuronyltransferase activity toward morphine.[984] Activity to 2-aminophenol and bilirubin is inhibited by certain monoamine oxidase inhibitors.[985]

Certain iodinated molecules in radiographic contrast media are glucuronidated,[4,339] and several of these compounds have been shown to inhibit 4-nitrophenol glucuronidation by microsomes.[986]

Anesthesia with ether, or to a lesser extent with pentobarbital may, from biliary studies depress glucuronidation of iopanoate in the rat. Urethane anesthesia, or decerebration did not affect glucuronidation.[987] A dose-related, noncompetitive inhibition of rat-liver microsomal transferase activity towards 4-nitrophenol followed diethylether, halothane, methoxyfluorane, and chloroform exposure.[988] Diethylether may also inhibit rat liver UDPglucose dehydrogenase.[988]

Various drugs inhibit transferase activity towards estrogens[112] and bilirubin.[4,339] Sulfisoxazole displaces bilirubin from plasma albumin and increases incidence of kernicterus;[989] it displaced bilirubin from albumin in the medium of cultured hepatoma cells, so that the bilirubin entered the cells and inhibited several functions there, including that of UDPglucuronyltransferase.[989] Glucuronidation of bilirubin and 2-aminophenol was inhibited by ethinylestradiol, but not by other oral contraceptives tried.[990] Very high doses of gentamycin (above therapeutic level) inhibited transferase activity towards bilirubin and salicylamide.[991] Novobiocin inhibits[870,992] transferase less than it inhibits biliary transport, as often suggested[4,339] for this antibiotic, which causes hyperbilirubinemia.[993] Novobiocin may diminish bilirubin glucuronidation by interacting with Mg^{++}, as it inactivates enzymes synthesizing bacterial cell-wall membranes,[993] but dependence on Mg^{++} may also reflect its influence on bilirubin solubility.[111]

Some apparent inducers of transferase inhibited it in vitro, perhaps by competition for enzyme or UDPGlcUA. Examples are piperonyl butoxide[133,994] on activities towards testosterone, 1-naphthol,[133] and acetaminophen;[995] eugenol[996] or phenobarbital metabolites[405] on activity to 4-nitrophenol; pargyline on activity towards morphine;[983,984] Δ^9-tetrahydrocannabinol on activities towards 4-nitrophenol and estradiol.[997]

Xenobiotic inhibitors and activators of other enzymes involved in glucuronidation are discussed in the relevant Chapters; e.g., ethanol, rotenone, and menadione potently inhibited glucuronidation of 4-nitrophenol by rat-liver cell suspensions, but not transferase activity in microsomes, because rotenone inhibits formation of UDPGlcUA[166] and ethanol increases the $NADH/NAD^+$ ratio and thereby lowers UDPglucose dehydrogenase activity.[166]

Inhibitors of glucuronide transport into bile will inhibit glucuronidation observed in vivo; several anabolic steroids causing little or no inhibition of UDPglucuronyltransferase or UDPglucose dehydrogenase probably decrease observed glucuronidation this way.[4]

C. Induction of Glucuronidation by Phenobarbital and Other Barbiturates

Administration of phenobarbital to animals and man was known early to increase glucuronide excretion[727] and bilirubin clearance[721,728,946] (for other references see References 339, and 639). Increased excretion could be due to increased uptake of aglycon (even though this uptake appeared not to be limiting[581]), to increased formation of glucarate and glucarolactones inhibiting microsomal β-glucuronidase,[311,619,998] or to increased activity of UDPglucose dehydrogenase.[311,320,502,570,736] In homozygous Gunn rats and Crigler-Najjar patients, no increased glucuronidation on phenobarbital treatment has been found early[821,932] or later.[203,639]

At least part of the enhancement by phenobarbital is from stimulation of UDPglucuronyltransferase. Phenobarbital had no effect in vitro,[342] and enhancement was gradual, suggesting induction. Rise in rat was slower than for Phase 1 enzymes and, with low dosage, not great,[139,311,728,737,932,999] despite considerable increase in glucuronidation observed in corresponding liver slices.[1000] However, perturbation of microsomes disclosed that the transferase had been increased in rat liver, but in the latent form. Perturbants such as digitonin or phospholipases A and C, releasing little microsomal protein,[168,252,1001] did not break this latency, but treatment in vitro with Triton® X-100[139,314] or trypsin,[1001] solubilizing more protein,[211,1001] made stimulation quite evident after phenobarbital pretreatment of rat[139,1001] or of mouse.[314] Stimulation then occurred almost as rapidly and to much the same extent as that of cytochrome P-450 in rat[1001] e.g., detectable after 1 day's treatment, and reaching four-fold after 5 days. Prolonged treatment with phenobarbital (e.g., over 5 days) itself changes constraint, because increase is then seen without any perturbation.[1001] In some species, e.g., chick embryo, enhancement is always rapid and obvious[342] (Chapter 10, Section IV.B).

Use of phenobarbital in increasing perinatal glucuronidation and transferase activity has been reviewed[639] (Chapter 10, Section IV.B).

Phenobarbital treatment of Gilbert's syndrome[639] became a subject of controversy. As the drug increases liver and possibly microsomal weight, total transferase activity towards bilirubin must increase,[1004] but increased specific activity per milligram microsomal protein was not always noted except after periods of high dosage.[1004] Albumin used in the assay binds bilirubin, and substrate concentration may fall too low for the increased $K_{bilirubin}$ accompanying phenobarbital treatment.[383] Again, activation is usually required before induction becomes apparent.[213] Specific increases in transferase activity to bilirubin after phenobarbital treatment are recorded in mouse,[314,377,728] rat,[336,881,935,937] cat,[383] but not in guinea pig,[902] and often not in man. In man, specific transferase activity to bilirubin is recorded as rising after phenobarbital treatment,[1002] sometimes rising,[962] or rising only with the increased microsomal protein.[1003] Phenobarbital may assist excretion of bilirubin conjugates,[933] and this itself could account for many observations of more rapid bilirubin clearance in vivo and of glucuronidation rate in slices following treatment.

Examples of recent reports of UDPglucuronyltransferase activities induced by phenobarbital concern chloramphenicol,[336] 4-methylumbelliferone,[314,324] morphine,[364,1005] 1-naphthol,[278] 4-nitrothiophenol,[206] phenolphthalein,[314] and serotonin.[4269]

Phenobarbital can act directly on the liver cell itself to induce activity from zero[735,750] or from low adult values,[1006] and on kidney tissue.[197,735] It enhanced 1.8-fold transferase activity to several substrates in liver nuclear envelope from pretreated rats,[147,149,150] this increase not being due to increased envelope surface, but to increased enzyme activity per unit surface area,[147] i.e., an induction. The enzyme it induces in embryochick liver resembles that of the adult bird in degree of latency and "kinetics",[193] and in rat liver induced and noninduced transferase possess the same molecular weight and charge on purification to homogeneity.[240,241]

Several barbiturates were examined[1007] for induction of transferase activity towards

4-nitrophenol in detergent-activated liver microsomes from pretreated rats. Substituents on the structure

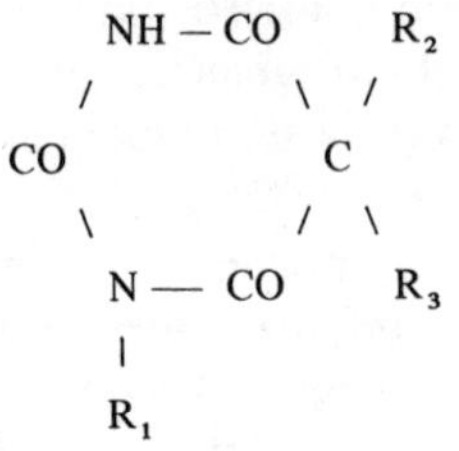

$$\begin{array}{ccc} NH-CO & & R_2 \\ / & \backslash \; / & \\ CO & & C \\ \backslash & / \; \backslash & \\ N-CO & & R_3 \\ | & & \\ R_1 & & \end{array}$$

are, for phenobarbital, H at R_1, phenyl at R_2, and ethyl at R_3; phetharbital, with phenyl at R_1 and ethyl groups at R_2 and R_3, was also an inducer. Others tested were not inducers. They possessed a phenyl group at R_1 and H at R_2, and either a butyl or an ethyl group at R_3, or a saturated cyclohexane ring at R_1 with ethyl groups at R_2 and R_3, or with H at R_2 and ethyl at R_3. Bucolome, with cyclohexane at R_1, H at R_2, and butyl at R_3, lowered this transferase activity on administration, but has been reported elsewhere to increase salicylamide glucuronide excretion and reduce hyperbilirubinemia.[1007] The long metabolic half-life associated with lipid-solubility enhances the inductive power; factors such as lipid-solubility, as well as structure, of the barbiturates must be studied.[1007]

Induction by barbiturates is accompanied by hypertrophy of liver endoplasmic reticulum, distinguishable from the hyperplasia occurring during regeneration. The two processes of hypertrophy and hyperplasia compete for protein synthesis,[1008] and hyperplasia normally receives priority, so that enzyme induction by drugs is less in regenerating than in normal liver.

D. Induction of Glucuronidation by Xenobiotics other than Barbiturates

The first examples of induction of UDPglucuronyltransferase were 3,4-benzo pyrene and 3-methylcholanthrene.[733] Liver microsomal transferase activity towards 2-aminophenol had increased within 18 hr by 40% in the injected males and by 250% (up to normal male levels) in the females; pretreatment with testosterone doubled this increase in females.

Since then, reports on effect of pretreatment with xenobiotics on transferase activity or on glucuronidation have been common. Many are listed below (Table 4). Some interesting examples are discussed in succeeding sections. Induction by aromatic hydrocarbons and by barbiturates appear to exhibit differences, though not yet so obviously as with the Phase 1 system,[1039] recently, a third group of inducers, with 2,3,7,8-tetrachlorodibenzo-*p*-dioxin (TCDD) as example, has been postulated.[389]

3-Methylcholanthrene has been used to prove *de novo* transferase synthesis[337] during its "induction"[336] by a xenobiotic. Pretreated rats were given ^{3}H-labeled leucine in an 8-hr pulse at the period of maximal increase of transferase activity (24 to 72 hr); control rats received ^{14}C-leucine. Livers of tests and controls were pooled and transferase activity to 1-naphthol purified[337] (Chapter 6, Section I), when ^{3}H:^{14}C ratios increased 2.5-fold; after isoelectric focussing of partially purified enzyme the highest ^{3}H:^{14}C ratio was associated with transferase activity and a band in SDS-gel of molecular weight 54,000. This was considered evidence that the enhanced transferase activity after 3-methylcholanthrene treatment was primarily due to *de novo* protein synthesis.[337]

Ethanol does not seem to induce UDPglucuronyltransferase to any extent, probably because of rapid metabolism and clearance. After heavy dosage, unoxidized ethanol is excreted as glucuronide.[1040] Such conjugation should be examined in man, even

TABLE 4

Some reports on UDPglucuronyltransferase activity or glucuronidation following pretreatment with certain xenobiotics

Xenobiotic	Ref.
Aldrin	1009
Aminopyrine	570, 732
Amphetamine	112
Aniline	1010, 1023
Arochlor 1260	1010
Atrazine	1010
Barbital	570
Benzene	1010
Benzo(a)pyrene	112, 733, 1000, 1011—1015
Chlorbutol	1019
Chlorcyclizine	732
Chlordane	619
Chloroquine	732
Chlorpromazine	112, 1016
Chrysene	216, 1017
Cinchophen	1000, 1016
Clophen	see polychlorinated biphenyls
Dexamethasone	319, 391, 698, 711, 764
1,2,5,6-Dibenzanthracene	732
Dieldrin	619, 1009
Diethylnitrosamine	234
3,5-Di-*tert*-butyl-4-hydroxytoluene	1018
Dimethoate	1010
Disulfiram	1010, 1019
Dodin	1010
Ethanol	1010
Eucalyptol	1021
Glutethimide	937
Heptachlor	1010
Heroin	724
n-Hexane	220
Hexachlorobenzene	1010
Hycanthone and chlorinated analogs	1020
Isosafrole	1009
3-Methylcholanthrene	112, 174, 175, 209, 285, 324, 336, 733, 794, 902, 910, 1014, 1016, 1017, 1022, 1044
Morphine	112
Nikethamide	1019
Nitrobenzene	1010
Pargyline	983, 984
Perthane	1010
Phenol	1023
Piperonyl butoxide	133, 994
Polychlorinated biphenyls	797, 1017, 1025—1028
Pregnenolone 16α-carboxynitrile	168, 945, 1018, 1029
Pyrazole	1030
Salicylate	1000
Spironolactone	1031—1033
Styrene	1034
2,3,7,8-Tetrachlorodibenzo-*p*-dioxin (TCDD)	334, 388, 389, 1035
Tetraethyl lead	1010
Δ^5-Tetrahydroxycannabinol	997

TABLE 4 (continued)

Some reports on UDPglucuronyltransferase activity or glucuronidation following pretreatment with certain xenobiotics

Xenobiotic	Ref.
1,1,1-Trichloro-2,2-bis (*p*-chloro-phenyl)ethane (DDT)	245, 874, 1018, 1019, 1036, 1037, 1048
Thiouracil	975, 976, 1038
Toluene	1010

though UDPglucuronyl transferase can be inhibited by ethanol.[157,1041] Ethanol, up to 1% (v/v) had no effect, stimulatory or inhibitory, on UDPglucuronyltransferase in cultured fetal, embryonic or adult hepatic tissue from rat or chicken;[319,711] or even when injected or infused *in ovo* (G. J. Dutton, unpublished results, 1977). Slight stimulation by ethanol was reported in rats within 8 hr of two oral doses each of 5 mg/kg[1019] but may have been due to activation, so soon after the dose; UDPGlcUA levels fell somewhat at that time, suggesting inhibition of UDPglucose dehydrogenase activity, but the latter was reported normal.[1019]

UDPGlucose dehydrogenase can be inhibited by ethanol[157,1041] and ethanol probably inhibits glucuronidation in intact cells this way. Glucuronidation in isolated rat hepatocytes of harmol, 2-naphthol, 4-methylumbelliferone and phenolphthalein was 50% inhibited by 10 mM ethanol and not further inhibited by 100 mM ethanol.[511a] Saturating concentration for ethanol dehydrogenase is 6 mM, and 10 mM ethanol had no effect on transferase in microsomes; high NADH/NAD ratios from the ethanol dehydrogenase action were probably responsible, as supported by lack of inhibition by ethanol when 4-methylpyrazole was present to inhibit ethanol dehydrogenase;[511a] also, UDPGlcUA level fell in hepatocytes exposed to ethanol.[511a]

A drug altering the UDPGlcUA pool (e.g. D-galactosamine) is less effective in changing acetaminophen metabolism than an inducer (phenobarbital) or an inhibitor (prednisolone) of UDPglucuronyltransferase.[1041a]

The many results on glucuronidation treatment with inducers are conflicting. However, such pretreatment requires, and directly varies with, amino acid incorporation into protein:[336,735] i.e., the process involves induction, probably accompanied by change in membrane environment. Enzyme constraint is therefore likely to be progressively changed, and variables of tissue, substrate, species and type of inducer would produce considerable scatter. These variations will be briefly discussed.

E. Activation Characteristics of UDPGlucuronyltransferase Induced by Xenobiotics

Mulder[139] and Winsnes[314] observed that induction of UDPglucuronyltransferase with phenobarbital was most evident when perturbants were present during assay. A greater percentage of total activity was therefore latent after induction than before, suggesting the unsatisfactory term "increased latency". With some substrates, no induction of transferase appeared before activation.[139] Because activation characteristics alter on induction,[314] the degree of induction observed depends on the activating procedure; and as many preparative procedures incur activation (Chapter 5), depends also on preparation of the enzyme. Limited microsomal digestion with trypsin,[1001] which may peel off the superficial protein layer of the membrane, suggests from the amount of protein released, that phenobarbital pretreatment of rats increases thickness of this layer;[167,1042] "increased latency" might arise from the transferase protein being buried deeper in the membrane.[1042] This phenobarbital-induced transferase activity towards 4-nitrophenol in rat liver, activated also by Triton® X-100,[139,336] could not be activated by other compounds able to activate the noninduced enzyme,[167,325] which, together with changed submicrosomal distribution,[139] suggests a change in membrane characteristics, and transferase constraint, following phenobarbital treatment.

Induction of transferase activity by 3-methylcholanthrene, differs from that seen by phenobarbital. (It should be noted[336] that 3-methylcholanthrene is only slowly absorbed from the peritoneal cavity on injection and persists in the liver after a single injection, whereas phenobarbital is rapidly "detoxicated" and requires multiple injections or prolonged feeding to achieve comparable hepatic concentrations in vivo.) Pretreatment of young rats with 3-methylcholanthrene enhanced specific activity towards 2-aminophenol and bilirubin within 6 to 12 hr, but pretreatment with phenobarbital or chlorcyclizine yielded no observable results after 3 days[285] possibly because of "increased latency"; kinetic parameters of the enzyme changed following pretreatment with 3-methylcholanthrene, but as the apparent K_m for 2-aminophenol decreased in rat and increased in guinea pig, the authors wisely limited their conclusions.[285] Again,[902,1022] transferase activity in unactivated microsomes to bilirubin specifically increased on a single injection of methylcholanthrene, whereas multiple doses of phenobarbital over 10 days increased only the total liver activity. "Increased latency" can result from treatment with 3-methylcholanthrene as well as phenobarbital; unactivated transferase from rats increased its activity to 1-naphthol by 10 and 70% respectively, while with the activated enzyme, the increases were fourfold in each case.[177]

Induction by 3-methylcholanthrene, not additive to that by phenobarbital in activated preparations on a microsomal protein basis,[324] is additive if measured per 100 g body weight.[324] Even if not additive, the two inductions may induce different membrane constraints on the enzyme. After phenobarbital treatment, trypsin "solubilized" the enzyme, but after 3-methylcholanthrene treatment, trypsin was not effective, and phospholipase A or detergents were best.[325] The fluorescent probe, 8-anilino-1-naphthalene sulfonate (ANS), showed no difference between liver microsomes after pretreatment of rats with either of the two inducers if measured at 25°C,[269] but a nonlinearity was observed at 40°C, in those from phenobarbital-pretreated rats.[1043]

"Increased latency" in liver transferase activity towards 4-nitrophenol occurs also after pretreatment of rats with DDT,[1048] styrene,[1034] aldrin, dieldrin, or isosafrole,[1009] trypsinization or digitonin addition being required before enhancement was evident.[1009] Polychlorinated biphenyls also increase the latency, sometimes to a high degree;[1026] dietary cholesterol lowers this increase, probably affecting membrane structure.[798] Polychlorinated biphenyls enhance biliary secretion of thyroxine by rats, at least partly by enhancing UDPglucuronyltransferase activity to thyroxine.[1025]

Transferase activity to 4-nitrophenol was stimulated by intraperitoneal administration of n-hexane to guinea pigs.[220] Although n-hexane is a typical activator in vitro,[220] its long-term stimulation was considered due to induction, being additive to in vitro activation.[220]

F. Kinetic Changes in UDPGlucuronyltransferase Activity on Induction by Xenobiotics

Some kinetic changes were mentioned in the preceding section, and they would seem to confirm that constraint of transferase changes on induction. Many other reports exist,[105,173,314] and in some, the changes are less[139,179,314] or absent[336] after perturbation, consistent with origin in the surrounding membrane. Inducer-specific changes are described; e.g., phenobarbital pretreatment (multiple injections) increased rat hepatic activity to testosterone without changing apparent K_m for this substrate, whereas methylcholanthrene (single injection) changed both K_m and specific activity.[1024]

G. Substrate Specificity of UDPGlucuronyltransferase Induced by Xenobiotics

If relative activity of UDPglucuronyltransferase towards different substrates concerns membrane constraint it might vary on induction and with different inducers.

Some observations suggest this. Bock et al.[336] found that phenobarbital pretreatment stimulated transferase activity towards chloramphenicol and bilirubin rather than towards 4-nitrophenol and 1-naphthol, whereas pretreatment with 3-methylcholanthrene stimulated activity towards the second pair and tended to decrease it towards the first pair. However, the difference persisted in solubilized,[336] partly purified[336] and very considerably purified[262] transferase preparations. Enzyme activity was more stable during purification from rats treated with 3-methylcholanthrene than from those treated with phenobarbital. Further, specific activity of peak transferase activity towards morphine from methylcholanthrene-treated rats was only 1/6 that found from phenobarbitaltreated rats, whereas specific activity towards 1 naphthol was almost fourfold higher.[262] Bock et al. conclude that the two inducers differentially stimulate separate transferase activities, and have recently[337] physically separated the activities. These observations of Bock's group are consistent with the divisions of functional heterogeneity observed by Wishart et al.,[145a,390,391,1004] where under standard conditions the six transferase activities in the "late fetal" cluster (Chapter 10, Sections III.E.1 and IV.D.2.b) are induced in adult rat liver by methylcholanthrene whereas the six in the "neonatal" cluster are not; the latter six are preferentially induced by phenobarbital.[1004] (For extension to further substrates, see Reference 145a). Again, transferase activities to 4-methylumbelliferone, 4-nitrophenol, and 1-naphthol are inducible by TCDD,[389] whereas activities to diethylstilbestrol, β-estradiol, estrone, testosterone and phenolphthalein are not,[389] this division also relating to onset of "natural" induction perinatally.

TCDD raised transferase activity to 4-nitrophenol sixfold[334] in crude, solubilized, or semipurified preparations, but not towards testosterone or estrone; the enhancement, blocked by Actinomycin D, was presumed induction.[334] Chlorpromazine increased specific transferase activity to estrogens[112] but not towards 4-methylumbelliferone.[1016]

H. Species and Tissue Specificity of UDPGlucuronyltransferase Induced by Xenobiotics

As will have been gathered, induction of UDPglucuronyltransferase activity by xenobiotics need not be similar among species or tissues. A few examples follow. When liver transferase activity towards 2-aminophenol is induced by methylcholanthrene, the apparent $K_{2-aminophenol}$ rises in guinea pigs but falls in rats.[285] Phenobarbital induces liver transferase activity to 4-nitrophenol in rat but apparently not in cat,[383] and induces activity to phenols in chick-embryo liver and kidney, but not in gut or skin.[342] Exposure to cigarette smoke (3,4-benzopyrene) doubles transferase activity to 4-methylumbelliferone in rat gut, slightly raises it in lung, and does not change it in kidney.[1045,1046] 3,4-Benzopyrene and 3-methylcholanthrene,[925,1014] but not phenobarbital,[209] induce transferase activity to 4-nitrophenol in rat duodenum; activity induced by 3-methylcholanthrene appeared highest laterally in villi, not at tips or in the crypts, and one stage in the cell's growth was optimal.[1047] Route of administration is important; liver transferase activity increased with DDT given intraperitoneally[1048] but decreased when it was given intragastrically.[245] 3-Methylcholanthrene induced gut mucosal transferase by administration intragastrically rather than intraperitoneally.[797] Several unexplained, anomalies exist with route and period of administration, e.g., phenobarbital injected for 3 days raised the (activated) specific transferase activity in rat liver to estradiol, but not to estrone; whereas if given in the drinking water over 20 days, it raised specific activity to estrone, but to estradiol only if on a total liver protein basis.[213]

I. Decreased or Unchanged UDPGlucuronyltransferase Activity Following Administration of Xenobiotics In Vivo

Certain xenobiotics decrease UDPglucuronyltransferase activity; or block stimulation due to others. When decrease is only observed in detergent-activated microsomes, the xenobiotic may have already activated the enzyme in vivo, by degrading the endoplasmic reticulum as reported after chronic administration of carbon disulfide,[221,222] and likely with carbon tetrachloride or tetrachloroethane.[1049] Carbon tetrachloride and hematin administered in vivo blocked induction by 3,4-benzopyrene of both Phase 1 and the transferase, suggesting some product of Phase 1 (an epoxide?) was necessary to induce the transferase. SKF 525-A® may block in a similar way,[1013] and also hemin.[781]

Phenylbutazone[1050] and ethinylestradiol[733] decrease glucuronidation, but the reason is unknown.

Several compounds do not stimulate rat-liver transferase activity to 4-nitrophenol, even after daily intraperitoneal injection for 2 to 3 weeks. These are hepatochlor, Perthane, toluene (in contrast to benzene), lead tetraethyl, Dodin, Atrazine, and dimethoate; Phenylmercuric acetate halved the transferase activity.[1010]

J. Effect of Xenobiotics on Enzymes other than the Transferase Affecting Overall Glucuronidation

As mentioned in other sections, phenobarbital exposure sometimes increases overall glucuronidation e.g., of paracetamol in infants[1051] and of 2-aminophenol in rat or chick embryo liver slices,[173,342] and sometimes has little effect, e.g., with 4-nitrophenol in rat liver cell suspensions.[166] We now examine effect of xenobiotic inducers on hepatic UDPGlcUA and UDPglucose dehydrogenase.

Liver UDPGlcUA concentrations may often appear rate-limiting even in adult animals (Chapter 9, Section III.F). It is not yet known whether the approximate doubling[177,1019] or significant increase[180] of UDPGlcUA concentration in perfused[177,180] or fresh[1019] liver of rats pretreated with phenobarbital or 3-methylcholanthrene is in vivo adequate for the occasional great increase in transferase activity following such pretreatment. Concentration of UDPGlcUA in vivo need not indicate its turnover or accessibility to the enzyme. As no increase of UDPGlcUA occurs in nonperfused livers of rats pretreated with phenobarbital or methylcholanthrene,[180] possibly (K. W. Bock, personal communication, 1977) the level of UDPGlcUA is strictly controlled in vivo and this regulation may be weakened in the isolated perfused liver.

Levels of UDPglucose dehydrogenase need not correlate with availability of UDPGlcUA for the transferase protein (Chapter 9, Section II.B). Further evidence for this comes from administration of xenobiotics. Administration of barbital, phenobarbital, nikethamide, chlorbutol or DDT raised UDPGlcUA levels within 8 hr in rat liver, e.g., from some 0.26 to some 0.69 μmol/g liver with phenobarbital,[1019] yet UDPglucose dehydrogenase activity did not increase.[1019] However, sometimes xenobiotics inducing transferase activity and glucuronidation also increase UDPglucose dehydrogenase activity. The evidence was at first conflicting. Increases in rat or guinea pig liver dehydrogenase activity up to twofold followed intraperitoneal administration of 60 mg barbital or 50 mg chloretone for 4 days, but not of 3-methylcholanthrene or 3,4-benzopyrene;[570] another report[736] found no increase after treatment with chloretone (150 mg/kg body weight) intragastrically. Injection (i.p.) of rats for 5 days with 20 mg phenobarbital twice daily increased it nearly two-fold[737] but 10.5 mg injected daily for 7 days increased it only 1.25-fold.[311] In mice allowed phenobarbital in drinking water, it rose two-fold.[502] Small rises in UDPglucose dehydrogenase activity have also been reported after pretreatment with chloroquine,[732] morphine,[1052] salicylamide,[1053] or eugenol,[1054] and a two-fold rise with cinchophen.[311]

Fragments or cell monolayers of chick embryo liver cultured with phenobarbital in the medium exhibited only two- to three-fold increase in the fragments and none in the monolayers, even though transferase activity and, notably, glucuronidation increased over 20-fold.[502] However, injection of 40 mg phenobarbital into eggs at 8 days' incubation increased liver dehydrogenase activity five-fold (dose-dependently) over saline-injected controls.[502] The induced enzyme appeared identical to the normal enzyme in all characteristics examined.[502] As no activation by phenobarbital or its metabolites could be demonstrated,[502] UDPglucose dehydrogenase, therefore, appears clearly stimulated, probably through induction, by pretreatment with phenobarbital in chicken. Salicyclamide inhibited the rat liver enzyme in vitro[1053] and tolbutamide and phenoformin[1055] decreased it when fed. Certain xenobiotics inhibit overall glucuronidation by interfering with UDPGlcUA production. Rotenone, menadione, and ethanol maximally inhibit glucuronidation in isolated rat-liver cells at concentrations having virtually no effect on microsomal transferase. The first two could inhibit by blocking mitrochondrial oxidation, but ethanol increases intracellular $NADH^+$, and inhibits UDPglucose dehydrogenase (Section II.D above).

Microsomal β-glucuronidase is inhibited by phenobarbital[998] and some other inducers of the transferase,[619] in rat liver, but lysosomal β-glucuronidase is either relatively unaffected by them,[619] or enhanced by phenobarbital.[620] 3,4-Benzopyrene[1056] and DDT[1057] increase total β-glucuronidase activity, while disulfiram diethylcarbamate,[1058] cinchophen,[311] and pyrazole[1030] decrease it.

UDPGlcUA acid pyrophosphatase is only slightly increased by some xenobiotics (e.g., chloretone, ethionine,[570] disulfiram, and diethyldithiocarbamate[1058]) and decreased by others (e.g., cinchophen,[311] DDT,[1057] pyrazole,[1030] and dimercaprol[1059]).

None of these effects appear responsible for, nor influence significantly, the enhancement of glucuronidation brought about by the major process, the induction of UDPglucuronyltransferase.

K. Excretion of Glucuronic Acid Metabolites Following Administration of Xenobiotics

Many xenobiotics stimulating activities of Phase 1 and 2 drug-metabolizing enzymes increase metabolic flux through the "glucuronic acid cycle" of glucose and galactose metabolism,[543,555] increasing the urinary excretion of its components, e.g.,L-gulonic, L-ascorbic, (conjugated) D-glucuronic acid, and D-glucaric acid.

Urinary glucarate excretion has, therefore, been proposed as an index of hepatic microsomal-enzyme activities,[945,1060] (for references see References 591 and 1061). Because many enzymes are involved, this index must be applied with caution, especially if a rise in UDPglucuronyltransferase activity is to be deduced. If UDPglucuronyltransferase is low, perhaps more UDPGlcUA is then available for breakdown through the pathway to glucarate. Then, a higher glucarate excretion could indicate lesser transferase activity, not greater.[106] For example, hypothyroidism increased UDPglucuronyltransferase activity, but decreased glucarate excretion;[975] increased excretion of D-glucarate did parallel a rise in transferase after *n*-hexane administration, but required a long interval.[220] Conversely, glucarate excretion rose markedly within 8 hr of administering barbiturates, during which time liver UDPglucuronyltransferase does not change.[1019] Stimulation of glucuronic acid pathway and of transferase cannot, therefore, be directly related.[1019]

Glucarate excretion has been studied: (1) to screen for induction of UDPglucuronyltransferase activity towards bilirubin following the neonatal administration of phenobarbital, and (2) to monitor an infant's capacity for glucuronidation, but results are equivocal. Whereas some correlation between lowered hyperbilirubinemia and increased glucarate exists,[1063] D-glucarate excretion could not predict ability of a neonate to glucuronidate acetaminophen.[1064] Notten and Henderson[1019,1065] consider that, because xenobiotics do not increase UDPglucose dehydrogenase activity before stimulating glucarate excretion, but do increase UDPGlcUA concentration[1019] within this first 8 hr, then UDPGlcUA availability must be increased, possibly by an inhibition of glycogen synthesis allowing UDPglucose to be diverted to UDPGlcUA formation. They cite phenobarbital pretreatment reducing particulate glycogen in smooth endoplasmic reticulum.[1066] They have themselves noted[1067] in isolated rat liver cells that phenobarbital pretreatment increases UDPGlcUA and UDPglucose, decreases glycogen, and lowers ^{14}C incorporation into glycogen. Their recent review,[1068] concludes that urinary D-glucarate level need not be related to hepatic drug-metabolizing capacity, but can indicate exposure to certain xenobiotics.

Flux along the "glucuronic acid pathway" might be more reliably, though less conveniently, monitored from the spectrum of excreted participants (e.g., xylitol, L-gulonic acid, as well as D-glucaric acid) studied by g.l.c. techniques (see Reference 1018 for references). Glucaric acid determination has been reviewed.[591]

III. ADDITIONAL NOTES

Section I.A., I.C: Urinary output of estrogen glucuronides in women has been followed in detail by direct radioimmunoassay.[1068a] No diurnal variation was noted, but a regular pattern occurred in the menstrual cycle and during pregnancy.[1068a]

Section I.B: Secretion infused into rats for 90 min increased transferase activity to bilirubin 1.5-fold, but not that to 4-nitrophenyl.[1068b] The authors suggest that this effect may be related to the similar low stimulation of this transferase activity observed with dibutyryl cyclic AMP.[976a]

Section II.A: It is now clear that "functional heterogencity" of the transferase extends to relative induction by different xenobiotics. Bock's GT$_1$ activities[336,392c] and Wishart's late-fetal cluster 1 activities[145a,1044,1068c] are preferentially induced by 3-methylcholanthrene; the corresponding GT$_2$ and neonatal cluster 2 activities are preferentially induced by phenobarbital. Both authors note that substrates for the first type are planar molecules, those for the second more bulky. An overlapping configuration — as in a compound existing in equilibrium between planar and nonplanar forms — may well exist; Bock et al.,[392c] suggest 2-hydroxybiphenyl and benzo(a)pyrene 7,8-dihydrodiol are overlapping substrates. Lucier's late-fetal "nonsteroid" cluster is inducible by TCDD, the neonatal "steroid" cluster is not.[389] This phenomenon of functional heterogeneity links perinatal development, inducibility by xenobiotics and glucocorticoids, tissue distribution and, to some extent, physical separation; it is of great interest.

Chapter 14

EXTRAHEPATIC GLUCURONIDATION

I. GENERAL

Early work[4,5] indicated that liver was not the only site of glucuronidation. Various organs, perfused or sliced, performed the process, and experiments with hepatectomized or further-eviscerated "whole" animals suggested low-level background glucuronidation.

The "hepatic obsession" in research has lessened by increasing need to monitor reception sites in the body for environmental pollutants; hence, for example, the growing interest in gastrointestinal (GI) tract and lung.

Extrahepatic glucuronidation might seem a more logical method of dealing with ingested toxic molecules than allowing them to be absorbed and circulate unmodified to the liver. Powell et al.[1069] elegantly confirmed this possibility by demonstrating that total radioactivity from labeled phenol fed to rats was no greater in liver than in blood; they found intestine the main conjugatory site of orally administered phenol in this animal.

However, even allowing for the large bulk of gastrointestinal tract and high activity in kidney, liver seems in guinea pig or mouse the major site of glucuronidation of 2-aminophenol or bilirubin because of its mass and high level of UDPglucuronyltransferase.[574] A detailed tissue survey[1070] with the sensitive assay for methylumbelliferone glucuronide came to the same conclusion in rat, whose gastrointestinal tract possessed only 15 to 20% of hepatic activity; the results[1070] (Table 5) unfortunately omit activity in skin, a large glucuronidating organ.[573]

Many examples are noted below of specific tissue distribution of transferase depending on substrate. Liver often contains the only activity demonstrable for some substrates (e.g., for 4'-hydroxyamphetamine, absent from kidney, gut, lung, spleen, brain, and heart of rat[408]), but occasionally it lacks an activity present elsewhere (e.g., in man for glucuronidation at the 3-α position of estriol,[1071] performed in the gut).

Use of activators, inducers or of glucarolactone (to inhibit β-glucuronidase) has revealed UDPglucuronyltransferase activity in several tissues (e.g., spleen and lung) previously thought to lack it. No quantitatively significant new site has been discovered, although specific local glucuronidation may be important, as in adrenal gland.

II. SELECTED EXAMPLES OF DIFFERENTIAL ACTIVATION AND INDUCTION OF UDPGLUCURONYLTRANSFERASE AMONG TISSUES

Because membrane environment of UDPglucuronyltransferase will differ among tissues, activation and induction characteristics of the enzyme may likewise differ. Digitonin activated rat hepatic transferase towards 4-methylumbelliferone and 4-nitrophenol eight- to ninefold but in lung only two- to threefold and a similar fall, from four to fivefold to two- to threefold, was found between liver and lung in guinea pig;[1070,1072] however, the one concentration of digitonin used might not have been optimal for lung. Confirming low degree of latency in lung, no activation occurred in rat lung and gut with 4-nitrophenol as substrate, unlike kidney and liver.[829] In gut, activation probably occurs during preparation (Section IV below), but this explanation is less likely for lung.

Many differences depend on how soon and in what amount the inducer reaches the

Table 5
TRANSFERASE ACTIVITIES IN
RAT TISSUES TOWARDS 4-
METHYLUMBELLIFERONE[1070]

	A	B
Liver	460	3200
Duodenal mucosa	260	96
Kidneys	150	120
Adrenal glands	170	63
Spleen	30	35
Lungs	28	34
Thymus	19	8.8
Heart	1.4	1.1
Brain	0.8	1.3

Note   Column A contains the nanomole substrate glucuronidated gram^{-1} wet weight. Column B contains the nanomol substrate glucuronidated per whole organ.

respective tissues: injected, 3-methylcholanthrene or 3,4-benzopyrene induces transferase activity in liver but not in kidney or alimentary tract,[1000] whereas when fed they induce first, and to greatest extent, in gut.[1014] Intragastric administration of salicylate induces transferase activity in rat kidney but not in gastrointestinal tract or liver, whereas intragastric cinchophen enhanced activity in all three tissues.[1000]

Species differ in tissue response. Phenobarbital injected into rat induced transferase to 2-aminophenol in liver but not kidney or gut,[1000] whereas when injected into chick embryo response occurred in liver and kidney but not gut or skin.[324]

Specificity of steroid glucuronidation at various C-atoms varies between liver, gut, and kidney among species, a complex question extensively studied by Breuer's group.[43]

III. GLUCURONIDATION IN KIDNEY

Glucuronidation in kidney was first shown[85] using slices and with borneol as substrate, being confirmed for other substrates at glucuronidation rates (glucose present in Reference 574) almost as high as liver.[4] Transferase activity in homogenates appeared very low (or was thought absent[4]), largely due to rapid UDPGlcUA breakdown by kidney UDPGlcUA pyrophosphatase.[559,572] Kidney cortex in guinea pig contained ten times more transferase activity to 2-aminophenol than medulla,[574] and in man was[1073] 20 times more active towards estriol and 17β-estradiol than medulla. Activity towards bilirubin is confined to the proximal tubular cells in dog and rat.[475,1074] The "microsomal" fraction contains highest transferase activity.[232,574] UDPGlcUA and UDPglucose dehydrogenase are present in kidney,[574] so that glucuronidation in vivo presumably occurs via UDPglucuronyltransferase; no direct synthesis of glucuronide from glucuronate or glucuronolactone was demonstrable in kidney.[574]

Activity to certain substrates seems low or absent in kidney. Lack of renal activity to bilirubin or 2-aminophenol in Gunn rat[462] or to 2-aminophenol in cat[886] resembles their hepatic deficiency, but liver contains activities which in the following species are absent from kidney: activity to bilirubin in man,[475] to phenolphthalein in pig[697,710] and rat,[389] to steroids in rat,[356,389] (but not in pig[43,232]) and to 4-hydroxy-3-methoxyphenylethanol[894] and chloramphenicol (K. W. Bock, unpublished work) in rat.

Estrogen glucuronide synthesis occurs in the kidney of many species, including man,[197,1075] and the kinetic properties[232] of the pig kidney enzyme suggest it glucuronidates estrone with a specific activity higher than the liver enzyme, possibly by a sequential catalytic mechanism,[232] which operates somewhat in reverse also.[232]

Transferase activity in kidney develops at similar rates to that in liver in a range of species (e.g., chicken,[851] guinea pig,[682] man[197]) and is increased in chick embryo by phenobarbital[342] or glucocorticoids,[711] and in mammalian fetus by glucocorticoids.[693] These inducers, act directly on the kidney cell in culture.[342,711]

The physiological role of renal conjugation has been discussed[1076] with the chicken as model. The kidney may contribute to the urinary conjugates of bilirubin in animals (not man) during hyperbilirubinemia and bile duct ligation[931] and to the 3α-glucuronidation of estriol in man (Section IV). Injection of hemoglobin or bilirubin into isolated dog kidney resulted in urinary excretion of conjugated bilirubin,[1074] and Wistar rat kidneys grafted into Gunn rats markedly lowered the congenital hyperbilirubinemia of this strain,[939] presumably because of the demonstrable transferase activity in Wister rat kidney.

Many compounds are cleared into urine by both filtration and active tubular secretion, e.g., organic anions of relatively low molecular weight and many glucuronides,[1076,1077] such as that of resorcinol. Glucuronides of higher molecular weight, e.g., of androsterone and aetiocholanolone in man[1078] and pregnanediol in chicken[1079] are excreted solely by glomerular filtration.

Sex differences occur in kidney, male rats possessed less transferase activity towards 4-nitrophenol in kidney than females, contrary to liver.[829]

IV. GLUCURONIDATION IN ALIMENTARY TRACT

A. General

A recent review exists.[1080] Glucuronidation in the alimentary tract was first reported using slices of intestinal mucosa and confirmed with glucose-fortified slices, strips, and sections in many species for various substrates including bilirubin.[4] Demonstration of UDPglucuronyltransferase activity in homogenates of this tissue was hampered by rapid destruction of added UDPGlcUA. The transferase appeared located in mucosa[574] and for 4-methylumbelliferone at least, in villi and crypts, not Brunner's glands,[21] being induced by 3-methylcholanthrene along the sides of the villi.[1047]

The whole length of tract forms glucuronides. Transferase activity towards 4-nitrophenol,[1083] 2-aminophenol,[1081,1082] and 4-methylumbelliferone[1070] is high at the oral and low at the aboral end, an exactly opposite distribution to that of β-glucuronidase activity.[1070,1083]

UDPGlcUA and UDPglucose dehydrogenase are found in the GI tract,[320,574] and no evidence for glucuronidation other than by UDPglucuronyltransferase is apparent. In intestine the formation of UDPGlcUA seems constant and high, a good rate of glucuronidation being maintained by a perfused rat intestinal loop.[1084] Although starvation did not lower glucuronidation in the rat intestine in vivo,[789] glucose added to the medium increases glucuronidation of intestinal mucosal strips.[574]

B. Latency of UDPGlucuronyltransferase Activity in Alimentary Tract

UDPGlucuronyltransferase activity to morphine as usually prepared from alimentary tract lacks latency, not being further activatable;[365] indeed, with 4-methylumbelliferone or 4-nitrophenol as substrates,[326] all perturbants, except cetylpyridinium chloride, inactivated it. These observations indicated that intestinal UDPglucuronyltransferase operated unconstrained in vivo or that it was activated, e.g., by gut proteases, during preparation and assay.

Josting et al.[1085] investigated this, finding that transferase activity to 1-naphthol was not enhanced by UDPGlcNAc or Triton® X-100 when assayed in microsomes from rat intestinal mucosa or in homogenates assayed a similar period (4 hr) after death of the rat. When fresh homogenate (30 min after death) was assayed, then activation by these agents was immediately apparent, activity rising to levels seen in the aged (4 hr) microsomes.[1085] This suggests that: (1) the enzyme is activated artefactually when measured in gut microsomes, (2) it exists in rat intestinal mucosa, as in liver (Chapter 5), in a constrained state, and (3) the low level seen by early workers with gut homogenates was not only due to destruction of added UDPGlcUA, but also to their care in using fresh, cold (and, therefore, unactivated) preparations![1085] Consistent with conclusion (2) was the finding that[1085] glucuronidation rates per gram mucosa were similar in perfused loops and in fresh homogenates containing the physiological concentration of UDPGlcUA found in liver (there is no reason to suppose concentration of UDPGlcUA to be greater in mucosa than in liver).

C. Substrate, Sex, and Species Differences in Glucuronidation by the Alimentary Tract

Because of variable and unknown activation of transferase in alimentary tract, we need not list the many comparisons made between activities in this and other organs nor the sex differences there.[797,830,904] Among species it seems that herbivores (possibly more exposed to dietary glucuronidogenic inducers) generally possess higher transferase activities to xenobiotics studied than do carnivores.[1086]

Transferase activity glucuronidating steroids in pig intestinal mucosa has been described in detail,[43] its existence being suggested by formation of estradiol glucuronide[1087] and estrone glucuronide[656] in rat duodenal slices and human intestinal loop, respectively.

Some conjugation of bilirubin is reported in the tract from guinea pig[574] and rat (K. W. Bock, personal communication, 1975) and suggested indirectly from work[1088] with hepatectomized and nephrectomized rats; which still formed a little bilirubin monoglucuronide; but when evisceration was additionally inflicted upon them, bilirubin was virtually unconjugated. The same drastic treatment accorded to dogs[1089] also suggested intestine a site of bilirubin glycosylation.

D. Significance of Development of Glucuronidation in the Perinatal Alimentary Tract

With some substrates and species (e.g., phenols in chicken[369,500]) alimentary track UDPglucuronyltransferase activity does not appear until birth, but in mice and guinea pig[574,682] remarkably high levels of activity towards 2-aminophenol was noted in whole organ and homogenate preparations of the fetal tract (Chapter 10, Section III.E.2.b). These experiments have not yet been repeated with optimally activated enzyme.

The well-documented estrogen glucuronides of human amniotic fluid, meconium, and fetal tissues (Chapter 10, Section II), may derive from the low glucuronidation in human fetal stomach or human gastrointestinal tract (Chapter 10, Section III.E.2.b).

Deconjugation in the gut of bilirubin already conjugated in liver and passed into the tract via bile could be of great importance in the neonate if, as is likely at that age[678] some free bilirubin could be reabsorbed for enterohepatic circulation. Evidence is discussed in Chapter 10, Section III.D. Treatment of neonatal hyperbilirubinemia has, therefore, recently included dosage with agar to bind liberated bilirubin in gut,[1090] although its efficacy has been questioned.[1091] β-Glucuronidase activity in the ileum of infant rats falls to the adult levels about weaning, possibly from action of hormones such as cortisone.[1092]

E. Significance of Glucuronidation in the Adult Alimentary Tract

Glucuronides of salicylate, salicylamide,[1093] and of certain steroids and thyroxine analogs[732] form in everted intestinal sacs and accumulate on the serosal side. Aglycon was thought to enter from the mucosal side, become conjugated in the mucosal cells, and the resulting glucuronide leave by the serosal side, the mucosal side being impermeable to glucuronides; in this way uptake and transfer of the less polar steroids and possibly of certain amino acids might occur.[918] Slower absorption after oral administration of salicyl glucuronide than free salicylate does suggest[1080] a barrier to glucuronide uptake. A report[1093] that oral administration of salicylate gave 100 to 200 times more free aglycon than glucuronide in blood is countered by another[1094] which incubated labeled Prontosil with everted rat intestinal sacs and found 60% of absorbed label as Prontosil *N*-glucuronide and only 32% as free drug; also, intraduodenal administration of Prontosil was more effective in forming biliary glucuronide than intravenous administration of the drug. This evidence and that of Powell et al.[1069] (see Section I above) suggests that considerable conjugation of orally administered aglycon occurs in gut, and recent work by Bock's group clearly shows its importance.

Bock and Winne[1084] infused 1-naphthol into the lumen of closed or single-pass perfused jejunal loops from rat and measured the appearance of 1-naphthol glucuronide both in the intestinal lumen and in venous blood. When 1-naphthol was infused into the lumen, 69% was metabolized within 30 min, almost all as glucuronide; 49% of this appearing in venous blood and 20% in the lumen. Only 19% free naphthol was found in venous blood, most of it bound to cellular elements. When glucuronide was infused into the lumen, 59% remained there after 30 min, whereas 20% had appeared in the venous blood. On single-pass perfusion with 1-naphthol, a steady 91% glucuronidation of absorbed aglycon was achieved after 10 min, of which 82% was released into blood, the rest into the lumen. Glucuronide did not accumulate in the mucosa.[1084]

These results indicate that glucuronides are excreted by epithelial cells of intestinal mucosa into blood and into intestinal lumen, their regulation between these two compartments being unknown.[1084] Further, glucuronides are absorbed to some extent from lumen to blood without hydrolysis, and this way contribute to enterohepatic circulation of drugs.[1084]

Uptake by gut is substrate dependent. Paracetamol was rapidly absorbed from lumen to blood, 75% appearing unconjugated in blood within 30 min and 51% within 10 min, only 4% being conjugated (largely as glucuronide).[1085] Morphine was absorbed slowly (only 1.4% within 10 min), about half being free and half conjugated (largely as glucuronide).[1085] Glucuronidation by gut, therefore, markedly reduces the amount of morphine and 1-naphthol reaching the circulation when administered by this route, whereas paracetamol uptake is scarcely affected.[1085] Work with rat intestinal "microsomes", also indicates that glucuronidation in rat gut lowers the analgesic effect of oral morphine;[303] this rat gut transferase conjugates the more lipophilic phenolic opiates most readily,[403,405] buprenorphine and etorphine at low concentrations in gut mucosa (1 µg/mℓ), being almost wholly glucuronidated, no free drug appearing in the portal blood. These findings largely explain the early reports.

F. Biliary Secretion and Enterohepatic Circulation
1. Biliary Excretion of Glucuronides

Smith[22] surveys the early history of the excretion of xenobiotics in bile. The general subject is critically discussed in "The Hepatobiliary System",[1095] A brief review exists.[1095a]

Smith[22] considers that biliary, rather than urinary, excretion of both endogenous and xenobiotic compounds or their conjugates is favored when the molecular weight is above 300 to 400. He lists[22] such high molecular weight glucuronides, ranging from 4-hydroxybiphenyl glucuronide (346 daltons) to iodopanoic acid glucuronide (747 daltons). Species vary in their moleculars weight threshold for biliary secretion, and the value differs between anions, the most common form, and cations, such as quaternary ammonium compounds. If the molecule is polar and large enough, conjugation is not necessary for its excretion into bile; but bilirubin (Chapter 1, Section V) indicates that other requirements exist for biliary excretion than size and polarity. Another example is the greater biliary excretion of the glucuronide of 4,4′-dihydroxybiphenyl than of the free compound.[1096] Glucuronides are more commonly excreted in the bile than are sulfates, largely because the glucuronyl moiety adds 176 daltons to the molecular weight.[22] Glucuronides excreted in bile tend to be more lipophilic than those excreted in urine because of the greater size of the usually strongly lipophilic aglycon than of the glucuronyl moiety,[22] and because it may be far enough away from the polar glucuronyl moiety for the conjugate to be accepted by transfer proteins excreting a compound into bile.[8] All major types of glucuronides are found in bile, but information for the C-linked glucuronides is not yet available.

Excretion into urine rather than bile is not necessarily because the smaller molecular weight glucuronides leave the body more efficiently that way; in rats with nonfunctioning (ligated) kidneys, those compounds normally only excreted in urine remained mostly in the body according to one report,[181] only a small proportion overflowing into bile,[181] but another found a more significant overflow into bile.[1097]

As much work suggests that low molecular weight anions which are poorly excreted in the bile (e.g., 2-aminophenyl glucuronide) are more readily reabsorbed from the rat biliary tree on retrograde infusion than larger molecules (e.g., phenolphthalein glucuronide) predominantly excreted in the bile,[1098] then possibly, as with renal tubular reabsorption, those compounds not appearing finally in bile may have been reabsorbed;[1098] pressures used in this retrograde infusion were, however, unphysiologically high. Bile collected experimentally is not the same as canalicular bile,[1095,1098] and only examination of the latter can indicate if such reabsorption actually occurs.

Another factor influencing excretion of a compound into bile is its binding to specific proteins. Ostrow in discussion to Reference 1098, pointed out that conjugates are usually less strongly bound to both ligandin and serum albumin than are free aglycons, and in cholestasis, for example, should be cleared via the urine more rapidly than free aglycons despite their greater molecular weight; he considered protein binding more important for biliary excretion than molecular weight. Taylor in discussion to Reference 1098, suggested that protein binding in canalicular membranes would be specific, especially with endogenous compounds, such as steroids; he considered carrier proteins in rat canalicular membranes relatively unselective in binding, so that almost any molecule of a given molecular weight will be transported through the membrane, whereas man's more selective carrier proteins would account for his more selective biliary excretion. The role of ligandin or similar binding proteins in liver cell transport, has been reviewed.[729]

In summary, molecular weight and charge largely determine whether glucuronides pass into bile or into urine, but probably other specific factors contribute.

Passage of glucuronides into bile itself has been discussed.[1099,1100] 4-Nitrophenyl glucuronide was more readily excreted into bile when newly synthesized by the liver cell than if already circulating as a conjugate, as if newly synthesized glucuronide enters a separate excretory pool.[1100] This pool may not be in the lumen of endoplasmic reticulum,[1100] because polar metabolites such as glucuronides enter the cytosol before excretion,[100,165,1101] and although that route is not yet proved for biliary glucuronides, sulfates — many of which are excreted in the bile — are largely formed in the cytosol. If the pool is in the cytosol, circulating glucuronide will not enter the hepatocyte and join it as long as new glucuronide is being synthesized;[1100] this new glucuronide would be actively transported from cytosol to canaliculi and sinusoids, as proposed for glucuronides of thyroxine[1102] and bilirubin.[1103] Synthesis completed, the intracellular glucuronide concentration falls, and circulating glucuronide enters the hepatocyte.[1100]

Whether excretion or conjugation limits clearance of a glucuronide via the bile is obscure. It must largely depend (G. J. Mulder, personal communication, 1978) on: (1) dose of substrate relative to K_m and V_{max} values for its glucuronidation in vivo, (2) rate of hepatic uptake of substrate, and (3) an intact or obstructed urinary excretion. At low enough dosage in normal animals, conjugation would be rate limiting.

Several observations, however, have suggested excretion being rate limiting. Phenobarbital administration raised transferase activity to bilirubin in vitro, but not the biliary excretion of bilirubin glucuronide in Wistar rats,[935] while in heterozygous Gunn rats, both enzyme activity and, significantly, excretion rate increased.[935] Mulder,[1097] studying mutual inhibition of biliary excretion of several transferase substrates and their glucuronides, concluded excretion rate limiting for the high doses then employed. Others agreed with this.[1005,1104] One group,[1005] measuring UDPglucose dehydrogenase and UDPglucuronyltransferase activities together in vitro with morphine as substrate found phenobarbital pretreatment enhanced this glucuronidation, whereas biliary excretion of the morphine glucuronide was retarded. Another studied biliary excretion of 4-nitrophenyl glucuronide, usually a urinary conjugate, in nephrectomized rats.[1104]

In contrast, many observations suggest that conjugation or hepatic binding can be rate limiting. Saturation was achieved sooner after injection of thyroxine than after injection of its glucuronide.[923] Iopanoate or its glucuronide, a conjugate only excreted in bile, was injected intraveneously into rats.[1105] Biliary clearance of iopanoate after injection of the conjugate was 20 times greater than after injection of iopanoate; phenobarbital pretreatment, increasing transferase activity, increased clearance in the latter, but not the former case.[1105] In rats with a portacaval shunt, transferase activity to 4-nitrophenol decreased, and the excretion of bilirubin glucuronide fell;[1106] phenobarbital pretreatment raised transferase activity and also excretion of bilirubin glucuronide.[1106] Controls with injected bilirubin glucuronide showed the operation did not delay excretion of conjugated bilirubin.[1106] Unfortunately, transferase activity to bilirubin itself was not measured. This latter activity, measured in 129 patients, yielded an asymmetric continuous distribution pattern,[1002] suggesting that transferase activity, and not excretion rate, is rate limiting in much of the population.[1002]

A possible reconciliation of these opposing conclusions come from the work of Bock,[178] who found that glucuronidation of 1-naphthol in perfused liver occurred at much the same rate as that in vitro if the broken cell preparations contained virtually latent transferase, low (physiological) levels of UDPGlcUA and physiological levels of the modulator UDPGlcNAc. Increasing aglycon load in the perfusate achieved "saturation" (i.e., secretion did not increase beyond a certain rate) but simultaneous measurement of conjugated aglycon in the tissue, in the perfusate, and in the bile revealed that the glucuronide did not accumulate within the cell;[178] secretion with 1-naphthol, therefore, was not rate limiting, but rather conjugation.[178] Mulder's warning[1097] regarding extrapolation from optimal in vitro conditions to the in vivo situation seems, therefore, justified; previous in vitro assays, not seeking to reproduce conditions in vivo, have boosted transferase activities artificially high, so that excretion appeared rate limiting.

2. Enterohepatic Circulation of Glucuronides

When biliary glucuronides reach the gut lumen, they can be hydrolyzed there by β-glucuronidase (Chapter 9, Section IV.D.2.b) and the liberated aglycons reabsorbed and recycled to the liver via the bloodstream. Enterohepatic circulation, understandably valuable for recovery of bile salts, seems less so in the case of glucuronides of endogenous compounds. However, enterohepatic circulation of steroids may serve some yet undiscovered purpose, and that of bilirubin is unlikely except in the neonate because of prior degradation by intestinal bacteria.[678]

The enterohepatic circulation of estriol in man well illustrates the role of glucuronidation in this process. Circulating estriol is converted by liver to estriol 16-glucuronide and estriol 3-sulfate-16-glucuronide. Almost all the estriol 16-glucuronide passes into blood and so into urine.[1107] Almost all the estriol 3-sulfate-16-glucuronide is excreted into bile, when part of it then passes intact through the mucosa of the upper small intestine,[1108] but most is hydrolyzed in the lower lumen and is then reglucuronidated in the mucosal cells at C-3 or C-16. These new glucuronides are then absorbed and transported to the kidney and so into the urine.[1109] As estriol 3-glucuronide is formed exclusively in the intestinal mucosal cells,[353] its appearance in urine depends on and is diagnostic of a functioning human enterohepatic circulation. Ampicillin treatment markedly lowers the amount of urinary estriol 3-glucuronide, probably by diminishing the bacterial hydrolysis of estriol 3-sulfate-16-glucuronide in the gut.[1110] A very full account of pre-1970 work on the enterohepatic circulation of estrogen exists.[1111]

Similar enterohepatic circulation has been shown in women[1112] for 17β-estradiol and estrone; biliary estradiol 3-sulfate-17-glucuronide is deconjugated in the gut with partial conversion of aglycon to estrone; then both estrone and 17β-estradiol are reconjugated at C-3 with glucuronic acid in the intestinal mucosa and excreted in the urine.[1112] The enterohepatic circulation of glucuronidated aryl sulfate esters likewise involves intestinal hydrolysis of the glucuronyl bond and in this case, urinary excretion of the sulfate.[1113]

An important result of treatment with alimentary tract antibiotics, such as ampicillin, has recently become apparent. Many steroidal oral contraceptives are normally enterohepatically circulated as glucuronides. Concomitant treatment with ampicillin markedly reduces their efficacy, presumably by removing gut bacteria and thereby much β-glucuronidase, so that the steroids do not undergo enterohepatic circulation and their normal dose becomes insufficient to prevent conception.[1113a]

If intestinal β-glucuronidase itself is inhibited, then the enterohepatic cycling of aglycons should again be markedly lessened, and if the aglycons are active drugs, pharmacological action should be diminished. One example has already been described (Chapter 9, Section IV.D.2.b).[627] Other reports note that glucarolactone added to intraduodenally infused, labeled stilbestrol monoglucuronide markedly decreased the label in bile, presumably by inhibiting its enterohepatic circulation,[1114] and that oral treatment with glucarolactone or with Neomycin diminishes enterohepatic circulation of the metabolites of orally administered vanillins, produced by bacterial action in rat gut.[1115]

Section IV.E above notes that glucuronides can be absorbed from lumen into blood without apparent hydrolysis, and so could to some extent contribute to the enterohepatic circulation.[1084]

V. SKIN

Glucuronidation occurs in skin, and the "negligible" levels reported in this large organ must add up to a significant contribution.

Harper[1116] and Zini[1081] first suggested that skin might form glucuronides, and this was confirmed for 2-aminophenol in strips and homogenates of shaved skin from mouse and guinea pig;[573] with strips, glucose doubled overall glucuronidation. Activity also occurs in human epidermis not dermis (unpublished results, quoted by Dutton[4]). UDPGlcUA present in skin of mouse and guinea pig,[573] is readily destroyed during transferase assay there (G. J. Dutton, unpublished results) and in rat skin.[558]

UDPGlucuronyltransferase activity and overall glucuronidation in skin towards 2-aminophenol is enhanced fourfold by prior painting with 3,4-benzopyrene,[1015] most obviously (within 4 days of a single painting) in a hairless strain of mouse.[1015] Normal shaved mice required 3 days of painting and stimulation then needed 5 to 11 days.[1015] Skin of benzopyrene-painted mice contains the glucuronides of 3,4-benzopyrene derivatives such as 8-benzopyrenol,[1117] indicating that carcinogens can be "detoxicated" by this tissue.

Guinea pig fetal skin[573] glucuronidates 2-aminophenol, and the low levels must contribute significantly to fetal glucuronidation in this animal. Skin from human fetuses (G. J. Dutton, unpublished results) or chick embryo[369] appears virtually devoid of activity to this substrate, but the problem needs more study. Human adult skin epithelial cells form glucuronides in culture[385] (Chapter 15, Section II).

VI. LUNG

Lung, earlier thought to lack UDPglucuronyltransferase activity or contain only a trace,[4] is now a well-accredited site. Activity in rabbit lung towards oxazepam[1118] was reported as 30%, and towards 2-aminophenol as 25%[368] that of the liver, but no activity occurred there towards 4-nitrophenol or phenolphthalein.[368] In rat lung, high activity to 4-nitrophenol has been noted,[829] and in pig lung, some activity to phenolphthalein.[710] Activity to 1-naphthol occurs in human lung prepared as fresh "microsomes" (K. W. Bock, personal communication, 1975) or "maintained" as an organ;[714] "hepatic level" of activity in lung to 1-naphthol has been claimed,[1119] but the low aglycon concentration employed may have been limiting for liver. With 4-methylumbelliferone as substrate, rat lung transferase activity to be 10% of liver by wet weight, but 30 to 60% on specific activity of the few "microsomes" present;[1072] other cells than alveolar macrophages may have contributed.[1072] Lung activity towards 4-methylumbelliferone was inducible by pretreatment with cinchophen or 3-methylcholanthrene,[1016] and in rats exposed to cigarette smoke it tended to remain constant or increase a little.[1045,1046] Maternal injection of dexamethasone evoked precocious transferase development towards certain substrates in the lung of fetal rats (Chapter 10, Section IV.D.2.b).

No activation of rat lung transferase to 4-nitrophenol could be demonstrated[829] and no sex differences,[829] the enzyme thus resembling that from rat gut, but in some other responses it resembled that from kidney.[1120]

The glucuronidating capacity of the lung as an organ may be less than its transferase activity suggests. For example, perfused whole lung glucuronidated 4-methylumbelliferone much more slowly than broken cell preparations;[1121] possibly access of aglycon (delivered by perfusate, not by inspiration) or availability of UDPGlcUA was limiting.

VII. OTHER TISSUES

A. Adrenal Gland

Adrenal gland, earlier reported negative except possibly positive in cockerel for steroids,[4] possesses a relatively high activity (similar to whole kidney, weight for weight) in rat for 4-methylumbelliferone.[717] Steroid glucuronidation there should be further examined.

B. Spleen

Formerly thought to possess "little if any UDPglucuronyltransferase activity",[4] the spleen exhibits it towards 4-methylumbelliferone[717] and 1-naphthol (K. W. Bock, personal communication). As β-glucuronidase is high in spleen,[594,717] doubling of transferase activity there with glucarolactone[717] is understandable even at neutral pH; but was still only 10% weight for weight of that in rat liver.[717]

C. Thymus

Thymus displayed activity towards 4-methylumbelliferone much as spleen, but added glucarolactone caused little increase.[717]

D. Other Organs*

Brain, heart, fat, and diaphragm showed only trace transferase activity towards 4-methylumbelliferone, with one questionable assay claiming high activities in the first three for 1-naphthol.[1119] Brain may possess some activity towards bilirubin[919] and oxazepam,[410] but no specific activity towards serotonin, 2-aminophenol, or 4-nitrophenol could be detected (J. E. A. Leakey, personal communication, 1975). Low activity towards estrogens was claimed in muscle and heart.[112] All these sources were earlier reported virtually negative.[4]

Mammary gland, lactating or nonlactating, from virgin or older rats possesses transferase activity to 1-naphthol; in nuclei and "microsomes" respectively the activity is 20 and 10% that in liver (G. Brüder, D. J. Fry, and E. Jarasch, personal communication, 1977). Glucuronic acid conjugates of xenobiotics, though not of steroids, are found in cow's milk[824,825] (see Reference 824 for early references), sometimes in considerable amount, and include those of n-aliphatic acids and of phenols, catechols, and other compounds from plant sources. The role of mammary gland UDPglucuronyltransferase in forming these conjugates, and any from administered drugs, requires investigation.

In rat, transferase activities were reported in uterus towards diethylstilbestrol, 4-nitrophenol, 4-methylumbelliferone, and 1-naphthol[389] and towards 17α-estradiol in rabbit.[442] Activities towards estrone, β-estradiol, testosterone, and phenolphthalein were not evident in rat uterus[389]

From the above, and earlier work[4] it seems that UDPglucuronyltransferase activity is detectable by sensitive methods in epithelial cells of most tissues.

Transferase activity towards 4-methylumbelliferone and testosterone exists in human lymphocyte lines RPM 1788, B-411-4, and SN-1014.[1121a] In B-22 it was very low (10% of these), and very low in fresh human lymphocytes, although measurable in lines cultured from donors;[1121a] culture appears to induce the activity.[1121a] For tumor tissue, see Chapter 15, Section II.B.

VIII. ADDITIONAL NOTES

Section II: More evidence is reported of the specific tissue distribution of the transferase as one aspect of its functional heterogeneity. Bock's group find their GT_1 activities in many tissues of the rat, but GT_2 substrates only in liver and intestine to any extent.[392c] Lucier et al.,[389] had found earlier more limited distribution of their steroid cluster of activities than of their nonsteroid cluster.

Section IV: A useful paper describes the metabolism of harmine to harmol and then to harmol glucuronide by isolated rat small-intestinal cells, and a rapid method of preparing these cells;[1121b] they were capable of only low sulfation.

Section VI: Glucuronidation is confirmed as playing a very minor role in conjugation of benzo(a)pyrene metabolites in perfused rabbit lung, and this deficiency is suggested to contribute to carcinogenesis there.[1121c]

---

* For placenta see Chapter 10, Section III.E.2.c.

Chapter 15

GLUCURONIDATION IN CULTURED TISSUE AND IN PATHOLOGICAL AND TOXICOLOGICAL CONDITIONS

I. GENERAL

Glucuronidation in cultured tissues and in pathological and toxicological conditions may be conveniently considered together because cell culture incurs an abnormal environment, often involves damage, and many culture studies employ tumor tissue.

The following subjects under this chapter heading are encountered elsewhere in this book: toxic effects of glucuronides themselves, Chapter 1, Section IX,B.2; tissue damage on UDPglucuronyltransferase activity, Chapter 5, Sections II. C,L,M, and N; toxic chemicals, Chapter 5, Sections II. F,G, and I; culture of fetal and perinatal tissue, Chapter 10, Sections IV. C,D, and E; the breast milk factor in neonatal jaundice, Chapter 11, Section IV; hyperbilirubinemic conditions such as Crigler-Najjar and Gilbert's syndromes, Chapter 12, Sections III. C,D, and E; recurrent jaundice of pregnancy, Chapter 13, Section I.C; and culture as an investigative technique, Chapter 17, Section II.C.6.

An excellent recent review on drug metabolism in cultured tissues exists,[1122] including critical assessment of procedures employed.

II. GLUCURONIDATION IN CULTURED AND NEOPLASTIC TISSUES

A. Culture of Nonneoplastic Tissues

Culture conditions allow environmental control and continuous monitoring of cell and enzyme response. For culture of perinatal tissues in which transferase is induced from zero to adult values spontaneously or on addition of xenobiotics or hormones, with use of cycloheximide pulsing to demonstrate dependence of induction on protein synthesis, see Chapter 10, Sections V. C. and D.

A successful approach with adult liver was that of Sandström,[1123] who cultured liver explants from a 52-year-old man for 5 days, sandwiching the tissue between perforated cellophane layers; after 5 days, transferase activity to 2-aminophenol was detected in the apparently healthy dividing tissue.[1123] Suspensions of adult rat liver cells[1006] retained some transferase activity to 4-nitrophenol for 6 days; phenobarbital in the medium retarded loss of activity.[1006] Cycloheximide blocked this effect of phenobarbital,[1006] possibly not by inhibiting any transferase induction, but by preventing the increased cell integrity[1122] brought about by phenobarbital. Phenobarbital preserves morphology of chick embryo liver cell monolayers, delaying invasion by fibroblast-like cells.[735,750] Dissociation of cells from tissue prior to culture may in some procedures incur membrane perturbation and protein loss,[182] adult rat liver cells losing much transferase activity to 2-aminophenol and 4 nitrophenol (G. J. Dutton and A. M. Goheer, unpublished work, 1976), regaining it within 24 hr or so and then slowly losing it again; transferase in these cells appeared already activated, for additional perturbation inhibited the enzyme. A similar, though partial, activation occurred in cultured cells from adult mouse liver.[182,692] This change in activation characteristics may reflect changed membrane composition and relate to the replacement of cytochrome P-450 by a cytochrome P-448 in liver cell lines.[1122]

Cell lines derived from adult rat liver by one method[1124] hydroxylate xenobiotics and steroids even after continuous subculture, but (P. Padieu, personal communication,

1976) do not glucuronidate these metabolites; this selective loss of the usually culturally more stable[1125] Phase 2 system requires study.

Kidney tissue cultured as fetal organ[197,369] or adult tubule fragments (unpublished work quoted by Fry and Bridges[1122]) glucuronidates well, but fibroblasts (derived from liver cultures[735,740] or human skin[385,424a]) do not. A positive result with rat thyroid fibroblasts[1126] is now considered artifactual (E. Dybing and H. E. Rugstad, personal communication, 1975). Human skin epithelial cells in culture glucuronidate 2-aminophenol and 4-nitrophenol but not bilirubin,[385] and transferase activity, as so often in cultured tissue apparently already activated, was demonstrable in homogenates of them;[385] 3,4-benzopyrene or benzanthracene added to the medium did not increase glucuronidation over 24 hr,[385] either because exposure was too short (benzopyrene needed several days to affect mouse skin transferase[1015]) or culture conditions had already induced the enzyme "to its limit".[385]

B. Cultured and Fresh Neoplastic Tissue ("Fresh" means in most cases tumor tissue fresh from a host, following implantation from a cultured source)

At first[4] UDPglucuronyltransferase seemed absent from several tumors, i.e., mouse carcinoma I.C. Res. Fund 2146S, mouse Crocker sarcoma, rat hepatoma Dennis LC-18, and Novikoff, solid and ascitic, 2-aminophenol and 4-methylumbelliferone were substrates offered. Arias (quoted results) later found 80% host liver activity in Morris 5123[1127,1128] and Reuber H 35 rat hepatomas.

Microsomal activity was later found towards 4-nitrophenol in several well-differentiated hepatomas at high level and detectable in certain poorly differentiated tumors.[979] In hepatomas 7794A, 7787, 7316A, and H 35, activity in homogenates was higher on a protein basis than in normal or host livers, and in 3924A and 3683 hepatomas, comparable to that in host liver. Some activity occurred in hepatomas H 139, HC, and LC, but little or none in Novikoff hepatomas.[979] Controls showed that in these instances activation by Triton® X-100 was broadly similar in host and hepatoma preparations.[979] As cultured monolayers of Morris 7288C hepatoma cells glucuronidated 4-nitrophenol present in the medium, overall glucuronidation also took place.[979] To find such detoxication in neoplastic cells was important, for hepatomas had been largely assumed deficient in all drug-metabolizing enzymes, and anti-tumor drugs might have been designed on that false basis.[979]   The clonal MH₁C₁ strain of rat hepatoma cells derived from Morris 7795 hepatoma glucuronidated 4-aminophenol added to their cultured monolayers,[1126] but Chang cells did not.[1126] Glucuronidation of both 4-amino and 4-nitrophenols by MH₁C₁ cells was noncompetitively inhibited in culture and in homogenates when chlorpromazine was present,[981] but only at high concentrations of chlorpromazine did morphological damage appear.[981] Other drugs (SKF 525-A,[1129] promazine, imipramine and thioridazine[982]) inhibited overall glucuronidation of 4-aminophenol, 4-nitrophenol and bilirubin by cultured intact MH₁C₁ cells more than they inhibited the transferase activity of the homogenized cultures; as three times higher concentrations were required for the transferase inhibition, these drugs (especially the psychotropes, which act as membrane stabilizers) may inhibit glucuronidation by slowing substrate entry into the cell.[982] These drugs also inhibited amino acid transport and incorporation into cell protein[982] and so could reduce glucuronidation by restricting protein synthesis;[982] propanol and quinidine did not affect glucuronidation, but their effect on amino acid uptake was not reported.[982] A serum-free culture medium potentiated inhibition by all the drugs, probably because serum binds the free drug.[982]

MH₁C₁ cells cultured as a tumor in Gunn rat lessened its hyperbilirubinemia[938] and, cultured in vitro, glucuronidated bilirubin;[1130] the latter process was inhibited, as in liver, by flavaspidic acids, suggesting uptake resembled that in liver.[1130] Specific transferase activity to bilirubin of tumor cell homogenates resembled that of normal rat liver, but culture of the cells with phenobarbital for 8 days did not increase this activity;[1130] however, tumor cells in a host might also be unresponsive to phenobarbital.[1130] Hydrocortisone in the medium also had negligible effect.[1130] These MH_1C_1 cells synthesize bilirubin as well as conjugate it, and are convenient for study of both processes. Using them,[1131] changes in the molar bilirubin:albumin ratio in the culture medium were shown to markedly influence rate of bilirubin conjugation, the optimal being 1:1. Sulfisoxazole added to a medium containing a 1:1 or greater bilirubin:albumin ratio appeared to displace bilirubin from albumin and inhibit conjugation; at lower ratios, addition of the drug seemed to enhance bilirubin conjugation.[1131]

Competition between glucuronidation of 4-aminophenol, 4-nitrophenol and bilirubin was compared in cultures and homogenates of MH_1C_1 cells.[592] Bilirubin did not inhibit glucuronidation of the other two, but they tended to interfere with its conjugation.[592]

"Native", fresh, homogenates of MH_1C_1 cells displayed some 56 to 77% of the maximal activity of digitonin-activated homogenates, whereas fresh homogenates of normal liver cells displayed only some 5 to 8% of their maximal activity.[514]

Moreover, diethylnitrosamine or UDPGlcNAc did not activate, and apparent $K_{UDPGlcUA}$ values were high,[514] all consistent with the enzyme in tumor cells being less constrained (Chapter 5) than in liver and so existing in a different membrane environment. Culture conditions could have contributed, but transferase from subcutaneous tumor tissue taken freshly from the host Buffalo rats also displayed low latency,[514] and the phenomenon appears due to their nature as neoplastic or at least nonhepatic cells. UDPGlcUA levels appeared, indirectly, much as in liver.[514]

Other workers[149,1132] found an up to ninefold higher transferase activity to 4-nitrophenol in Morris 5123 C and 7800 hepatoma membrane fractions than in host or normal rat liver microsomes, and some three- to five-fold higher activity in similar preparations from 9618 A and 7777 Morris hepatomas, but did not correlate these high enzyme levels with overall glucuronidation. The fractions from the high-transferase hepatomas showed an increased amount of polypeptide of 59,000 daltons,[1132] possibly the transferase itself. No correlation of tumor growth rate with transferase content was apparent.[149,1132]

The Reuber H-35 hepatoma possesses transferase activity to 4-nitrophenol[1133] and estradiol,[1024] but not to testosterone.[1024] Consistent with the findings reported above,[514] it is not activated by Triton® X-100[1133] and unlike that in liver, the hepatoma transferase activity to estradiol is not enhanced by pretreatment of rats with phenobarbital or 3-methylcholanthrene;[1024] the hepatic transferase induced by 3-methylcholanthrene resembled in kinetic parameters that existing in the hepatoma and not that in untreated animals.[1024] Triton® X 100 activated, however, transferase to several substrates in Morris hepatomas 7777, 5123C, and 9121,[1133a] in the former changing K_m values markedly for 1-naphthol and UDPGlcUA. Compared with host-liver transferase, activity in these hepatomas to chloramphenicol was below 20%, to bilirubin some 100%, and to 4 nitrophenol and 1-nitrophenol, 300%.[1133a]

UDPGlucuronyltransferase activity (to 2-aminophenol) occurs in rat hepatomas resulting from administration of diethylnitrosamine (I. H. Stevenson, personal communication, quoted in Reference 4).

The contribution to bodily glucuronidation by a tumor in vivo[938] was demonstrated by the greater excretion of 4-nitrophenylglucuronide by rats bearing the intramuscular H-35 hepatoma (possessing transferase activity to 4-nitrophenol) than by those bearing the s.c. Walker-256 carcinoma (not possessing transferase activity).

Recently,[1134] high concentrations of dopamine 4-O-glucuronide have been found in a transplantable islet cell tumor of golden hamster. Conjugation of this substrate in nonneoplastic islet cell tissue should be investigated. MH_1C_1 cells glucuronidate dopamine metabolites.[424a]

III. GLUCURONIDATION AND CARCINOGENESIS

The active role played by *O*-glucuronides of certain *N*-hydroxy compounds in carcinogenesis has been discussed, and their possible contribution to tumorigenesis in kidney and bladder pointed out (Chapter 1, Sections IX.B.2 and IX.B.3). Nemoto and Gelboin[20] consider that the 7,8-diol of 3,4-benzopyrene is potentially more readily convertible to the 7,8-diol-9,10-epoxide, the presumed carcinogen, than to the 7,8-diol glucuronide; therefore, UDPglucuronyltransferase is a key enzyme and glucuronidation is a key process in carcinogenesis by 3,4-benzopyrene. In particular, transport of this compound or its metabolites to the nucleus where the transferase is active (Chapter 4) is being intensively studied[1135]

Several recent reports note the "scavenging" action of UDPGlcUA when added to systems forming reactive intermediates from Phase 1 reactions; e.g.,[1135a] 3 m*M* UDPGlcUA lessened the binding to microsomal protein of secondary metabolites of 2,2' di-hexachlorobiphenyl by 30% and when 3 m*M* UDPGlcNAc (itself ineffective) was also added, 75% of the binding capacity disappeared: a dramatic confirmation of synergism by the latter nucleotide. Similarly UDPGlcUA (2 m*M*) plus UDPGlcNAc (4 m*M*) completely removed the mutagenicity of benzo(a)pyrene phenols (except the 9-hydroxy compound) in the Ames test,[910a] UDPGlcUA alone being 85% effective, PAPS 50% effective, and glutathione of very little effect;[910a] UDPGlcUA addition decreased all the binding peaks to DNA except one.[910a] Nemoto,[1135b] however, using whole hamster embryo cells (13th day of gestation, grown for 3 to 4 days), found that although UDPGlcUA addition increased the glucuronidated benzo(a)pyrene phenols, it resulted in increased amount of benzo(a)pyrene metabolites bound to rat liver DNA nucleosides. He suggested this arose because phenols were removed from the microsomal matrix by glucuronidation, and dihydrodiols consequently had greater freedom there; these dihydrodiols were then readily metabolized by Phase 1 enzymes, producing more of the 9,10 oxide and hence, more binding.[1135b] UDPGlcUA would thus potentiate mutagenicity!

Another explanation, again incurring increased carcinogenicity through glucuronidation, would be hydrolysis of benzo(a)pyrene 3-glucuronide by β-glucuronidase, which liberates a derivative binding more strongly than the aglycon or its glucuronide to DNA.[1135c]

Goldman[1135d] mentions recent work on the hydrolysis of *N*-glucuronides in gut, with consequent carcinogenicity there.

Transferase activity of several species to the carcinogen 4-amino-3-biphenyl did not correlate with their susceptibility to tumorigenesis by this compound,[880] but previous steps could have been determining.

An ingenious suggestion for utilizing the β-glucuronides of anticancer drugs was that of Bicker.[59] As the largely anaerobic metabolism of tumor tissue gives rise to a lower local pH (ca.6.0) than normal tissue (ca. 7.3) in the presence of glucose, the local β-glucuronidase should be more active in tumor tissue. Bicker implanted a sarcoma into rats[59] and injected them intraperitoneally after 4 weeks with β-(hydroxy)-quinoline glucuronide in glucose solution and then repeatedly with glucose. After extracting tumor and surrounding tissue, he found in the three rats used (and after measurement by different techniques), free β-(hydroxy)-quinoline in tumor tissue, but none in normal surrounding tissue; controls, untreated or receiving the glucuronide without the glucose, possessed no free aglycon either within or about the tumor tissue;[59] a similar result occurred with 2-naphthyl glucuronide. He suggests[59] glucuronides of drugs such as 4-hydroxyaniline mustard should be tried therapeutically.

IV. GALACTOSAMINE HEPATITIS

D-Galactosamine, specifically among sugar analogs, evokes liver damage in rats resembling that from human viral hepatitis.[188] Both D-galactosamine and D-glucosamine are cytotoxic, inhibiting biosynthesis of RNA, DNA, and protein, especially in tumor cells.[188] Keppler et al.[188] provided an explanation. On administration of galactosamine to rats, free uridine phosphate content fell to below 10% normal within 3 hr, uracil nucleotide content increased and UDP-sugar pattern markedly changed, UDP-*N*-Acetylhexosamine content increasing ninefold, and UDP-hexosamines, formerly undetectable, rising far above UDPglucose-UDPgalactose levels.[188] D-Glucosamine and 2-deoxy-D-galactose also trapped uridine phosphates, but were less effective, did not lead to formation of UDP-hexosamines, and did not provoke hepatitis.[388] D-Glucosamine slightly increased UDPGlcUA, but D-galactosamine reduced its level from 0.28 to 0.09 μmol/g fresh liver.[388] This fall was attributed to inhibition of UDPglucose dehydrogenase by galactosamine 1-phosphate and UDPgalactosamine;[1136] inhibition of UDPglucose pyrophosphorylase[1137] could also contribute.

Galactosamine, as expected, inhibited glucuronidation in liver slices[305] and so did glucosamine,[305] despite its lack of effect on UDPGlcUA levels[305] and its great enhancement of the levels of UDPGlcNAc, the positive effector of the transferase.[305] Addition of UDPGlcUA to the medium partially relieved this inhibition.[306]

Galactosamine did not inhibit glucuronidation in rat jejunum slices, attributed to its slower conversion there to UDPhexosamines or to the different metabolic pools required for synthesis of mucins.[306]

With liver perfused in vivo, previous exposure to galactosamine lowered the UDPGlcUA content almost half;[179] after in vivo exposure to glucosamine, glucuronidation of 1-naphthol fell in these perfused livers, but not that of bilirubin.

V. DIABETES

In 1926, Quick[1138] noted that insulin administered to phlorrhidzinized dogs increased urinary excretion of glucuronide. Confirming this, administration of insulin to normal rats increased excretion of bilirubin glucuronide, hepatic conjugation of bilirubin, and level of hepatic UDPGlcUA, but did not change dehydrogenase or transferase activities.[1139] Liver of diabetic rats possessed an apparently unchanged transferase activity to testosterone and 2-aminophenol, but notably lower UDPglucose dehydrogenase activity, lower UDPGlcUA content, and lower overall glucuronidation capacity; they excreted much less administered bilirubin as biliary glucuronide in vivo and during liver perfusion than normal rats.[1139] In the insulin-deficient animals, decreased dehydrogenase was suggested the limiting factor, UDPGlcUA breakdown by pyrophosphatase being only slightly increased, and insulin operating by activating the dehydrogenase rather than by diverting UDPglucose to it.[1139] These interesting studies should be extended with recent knowledge.

Miettinen and Leskinen[543] reviewed earlier evidence for increased flow through the glucuronate pathway in diabetes. Neither glucuronide nor free glucuronate appears elevated in the serum of diabetics.[4]

Added insulin is required for glycogen deposition and full morphological maturation of cultured chick embryo liver cells,[1140] but not for their spontaneous development of very high transferase and glucuronidation activities.[735,738,740]

VI. PORPHYRIA

Steroids provoke attacks in hepatic porphyria and in porphyria cutanea tarda. Free steroids seem necessary, their glucuronides being relatively inactive even when present at the living cell; porphyrin synthesis in chick embryo liver—cell culture being stimulated by steroids added to the medium except when UDPGlcUA was also added (which may enter intact cells). Any condition, therefore, of diet, disease, or injury leading to impaired glucuronidation of steroids could lead to intracellular accumulation of steroidal inducer and so provoke an attack. Concomitant inhibition of UDPglucuronyltransferase by these steroids would accentuate their effect. Fed glucose may relieve attacks of acute intermittent porphyria by increasing availability of UDPGlcUA;[1141] it did not decrease the porphyria-inducers circulating in plasma.[1141] Starvation may provoke attacks by lowering hepatic UDPGlcUA levels and, possibly, by increasing β-glucuronidase levels.[1142]

Barbiturates and other inducing xenobiotics can provoke porphyric attacks even though they induce UDPglucuronyltransferase, because they induce δ-aminolevulinic acid synthetase and hence porphyrin biosynthesis.

VII. GASTROINTESTINAL (GI) TRACT

The attractive idea that gastrointestinal ulcerogenesis arose from competition of glucuronidogenic ulcerogens for UDPGlcUA needed in mucin synthesis[21] has been relinguished, because Brunner's glands, where mucin formation is diminished during ulcerogenesis, do not contain transferase activity.[21] Animals susceptible to experimental ulcerogenesis possess lower transferase activity in the tract mucosa than the resistant species.[1086]

Hypertrophic pyloric stenosis is frequently associated with hyperbilirubinemia and with an absolute decrease in hepatic UDPglucuronyltransferase activity to bilirubin.[1143] Transferase activity could be aggravated by concomitant lowered caloric intake, but as pyloric stenosis with caloric restriction does not always incur lowered transferase, another explanation is needed.[1143] The slight hyperbilirubinemia associated with Gilbert's syndrome can be aggravated by decreased food intake due to achalasia,[960] but the influence of diet on Gilbert's syndrome has been more fully discussed in Chapter 12, Section III.D.

VIII. CHOLESTASIS AND LIVER DAMAGE

A. General

Cholestatic conditions probably damage the liver cell membrane and endoplasmic reticulum,[1095] and may be considered here. Because UDPglucuronyltransferase activity is closely related to membrane composition, apparently contradictory observations on the activity of the enzyme are to be expected in damaged liver due to other causes than erratic sampling. At the periphery of damage, transferase could be activated (e.g., by bile salts, lipid peroxides, phospholipases, proteases, and accumulated glucuronides) or induced, whereas in more necrotic regions it could be inactivated; also, transferase activities could be differentially affected among substrates. Tissues have presumably evolved metabolic "priorities" by which the less necessary expenditure of UDPGlcUA and enzyme function is sacrificed during crises to keep vital processes working. These priorities may be upset as tissue damage progresses, so that even the "abnormal" pattern of transferase activity could change with duration of insult or area of damage sampled.

Contradictory reports have appeared[4,11,339] and must continue. The reasons for them should be evident from the present review. It should also be evident that medication of damaged tissue can stress glucuronidation pathways by inhibiting, competing for, or haphazardly inducing transferase activities and so disordering conjugation of both routine and xenobiotic metabolites.

B. Cholestasis

Normally, almost all the bile acid pool participates in enterohepatic circulation. In cholestasis, excretion of bile salts into bile falls, while bile acids rise in plasma and tissues other than liver and intestine. Cholestasis also results in the glucuronidation of bile salts and their excretion into urine;[428] structure of these glucuronides and properties of their transferase are discussed in Chapter 7, Section II.B.2.d. Urinary bile salt glucuronides occurred in 19 out of 20 patients studied with intra- and extrahepatic cholestasis, but only a trace in healthy subjects.[432] Although bile acid glucuronides are a small percentage of the plasma bile acids, they may be useful in eliminating mono-hydroxy bile acids (e.g., 3β-hydroxy-5-cholen-24-oic acid) which themselves exert cholestatic effects.[429] As phenobarbital pretreatment enhances activity of UDPglucuronyltransferase towards bile salts,[433] it could reduce the toxic effect of, for example, chenodeoxycholic and lithocholic acids by assisting their clearance from plasma via urine as well as by increasing bile flow; the lowering of plasma bile salt levels and disappearance of pruritus reported after phenobarbital treatment might be partially effected in this manner.

Abnormal conjugates of bilirubin containing glucuronic acid and several other sugars have been described in Chapter 7, Section III.D as arising in conditions of cholestasis.[1144] A histochemical study of conjugated and unconjugated bilirubin during cholestasis[1145] suggests that the fine liver cell granules corresponded to conjugated bilirubin and the coarse corresponded to unconjugated bilirubin; extracellular thrombi mostly contained the conjugate, and the Kupffer cell pigment mostly contained the free bilirubin. The role of deconjugation by β-glucuronidase at these sites is discussed in Reference 145.

Cholestasis, of course, blocks enterohepatic circulation, and deranges steroid metabolism depending on that process, e.g., that of estriol 3-glucuronide;[977] altered estrogen conjugation is an effect, not a cause, of the cholestasis.[978] Glucuronidation in pruritus gravidarum and in recurrent jaundice of pregnancy and in the Crigler-Najjar, Gilbert's, and Dubin-Johnson syndromes have been discussed (Section I of this chapter gives references). It is unfortunate that lack of transferase activity towards bilirubin in human kidney[475] prevents use of kidney grafts (so promising in Gunn rats[939]) to relieve the usually lethal Crigler-Najjar syndrome in man.

In obstructive jaundice,[1146] Gilbert's syndrome, or generally wherever transferase activity to bilirubin falls,[678] monoconjugates of bilirubin increase over diconjugates in man. Bilirubin glucoside, not glucuronide, occurs in urine of Crigler-Najjar patients and may be diagnostic there.[931] Glucosidation increases proportionally to glucuronidation in pathological bile, and some evidence suggests[181] that cholestasis releases constraint of UDPglucuronyltransferase so that it becomes inhibited by UDPglucose. Cholelithiasis may not affect transferase activity to bilirubin,[1147] or lower it slightly.[1002]

C. Liver Disease and Damage

Earlier reviews[4,11,339] noted that observed UDPglucuronyltransferase activity fell less rapidly in liver disease or damage than Phase 1 enzymes. This is probably due to its initial activation during membrane perturbation, the Phase 1 system being progressively inactivated. Parallel behavior is seen during the aging of tissue preparations. When chlorinated hydrocarbons provoke liver damage, the pattern is an initial rise, followed by progressive fall in transferase activity towards 4-nitrophenol and 4-methylumbelliferone[216] and estrogens[112] (for earlier work see References 11 and 339). Within 24 hr of administration to the animal of carbon tetrachloride,[180] activity to 1-naphthol increased eightfold in homogenates. As constraint was lost, UDPGlcNAc no longer activated, and UDP inhibited, thus theoretically rendering glucuronidation less effective in vivo, as confirmed in perfused liver, where net glucuronidation fell within 24 hr after administration.[180] Sulfation was much less affected. UDPGlcUA content did not fall even though 45% of the liver cells were necrotic or prenecrotic,[180] surprising in view of the fairly rapid decrease in UDPglucose dehydrogenase activity in aging tissue preparations (Chapter 9, Section II), and damaged liver can still form glucuronides, especially of bilirubin, whose conjugation seemed little affected;[179,180] perhaps glucuronidation of bilirubin is one of the suggested "priority" reactions in detoxication, although it eventually fell with CCl_4 poisoning[1147] and in ethanolic fatty liver,[380] where activity to 4-nitrophenol increased.[380] Subacute exposure to carbon disulfide fumes also raises transferase activity initially before a fall.[221] The active agents for activation and induction are probably not the administered substances (e.g., CCl_4 itself does not activate[216]) but their Phase 1 metabolites. Much recent work on the effect of administered aliphatic chlorohydrocarbons on the transferase and on Phase 1 enzymes, the latter being clearly more susceptible, has appeared.[1013,1049]

Chelating agents have been used in cases of heavy metal poisoning and recommended prophylactically for workers in certain industries; rats treated for 1 week with dimercaptrol, EDTA, or penicillamine showed no significant alteration in the glucuronidation system.[1059]

Ischemia of rat liver cells by temporary clamping causes some microsomal dysfunction; transferase activity (to 4-nitrophenol) was much less affected than the other enzymes studied.[204] It fell a little after 30 min ischemia, but in livers suffering 2 hr ischemia it was as high as in "normal" microsomes.[204] Part of the effect observed was probably due to activation, as the livers were partially necrotic;[204] transferase activity might have increased to a peak as in whole-tissue "aging" experiments[183] while glucuronidating ability would have been much lower.[183]

Liver disease usually involves some degree of cholestasis, but reports of transferase activity in these conditions are understandably conflicting. Rises,[917,1148,1149] normal levels,[808,1147] and falls[956,1150,1151] continue to be reported, and the enzyme activity clearly has little diagnostic significance; e.g., the decreasing excretion of urinary salicylamide glucuronide in increasing severity of cirrhosis is not unexpected, but apparently random rises or falls in excretion of glucuronides occur.[543] One reason is that extrahepatic glucuronidation, particularly renal, may compensate for much of the reduced liver function. Loading tests are not very informative;[543] even in advanced liver damage, transferase and overall glucuronidation will respond to xenobiotic inducers.[1152]

β-Glucuronidase is liberated in and from damaged liver cells,[1153] incurring high serum levels, and this might further embarrass glucuronidation. A better inhibitor of β-glucuronidase than glucaro-1,4-lactone, i.e., glucaro-1 → 4,6 → 3-dilactone, has been used in prophylaxis and therapy of bladder cancer (Chapter 9, Section IV). Meittinen and Leskinen[543] have reviewed pathological disturbances of the glucuronic acid pathway and their treatment. In regenerating liver, β-glucuronidase is especially active,[594,1154] but the transferase, bilirubin as substrate,[956] appears at normal levels; partial hepatectomy decreases UDPglucose synthesis and UDPglucose levels by some 70%.[1155]

Liver β-glucuronidase activity did not vary much in 113 subjects with various acquired and inherited liver diseases and did not correlate with transferase activities.[382]

X-Irradiation of the isolated organ at 400 to 1200 rd halved synthesis of 2-aminophenyl glucuronide by rat gastric mucosa or liver; whole-body irradiation of rats at 1000 rd decreased their output of urinary glucuronide (xenobiotic and steroid) and increased susceptibility to naphthalene poisoning.[4,339] The transferase itself may be relatively resistant to X-irradiation, for exposure of isolated chick embryo liver to 600 rd prevented mitosis, but allowed subsequent rapid development of transferase;[740] and over the 3rd to 6th day after 600 rd irradiation, rat liver UDPglucuronyltransferase

(4-nitrophenol as substrate) remained unchanged.[1156] The fall in glucuronidation after radiation exposure may be due to the fall in liver UDPglucose dehydrogenase and the rise in liver β-glucuronidase and UDPGlcUA pyrophosphatase noted after 600 rd irradiation of rats.[1156]

Chapter 16

RELATION OF OTHER DRUG-METABOLIZING PATHWAYS TO GLUCURONIDATION

I. GENERAL

The pathways of xenobiotic metabolism have been reviewed frequently. A useful summary is that of Williams.[2] As described in Chapter 1, they can be divided into two phases. Phase 1 concerns changes in the xenobiotic molecule itself, and Phase 2 concerns the conjugation of the xenobiotic or its metabolite with an endogenous molecule.

The metabolic and topologic relationship of Phase 1 to glucuronidation has been investigated, and is treated below. Of Phase 2 systems, glycosylations and sulfation especially have been studied in relation to glucuronidation, and are also treated below. Studies on the compensatory effect of other Phase 2 systems during deficient glucuronidation in neonates, referred to in Chapter 10, have been briefly reviewed.[636]

II. GLYCOSYLATIONS OTHER THAN GLUCURONIDATION

A. General

Glycosylations other than glucuronidation proceed by the respective nucleotide sugars and nucleotide glycosyltransferases, although some steroid glucoside formation independent of added UDPglucose has been reported in mammals,[113,411] suggesting involvement of a lipid acceptor, as in glycosylation of proteins. In plants, biosynthesis of phenolic glucosides in the absence of added UDPglucose has been reported,[1157] and a transglucosylase has been isolated from *Aspergillus niger*[1158] which will transfer α-D-glucose residues from maltose and similar α-D-glucopyranosides to phenolic and alcoholic hydroxyl groups and to carboxylic acid groups.

B. Glucosidation
1. Glucosidation in Plants

Glucosidation in plants of xenobiotics, known for many years[838,1159] is analogous to glucuronidation in mammals. The C—C-link, only recently observed in glucuronides, is widespread in plant glucosides.[76] UDPGlucosyltransferase forming β-D-glucosides exists in a variety of plants;[840-842,1159-1162] UDPglucose could often be replaced by other nucleotide sugars, e.g., thymidine diphosphate glucose.[1162]

UDP-Dependent reversibility of the UDPglucose:flavonol 3-O-glucoside system has been demonstrated[1163] in cell-suspension cultures of parsley with kaempferol- or quercetin-3β-O-glucoside and UDP as substrates; it could liberate the aglycons for further, directed, metabolism while conserving the conjugates' bond-energy in UDPglucose.[1163]

Glucosidation of small endogenous molecules in plants is well known; e.g., a particulate UDPglucosyltransferase accepting sterols[1164] and utilizing UDPglucose, but not ADPglucose, CDPglucose, or GDPglucose;[1164] its action with UDPGlcUA was not studied.

2. Glucosidation in Animals; General Aspects

Glucosylation in animals was earlier thought to be restricted to invertebrates;[4] e.g., in various insects, a UDPglucosyltransferase accepting phenols had been demonstrated.[840,848-851,1165] In insects as in plants, glucuronidation of foreign compounds seems not to occur. In molluscs, UDPglucosyltransferase, but not UDPglucuronyltransferase, has also been demonstrated.[206,845,846]

3. Glucosidation of Xenobiotics in Mammals

The first xenobiotic glucosylation in mammals was of 4-nitrophenol,[1166] arising from UDPglucose in mouse liver microsomes; the product was well characterized.[1167] Some 0.9 μmol of the glucoside per 100 mg microsomal protein per hour were formed under conditions (5 mM 4-nitrophenol and 5 mM UDP-sugar) yielding 5.1 μmol of the glucuronide. The enzyme resembled UDPglucuronyltransferase in several properties.[1167] UDPGlcUA inhibited formation of the glucoside,[1167] which in mouse urine accounted for 1 to 2% of the total metabolites, compared with 50% glucuronide.[1168] Glucosidation might become more important as aglycon load increased relative to available UDPGlcUA,[1167] but no firm evidence has yet appeared. As certain steroids (e.g., pregnane-3α,20β-diol, and pregnane-3α-ol-20-one) inhibited glucosyltransferase, but not glucuronyltransferase activity, and as these steroids were claimed to hinder excretion of glucuronides,[1167] a temporary glucosidation might assist excretion of glucuronides through the cell wall;[1167] this is consistent with accumulation of glucosidated glucuronides in biliary obstruction (Chapter 7, Section III.D, and below), but as these nonphenolic steroids are not glucosidated by mouse or rabbit liver,[454] the hypothesis is weakened. Labow and Layne[454] found 4-nitrophenol a better substrate in mouse liver than phenolic steroids, the latter being a better substrate in rabbit liver; their earlier failure[411] to find glucosidation of 4-nitrophenol by rabbit liver they considered[454] due to use of pH 8.0, optimal for steroid glucosidation, but inhibitory with 4-nitrophenol. The lack of success reported by Vessey and Zakim[1181] with activated guinea pig liver microsomes and 4-nitrophenol could similarly have been due to nonoptimal assay conditions (e.g., pH 7.5 and phospholipase-A-treated enzyme), as also the negative reports for other conjugates looked for by these workers (Sections II.C, II.D, II.E.1 below).

Isoflavones were both glucuronidated and glucosidated by rabbit liver microsomes, glucosides being identified by the ^{3}H from UDP-D-[6-^{3}H] glucose,[411] and by hydrolysis with almond emulsin and inhibition of this hydrolysis by the specific β-glucosidase inhibitor glucono-1,5-lactone.[594] Formation of glucosides was 3% or less of that of glucuronides; sheep liver and rabbit intestine form no detectable glucosides.[411] Other xenobiotic acceptors for the glucosyltransferase were diethylstilbestrol, 4-nitrophenol, phenolphthalein, 2,4-dihydroxyphenyl p-methoxybenzyl ketone, and the corresponding p-hydroxy compound.[411,454]

Glucosides seem unimportant in fetus and adult,[197] but they have not been searched for, and certain compounds may be extensively excreted in some species; e.g., 3-(4-pyrimidinyl)-5-(4-pyridyl)-1,2,4-traizole is excreted in dog bile at 60% of the dose.[1169]

4. Glucosidation of Endogenous Compounds in Mammals
a. Steroids

Microsomal preparations from rabbit liver[1170] possess UDPglucosyltransferase activity towards the 17α-hydroxyl group of estradiol and possibly of estradiol 3-glucuronide. This enzyme forms the mixed conjugate estradiol 3-glucuronide-17α-glucoside. In liver of rabbit,[113,1171] sheep,[453] and man,[1172] UDPglucosyltransferase activities exist towards steroids, the substrates preferentially and sometimes necessarily possessing a phenolic A ring characteristic of mammalian steroid estrogens.

In rabbit, two distinct glucosyltransferase fractions exist,[264] one transferring glucose equally well to the 3-hydroxyl group of estrone, estradiol-17α or estradiol-17β, the other to the 17α-hydroxyl group of estradiol-17α,3-glucuronide; in sheep, only the latter enzyme occurs.[453] In man, glucose is transferred to the 17α-hydroxyl group of estradiol-17α and less readily to the 16α-hydroxyl group of estriol,[1172] the 3-glucuronides of the estradiols and of estriol not being substrates.[1172] Use of tritium-labeled glucose in UDPglucose[454] to obviate interference by the non-UDPglucose-dependent steroid glucosyltransferase,[113] confirmed the two UDPglucosyltransferase fractions in rabbit liver and their differences from activity in mouse liver.[454] These steroid-glucosidating enzymes also appear less active than the glucuronyltransferases, rabbit liver forming 20 to 300 times more steroid glucuronide than glucoside.[331]

Separate enzymes probably exist for glucosylation and glucuronidation of xenobiotics; UDPglucosyltransferase activity for 4-nitrophenol was lost during purification of the UDPglucuronyltransferase activity accepting that substrate.[239,241]

No glucosyltransference occurred to estradiol-17α or its 3-glucuronide in solubilized rabbit liver microsomes if ADP-, CDP-, GDP-, or TDP-glucose replaced[1171] UDPglucose, contrasting with the relative lack of specificity in plants,[1162] (but see Reference 1164), but consistent with the nucleotide specificity for bilirubin UDPglucosyltransferase.[1173] With unsolubilized rabbit liver microsomes, UDPglucose, though enhancing glucosidation, is not essential (Section II.E below).

Evidence of separate UDPglucosyl- and UDPglucuronyltransferases for these steroids is:[411] estriol forms a glucuronide in rabbit liver, but not a glucoside; glucosidation, but not glucuronidation, proceeds for certain estrogens without added UDPsugar; and in rabbit liver, the 17α group of estradiol 17α-3-glucuronide is a substrate for glucosidation, but not for glucuronidation. Unpublished work is quoted[411] on partial separation and on characteristics of the two activities.[411] As with the various UDPglucuronyltransferase activities, boundaries of heterogeneity are not yet settled.

Steroid glucosides may be metabolic intermediates like the sulfates;[411] steroid monoglucosides (i.e., with no additional glucuronic acid moiety) are poorly soluble and understandably absent from urine or bile. A recent report of the transfer of glucose from UDPglucose to the 3-OH of estrogens in rabbit liver involves an isoprenoid lipid-soluble intermediate;[1173a] only strong estrogens are accepted — not, e.g., estriol. Injected estrogen glucosides disappear, but injected glucuronides are excreted.[1173a] Some evidence[1173a] suggests that estrogens are readily transported through the cell wall as glucosides, hydrolyzed in the cytosol, and taken to the nucleus as free steroid.

b. Bilirubin

Glucosidation of bilirubin was first indicated by identification of bilirubin glucoside in dog bile.[1174,1175] UDPGlucosyltransferase activity towards bilirubin was then shown in rat liver,[479,875,1173] the activity much resembling that of UDPglucuronyltransferase: distribution in cell fractions, stimulation by digitonin and metal ions, complete absence from the Gunn rat,[479] enhancement in rats by pre-treatment with phenobarbital,[881] activation by lecithin, and partial copurification.[257] However, pH-activity curves show discrepancies,[479] selective inhibition[181] or induction occurs,[244] the ratios of digitonin-activated to untreated enzyme activity are different,[472,475] and, significantly, there was no correlation in the ratio of relative liver glucuronidation:glucosidation. Activities varied between species, ranging from 1.0[475] or 2.2[881] in the cat to 10.5 in the mouse,[881] and 2.0 in sheep;[475] and cat, with lower glucuronyltransferase activity than the other animals, possessed specifically their amount of glucosyltransferase.[881]

In mammalian bile, the IXα-glucuronide is much more important than the IXα-glucoside; only in cat, mouse, rabbit, and dog did the glucose and xylose conjugates add up to 12 to 41% of the total;[475] in chicken bile, both glucuronide and glucoside were equivalent, but of very minor significance, presumably because biliverdin is the major product of heme breakdown in chicken.[475] Man, dog, cat, and rat excrete bilirubin largely as diconjugates, and the biliary glucuronide to glucoside ratio varied among species directly with the glucuronyl to glucosyl transferase activities of broken cell preparations.[475] Species varied in pH-optima and activation characteristics of the glucosylating enzyme.[475]

Extrahepatic glucosidation of bilirubin in dogs was suggested by an early report[1089] and detected[316] in kidney cortex of dog and rat, but not in that of man, cat, or ox.

Snake bile[875] contains only the monoglucoside as a bilirubin conjugate.

UDPGlucosyltransferase is deficient towards bilirubin in infant[708,881] and perinatal rats,[678] but developmental pattern differs from UDPglucuronyltransferase.[678,708] In Crigler-Najjar patients, some glucoside exists in urine and bile, but not in serum, possibly from selective glomerular filtration or from renal synthesis.[678] Noir[491] discusses glucoside and xyloside occurrence in cholestatic bile, where bilirubin glucoside increases proportionally over glucuronide, and this may be due to loss of constraint of UDPglucuronyltransferase.[181]

"Endogenous" formation of bilirubin glucuronide (i.e., non-UDP-sugar-dependent) is higher than of the glucoside,[316] in contrast to the situation with estrogens.[411]

c. Other Endogenous Compounds

UDPGlucose-dependent glucosyl transfer to ceramide occurs in chicken brain preparations, other nucleotides not substituting for UDP in the glucose donor.[1176]

C. N-Acetylglucosaminidation

Steroid N-acetylglucosaminides exist in urine of rabbits and man and in human bile.[436] They are double conjugates, with N-acetylglucosamine at alcoholic groups on the D ring; either glucuronic or sulfuric acid is conjugated at C-3.

Rabbit liver homogenates added N-acetylglucosamine to the 17α, but not to the 17β, hydroxyl group of estradiol; UDPGlcNAc or UDPGlcUA were both required unless the preformed 3-glucuronide was substrate, when UDPGlcNAc sufficed.[442,1177] UDP-N-Acetylglucosaminyl transferase activity was higher than glucuronyltransferase in rabbit kidney, but lower in liver and intestinal tract.[442]

Properties of this enzyme have been reviewed[436] and its easier inactivation than glucuronyltransferase by crude phospholipase C and protease noted;[264] it tended to resemble[264] (or may be identical with[454]) the steroid 17-glucosyltransferase, whereas the glucuronyltransferase tended to resemble steroid 3-glucosyltransferase.[264] The N-acetylglucosaminyltransferase and the glucuronyltransferase differed also in distribution and purification, activity of one disappearing during separation of the other.[454]

Steroid N-acetylglucosaminides may be concerned in the effect of steroids on synthesis or functioning of the mucopolysaccharide matrix;[436] N-acetylglycosamines are incorporated,[1178] independently of glucuronic acid,[1180] into oligosaccharides.[1179]

Identification of N-acetylglycosaminides in excretory fluids is unlikely for they are poorly soluble in water, and if formed at all with isoflavones, are undetectable with present techniques.[411]

N-Acetylglucosaminidation of 4-nitrophenol in vitro could not be detected.[1181]

D. Xylosidation

The D-xyloside of bilirubin was first found in dog bile along with the glucoside,[1174,1175] and its mechanism of formation from UDPxylose was demonstrated in rat liver.[316] Its structure is a D-xylopyranoside determined from its azodipyrrole,[479] probably with a β-linkage to bilirubin.[473] In vitro, digitonin caused less activation of UDP-xylosyl (and UDP-glucosyl) transferases than of the glucuronyltransferase, especially in rat.[475] The xyloside was found significantly in bile only of chicken, dog, and rabbit in a ratio of 1:3 with the glucoside when calculated as azopigments.[475] The dixyloside of bilirubin, like the other diconjugates, predominates over monoconjugate in bile from man, dog, cat, and rat, and in vitro is readily biosynthesized;[475] species (e.g., chicken, guinea pig, mouse, ox, pig, rabbit, and sheep) excreting largely monoconjugates formed only the monoxyloside in vitro.[316,475] In rat kidney, dixyloside, but not diglucuronide or diglucoside, is formed.[475]

Perinatally, rise of UDPxylosyltransferase resembled that of the glucuronyl rather than that of the glucosyltransferase.[678] Gunn rat is deficient in bilirubin UDPxylosyltransferase as well as in the other two enzymes.[476]

Rabbit liver microsomes transferred xylose from UDPxylose to estrone, 17α- and 17β-estradiol but not to estriol, testosterone or 17α-estradiol 3-glucuronide.[1182a] UDPXylose inhibited glucosyl transfer from UDPglucose to estrone, but no evidence of a lipid-soluble intermediate was seen, unlike that noted with glucosyl transfer. Only "poor" transfer of xylose to diethylstilbestrol or to 4-nitrophenol was noted.[1182a]

No evidence of xyloside formation was detected with UDPxylose, 4-nitrophenol, and an activated guinea pig liver microsomal preparation.[1181] For references to isolation of ribose conjugates, see Reference 1182a.

E. Other Glycosidatons

1. Galactosidation

A rabbit liver microsomal pellet or solubilized preparation formed galactosides at the phenolic 3-hydroxyl positions of estrone, 17α- and 17β-estradiol.[113] The sugar donor was UDPgalactose, and the estrone conjugate appeared more polar on thin layer chromatography than the glucoside conjugate. Radioactive estrone galactoside crystallized to constant specific activity with authentic chemically synthesized estrone-3β-D-galactoside.

As with UDPglucose in the formation of the glucoside (Section II.B.4.a above), UDPgalactose was essential only with Triton® solubilized microsomes.[113] A repeatedly washed microsomal pellet without UDPgalactose or UDPglucose formed both galactoside and glucoside but not glucuronide.[113] Added UDPgalactose or UDPglucose enhanced this formation of galactoside or glucoside but was not necessary.[113] UDPGlucose pyrophosphorylase, UDPglucose dehydrogenase, or UDPgalactose-4-epimerase, added to remove any possibly bound UDPgalactose or UDPglucose, had no effect.[113] However, addition of UDPgalactose 4-epimerase to added UDPgalactose diminished the amount of galactoside and increased the amount of glucoside formed.[113] The galactosyl and glucosyl donors therefore appear membrane bound within the pellet and are possibly involved in the glycosylation of endogenous protein there.[1182]

As steroid galactosides seem not to be excreted, they may be intermediates in a specific metabolic pathway.[113]

No decrease in 4-nitrophenol occurred when incubated with activated guinea pig liver microsomes and UDPgalactose.[1181]

2. Galacturonidation

After 4-nitrophenol had been incubated with activated guinea pig liver microsomes and UDPgalacturonic acid, the conjugate was detected on paper chromatography, eluted, and treated with β-glucuronidase.[1181] The residue was rechromatographed and on spraying with silver nitrate reagent gave a sugar spot with the R$_f$ typical of glucuronic and galacturonic acids; the additional lactone spot characteristic of glucuronic acid in acid solvents being absent, galacturonic acid must have been present in the conjugate.[1181] This UDPgalacturonyltransferase activity resembled UDPglucuronyltransferase in activation by UDPGlcNAc and in possible cooperativity of sugar nucleotide-binding effects,[1181] but was less activated by Triton® X-100 and phospholipase A and more readily inhibited by mersalyl;[1181] parallel studies with the two conjugates were not described, and the reproducibility not stated. Product inhibition studies indicated that UDPgalacturonic acid, unlike UDPGlcUA, did not inhibit the reverse reaction of UDPglucuronyltransferase; 4-nitrophenyl glucuronide inhibited the glucuronidation, but not galacturonidation, of 4-nitrophenol, increasing galacturonidation at some concentrations.[1181] Several glucuronides increased this galacturonidation, that of 4-methyl-umbelliferone increasing it ten-fold; no effect on the glucuronidation of 4-nitrophenol was observed.[1181]

The UDPgalacturonyltransferase and UDPglucuronyltransferase activities may be from different enzymes, but physical separation is required. No UDPgalacturonyltransferase activity existed in a highly purified UDPglucuronyltransferase preparation glucuronidating 4-nitrophenol.[241]

III. SULFATION

A. General

Sulfation, in a pharmacological context, yields biosynthesized sulfate esters of the structure:

$$R.O.SO_3^-$$

The linkage can be C–O–S, N–O–S, N–S, P–O–S, and S–S.

These ionized half-esters of sulfuric acid are formed more readily with phenolic hydroxyl groups than with alcoholic hydroxyl or primary amino groups. Sulfates, particularly of steroids are significant as metabolic intermediates (Chapter 1).

Sulfate is activated in two stages before transfer to an acceptor. First, ATP-sulfurylase (ATP-sulfate adenylyl transferase, EC 2.7.7.4) catalyses the reaction:

$$ATP + SO_4^{-2} \rightleftharpoons \text{adenosine-5'-phosphosulfate (APS)}$$
$$+ \text{ pyrophosphate}$$

Second, APS-kinase (ATP-adenylylsulfate 3-phosphotransferase, EC 2.7.1.25) catalyzes formation of "active sulfate":

$$ATP + APS \rightarrow \text{3-phosphoadenosine 5'-phosphosulfate (PAPS)}$$
$$+ ADP$$

Finally, one of several sulfotransferases (e.g., EC 2.8.2.1) catalyzes the conjugation itself:

$$PAPS + R.OH \rightarrow \text{3-phosphoadenosine-5-phosphate}$$
$$+ R.O.SO_3^-$$

Unlike the UDPglycosyltransferases and the Phase 1 system, the first two enzymes and the majority of the sulfotransferases which accept small molecules are cytosolic in mammalian cells.[1184,1185]

Sulfation, as a conjugation reaction, is relatively expensive in ATP and sulfate; in higher organisms, this sulfate was thought to come not from inorganic sulfate, but via thiosulfate, from bodily reserves of organic sulfur, and so sulfation is likely to be more rapidly limited by aglycon loading than is glucuronidation. It may also be a more primitive process than glucuronidation. Recent conclusions are well summarized by Dodgson.[1183] Other useful reviews exist,[1184-1186] and a short comparison of sulfation and glucuronidation in detoxication and "toxification."[1187]

However, Mulder and Meerman[1187a] believe that plasma inorganic sulfate is immediately available for drug sulfation by liver. Plasma sulfate falls as ester sulfate appears, and is rapidly replenished from other bodily sulfate pools.[1187a] Pretreatment with phenol did not result in sulfate shortage when the animal was again challenged with phenol.[1187a] Starvation of sulfate is not the reason for the decreasing proportion of sulfate and increasing proportion of glucuronide as phenolic substrate concentration is raised: the reason lies, they consider, in enzymic affinity (Section III. D below).[1187a]

Double conjugates (sulfo-glucuronides), are treated in Chapter 7, Section II.D.3.b. They can be formed both in vivo[456,1188,1189] and in vitro.[457] Their formation in vivo without prior hydrolysis of sulfate indicates that ester sulfates can enter cells and reach UDPglucuronyltransferase on (or in) the endoplasmic reticulum. This is especially surprising in the case of 2-hydroxy-5-nitrophenyl sulfate, a very polar conjugate[1188] whose glucuronide, 2-β-glucuronoside-5-nitrophenyl-sulfate, is notable in its juxtaposition of two acidic groups. Demonstration of UDPglucuronyltransferase specificity for this reaction would be valuable.

B. Specificity of Sulfotransferases: Relevance to Specificity of UDPGlucuronyltransferase

Specificity of sulfotransferases can be briefly considered here because they are physiologically analogous to UDPglucuronyltransferases in conjugating both endogenous and xenobiotic molecules, and so could suggest boundaries of substrate heterogeneity for such dual-function enzymes, and because they are soluble enzymes free of interpretative complexities. Even so, separation of activities is rarely convincing. Topological separation in vivo occurs between two classes of the enzyme; sulfotransferases responsible for sulfation of large molecules (e.g., chondroitin, dermatan, heparin, cerebrosides, glycoproteins, and polypeptide tyrosines) are found, probably as part of complex systems, in the Golgi membranes or endoplasmic reticulum, whereas those sulfating small molecules (e.g., drugs, steroids, sugars, etc.) are cytosolic.[1183] (The UDPglucuronyltransferase activities towards these two types of molecule are also distinct, though both are membrane bound[597,1180]). There seem to be five different sulfotransferases sulfating respectively of aetiocholanolone, androsterone, testosterone, deoxycorticosterone, and estrogens.[1183] The "multiplicity" of phenosulfotransferases is almost as confused as that of UDPglucuronyltransferases.[1183] Sulfation of 4-nitrophenol is distinguished from that of L-tyrosine methyl ester by selective loss of the latter activity from a purified enzyme preparation; but it appears that the enzymic redox state and hence conformation is important[1183,1190] in determining which substrates it accepts and therefore only one enzyme may be involved after all: a salutary example for workers on UDPglucuronyltransferase "heterogeneity".

C. Development of Sulfation: Relevance to that of Glucuronidation

Little is known on sulfation development,[636] but it seems more important fetally and perinatally than glucuronidation, especially for steroids.[1191] Possibly relevant to the study of glucuronidation enzymes is the finding of differential inhibition of phenol sulfotransferases in 1- and 28-day-old mice, suggesting a difference in protein structure (for these are soluble enzymes)[1192] possibly arising from conformational changes connected with their redox state (Section III.B above). A nondialyzable perinatal inhibitor of phenosulfotransferase in rat gut homogenates accounting for a transient fall in rapidly developing enzyme,[1183] could have been a PAPS-destroying protein analogous to UDPGlcUA pyrophosphatase (which does not increase over birth), but some PAPS may still remain after incubation (G. Powell, personal communication).

As with glucuronidation, a higher-than-adult phenol sulfotransferase activity exists in neonatal or infant mouse liver, falling to adult values at 35 days and developing sex differences after 20 days.[502] Sex differences were not altered in males by castration or testosterone,[1193] and may be maintained by a pituitary factor as are monooxygenases.[835] Powell and Curtis[1193a] quote unpublished work of Powell and

Jones in which phenolsulfotransferases rose from virtually zero over the 5 prenatal days in rat liver and gut.

D. Sulfation and Glucuronidation: The Effect of Substrate Loading

Examples have been encountered in this review of the greater role of glucuronidation as substrate loading increases. Bray et al.[1194] first clearly showed that sulfate supply was rate limiting in phenol sulfation, increasing load diminishing it to a zero order reaction. The fraction of salicylamide sulfated in man fell sharply with rising dose but increased again on feeding of cysteine.[1195] In intact rat, increased harmine administration decreased the ratio of harmol sulfate to harmol glucuronide, unless SO_3^- was simultaneously given;[1196] however, Mulder (personal communication) points out that the data indicate that SO_3^- merely decreased glucuronidation and did not increase sulfation. In perfused rat liver, the more 4-nitrophenol[175,1197] or paracetamol[1197] added, the greater the increase of glucuronidation over sulfation[1197] (G. Kahl, 1975, personal communication), and 4-nitroanisole largely produced 4-nitrophenyl sulfate because it only slowly released 4-nitrophenol into the system.[1197] Similarly, increasing harmol load on perfused rat liver raised, from a 1:1 basis, glucuronidation 11-fold and sulfation only threefold;[1198] apparent K_m for harmol with rat liver UDPglucuronyltransferase was ten times that for sulfotransferase.[1199] The K_m values generally suggest these competitive conjugations; for 4-hydroxy-3-methoxyphenyl-ethanol in rat, the K_m value with the UDPglucuronyltransferase is nearly 40 times that with the sulfotransferase,[894] and when the two conjugates were formed simultaneously in vitro, the sulfate clearly predominated.[894] However, the relative secretion of harmol conjugates by perfused liver could have also reflected slower biliary excretion of sulfate than of glucuronide.[1198] With harmalol as loading substrate, even small amounts are excreted almost exclusively as glucuronide in vivo,[1199] explicable by activity of sulfotransferase for harmalol being much lower than for harmol, whereas with UDPglucuronyltransferase the values are the same for both substrates.[1199]

Mulder and Meerman,[1187a] who found that pentachlorophenol selectively inhibits sulfation, conclude that a falling sulfate to glucuronide ratio with substrate loading does not result from limitation of sulfate — a likelihood they consider exaggerated — but from lower K_m values for phenols of sulfotransferases compared with glucuronyltransferases. Infused sulfate did not change the ratio markedly, the ratio being also unchanged following pretreatment with phenol to attempt to exhaust the sulfate pool,[1187a] but sulfate limitation with acetaminophen in man is to some extent overcome by concomitant administration of inorganic sulfate.[1201a] Others working with hepatocytes and benzo(a)pyrene metabolites, conclude like Mulder's group that glucuronidation has the greater capacity, but sulfation the greater affinity.[181a,1201b]

Selectively inhibiting sulfation demonstrated that biliary excretion of harmol sulfate is also rate limiting in vivo. Because of the easy saturation of sulfate excretion in bile and rapid excretion of harmol glucuronide, a marked and rapid appearance of the glucuronide in bile was better for detecting sulfation inhibition in vivo than waiting for eventual fall in excreted sulfate.[1200] 2,6-Dichloro-4-nitrophenol, at 1 μM, inhibited sulfation of harmol 100% and did not affect its glucuronidation when both processes were measured simultaneously in vitro[1200] by an elegant method.[1201] Detailed reports of the competition of the two transferases for harmol and of increase in one conjugation resulting from inhibition of the other in vivo[300,1198-1200] and in vitro[1200,1201] should be referred to for further information. With phenolphthalein, glucuronide and sulfate may compete for the same carrier during biliary excretion,[300] and phenolphthalein, phenolphthalein monoglucuronide, diethylstilbestrol monosulfate, and diethylstilbestrol disulfate each reduce biliary excretion of phenolphthalein disulfate in rats receiving constant infusion of the latter, also suggesting that these compounds compete for excretion.[1202]

The fate of estrone and of estrone sulfate in once-through and cyclically perfused rat livers, suggests[1203] different metabolism of free estrone taken up into the cell and of estrone derived intracellularly from hydrolysis of estrone sulfate; the former estrone is preferentially conjugated with glucuronic acid and excreted in bile.[1203]

E. Species and other Differences in the Sulfation to Glucuronidation Ratio

Whereas rat, rabbit, and guinea pig liver conjugated 4-hydroxy-3-methoxylphenylethanol by both pathways, human liver only formed glucuronide, and mouse and cat liver only sulfate;[894] the two enzymes were assayed separately and together.[894] Tyramine was only sulfated, not glucuronidated, by monkey liver;[422] 4'-hydroxyamphetamine[408] was glucuronidated, not sulfated, by the rodents used, but only sulfated in man and monkeys. Differences in excretion of diethylstilbestrol monosulfate, disulfate, and mixed sulfate-glucuronide between guinea pig[1204] and rat[456] arise from greater disulfation of disulfate in rat and glucuronidation of the resulting monosulfate; they can be explained by the relative activities of hepatic sulfatase, sulfotransferase, and UDPglucuronyltransferase systems, and not by differences in biliary transport.[1186]

Phenol administered intraperitoneally to hens yielded similar amounts of urinary sulfate and glucuronide, but when orally administered, more sulfate, possibly because of dilution and of predominance of hepatic sulfation following a low dose;[1205] when administered intramuscularly into the leg, 20 times more sulfate than glucuronide was formed, again possibly by slow release of aglycon, although preferential renal sulfation of circulating phenol from gut or muscle is also an explanation.[1205] Other species differences have been reviewed.[883]

IV. THE CYTOCHROME P-450-DEPENDENT MONOOXYGENASE SYSTEM

A. General

Much literature exists on the relation of glucuronidation to the Phase 1 system preceding it in the metabolism of xenobiotics,[168,635,1042,1206] but nothing certain is known.

B. Topological Relationship

Distribution of the two processes is similar among species[904] and tissues,[410] but not identical: in Gunn rat, the Phase 1 system appears normal and inducible,[209,925] and in human fetal liver, the cytochrome P-450-dependent hydroxylation develops much earlier than in nonprimate animals;[637,701,1207,1208] in both cases transferase is deficient (Chapter 10). Endoplasmic reticulum vesicles containing monooxygenase complex and UDPglucuronyltransferase are centrifuge differentially[145] and during purification members of the monooxygenase complex are clearly separable from solubilized transferase chromatographically and without causing obvious loss of transferase activity.[240,332,336,1209]

Many workers believe that the monooxygenase system is more superficial in the membrane than the transferase, e.g., the model of Vainio.[168,1043] It is more susceptible to perturbation than the transferase, being inactivated when the latter is activated, as by aging,[183] trypsinization,[160,167] chaotropes,[217] lipid peroxides,[168] surfactants,[211] carbon tetrachloride,[216,1009,1049] and protein deficiency;[782] methylmercury hydroxide inactivates the monooxygenase system before the transferase.[327] This readier susceptibility occurs with both endogenous[213] and xenobiotic substrates and is still evident following induction by phenobarbital or 3-methylcholanthrene.[325]

The idea that UDPglucuronyltransferase is situated below the cytochrome P-450 complex is consistent with the hypothesis noted earlier (Chapter 4, Section II) of the transferase enzyme being sited on the intraluminal surface of the vesicle. Xenobiotics would reach Phase 1 direct from the cytoplasm, be transformed, and the metabolite shuttled to the neighboring UDPglucuronyltransferase which would release the final products into the lumen. However, as noted earlier in this book, there is no evidence that glucuronides are excreted into the lumen, and they probably reach the cytoplasm first.

C. Metabolic Relationship

In vivo, Phase 1 and glucuronidation seem coupled, for only small amounts of hydroxylated intermediates occur in plasma, compared with glucuronidated products.[1210,1211] Evidence of some link between the two processes continually accumulates but the nature of the link, which must be indirect, is hard to define.

Both monooxygenase activity and UDPglucuronyltransferase are inducible by xenobiotics, such as barbiturates and methylcholanthrene,[1125,1206] or hormones, such as glucocorticoids.[755,759] Usually, the monooxygenase is induced the more rapidly;[168,759,1014] in chick-embryo liver exposed to corticosterone, aminopyrine *N*-demethylase and aniline hydroxylase activity reached their peak after 2 or 3 days,[759] when transferase activity was just beginning to rise, and it continued to rise when the Phase 1 activities were falling from their peak.[759] However, differential induction is apparent: monooxygenase, but not transferase, is induced by chrysene in rat liver,[216] and transferase, but not monooxygenase, by *n*-hexane in guinea pig liver.[219] Transferase can be induced to high levels from zero while monooxygenase activity never appears or declines to zero, as in culture of chick embryo liver[1125] or after injection of aminotriazole plus phenobarbital into eggs.[342] Induction of the two systems can be blocked to different degrees by various inhibitors of transcription or translation.[1212] All this suggests control by different operons. In various mouse strains, throughout crosses and back-crosses, induction of transferase segregated exactly as induction of monooxygenase; however, inheritance of transferase was an additive (codominant) trait, differing distinctly from inducible monooxygenase activity which behaved almost exclusively as a single autosomal dominant.[909,910] A single genetic locus for the two systems is, therefore, not proved, but both may be regulated by a common receptor.[910,1213]

Dependence of glucuronidation on Phase 1 has been much investigated. The total amount of hydroxylated metabolites, unlike that of glucuronide, formed from demethylimipramine by a coupled two-step process in vitro was independent of the concentration of UDPGlcUA, and no link seemed to exist between the two processes; but rate of removal of these metabolites from cytochrome P-450 did depend on the UDPGlcUA level. The coupled system 4-nitroanisole → 4-nitrophenol → 4-nitrophenyl glucuronide has been used frequently. It confirmed that ongoing conjugation does not increase the Phase 1 reaction[1214] and suggests, with perfused rat liver,[1197] or isolated hepatocytes[166] that "leakage" into cytoplasm of Phase 1 metabolites readily occurs, for sulfate of 4-nitrophenol is formed rather than glucuronide, at low 4-nitroanisole concentrations. A similar "two-step" system in isolated hepatocytes with biphenyl as primary substrate[1215] suggested that the Phase 1 inhibitor carbon monoxide did not inhibit glucuronidation (as also found in Reference 356 for steroids), again implying conjugation not be linked to Phase 1; SKF 5295A, another inhibitor of Phase 1, did inhibit conjugation, but has multiple effects.[4] A lag occurring between 4-hydroxylation of biphenyl and its glucuronidation was not observed if 4-hydroxybiphenyl itself was added to the hepatocyte suspension, suggesting that 4-hydroxybiphenyl at sufficient concentration "activates" the transferase.[1215] (A similar unsupported stimulatory role for a benzo(a)pyrene metabolite was suggested earlier[1013] to explain findings on "linkage"). Almost all the 4-hydroxymetabolite remained within the cells for 5 to 15 min,[1215] afterwards appearing in the extracellular medium as glucuronide or sulfate; these conjugates were scarcely ever detectable within the cell.[1215] Coupling of the two processes therefore seemed temporally inefficient, but only the conjugates, not the Phase 1 metabolites, could leave the cell.

Metabolism from the monooxygenase substrate to its glucuronidation is more than "two-step", the transient existence of an epoxide being necessary in formation of a hydroxylated derivative, and this reactive epoxide may itself be "detoxicated" to a diol through epoxide hydratase.[1216-1218] The phenol and the diol are then both substrates for subsequent glucuronidation. Coupling of epoxide hydratase and monooxygenase has been suggested.[1216]

This "three-step" system, from naphthalene, has been studied in isolated rat hepatocytes and in microsomes.[278] Naphthalene is first oxidized by the monooxygenase system to its 1,2-oxide, which can rearrange to 1-naphthol, be conjugated with glutathione, or be converted by epoxide hydratase to the 1,2-dihydrodiol; the latter can be further converted to 1,2-dihydroxynaphthalene and the toxic 1,2-naphthoquinone.[278] Linkage here between the transferase and the preceding step would help to prevent build-up of the reactive quinones.[278] In isolated hepatocytes, naphthalene was largely metabolized to the dihydrodiol (15% of total metabolites) or the dihydrodiol glucuronide (26%), 1-naphthyl glucuronide (1%) being relatively low.[278] In microsomal preparations, the same proportion of metabolites was achieved if UDPGlcNAc was present; if it were absent, the free dihydrodiol predominated over its glucuronide.[278] Phenobarbital or 3-methylcholanthrene induced transferase activity to 1-naphthol, but not to the dihydrodiol;[278] surprisingly, because these inducers stimulate activities of the monooxygenase and epoxide hydratase.[278] However, even the coupled systems monooxygenase and epoxide hydratase are themselves under separate genetic control.[278,1217]

Further pros and cons for linkage of glucuronidation and the Phase 1 could be quoted. NADPH enhances[1219] or inhibits[1214] transferase activity in vitro. Both hydroxylation and glucuronidation were depressed by hemin, a modulator of Phase 1 reactions and in itself an unlikely modulator of transferase;[781,1013] high concentrations of glucuronides inhibit the transferase and eventually also the monooxygenase.[168,332] The transferase and any of several NADH- or NADPH-dependent activities in the electron-transport systems in various rat hepatomas did not correlate. Use of the 2-stage 7-ethoxycoumarin → 7-hydroxycoumarin → conjugation pathway in hepatocytes, suggested[181a] that no direct linkage exists. Lind et al.[1219a] noted an indirect connection, finding, in a rat-liver microsome system forming the 3,6-quinone from benzo(a)pyrene, glucuronide produced only if NADPH or NADH were present. They suggest diaphorase is responsible, for its inhibitor dicoumarol prevented the glucuronidation, while trypsinised microsomes (with virtually no NADPH-cytochrome c reductase but unharmed diaphorase) could still perform it. As transferase activity to 3-hydroxybenzpyrene was inhibited by the 3,6-quinone, and 3-hydroxybenzpyrene inhibits glucuronidation of the quinone, they suggest a common pathway exists to the glucuronide; and as dicoumarol itself inhibits transferase activity to 3-hydroxybenzpyrene but not to other substrates, they also suggest diaphorase may have to rearrange this phenol before the transferase accepts it.[1219a] By analogy with epoxide hydratase,[1216] UDPglucuronyltransferase might exist in two forms, closely coupled to, or relatively free from, monooxygenase, the proportion depending on the endoplasmic reticulum structure, and this structure in its turn depending on the agent (e.g., Type I or Type II monooxygenase inducers) operating on the membrane at the time.[203,635] This scheme is not compatible with centrifugal separation of transferase from cytochrome P-450 unless the membrane is sheared horizontally as well as vertically during homogenization.[203]

V. ADDITIONAL NOTES

For recent work on the relationship of the perinatal development of other drug-metabolizing enzymes, especially of Phase 1, to that of UDPglucuronyltransferase, see Reference 636a.

Part III
Practical Aspects

Chapter 17

PRINCIPLES OF ASSAY OF GLUCURONIDATION IN BIOLOGICAL TISSUES OR FLUIDS AND SELECTED PRACTICAL PROCEDURES

I. INTRODUCTION

Section II outlines the principles of glucuronide assay commonly used at present, together with general precautions. Section III describes some assays of glucuronidation enzymes from a practical and critical viewpoint.

II. PRINCIPLES OF ASSAY METHODS

A. Introductory

This section will briefly deal with assay of glucuronides for identification and quantitation. Assay of glucuronide formed enzymically is discussed in Sections II.C.7 and III.C. below.

Earlier procedures are described or referred to by Williams[5] and by various authors quoted in Reference 4, especially for steroids.[80] A recent general review is by Tomasic.[1219b]

Conventional methods employed differential extraction of glucuronide and unconjugated aglycon. Quantitative release of conjugated aglycon, either enzymically or chemically, was difficult on a small scale, and risked a chemical change of aglycon. Certain newer methods directly measure specific glucuronides sensitively in biological samples without cleavage of the glucuronide link.

B. General Procedures
1. Radioimmunoassay
The potentially very sensitive and specific radioimmunoassay techniques have been applied to steroids for some time. Steroids are not immunogenic, but are made so by covalent linkage to, for example, bovine serum albumin. In order to bind steroid to the protein carrier in the first place, it is usually converted to a derivative with a carboxylic acid group, which is then linked covalently to the macromolecule by a mixed acid anhydride reaction whereby a peptide bond is formed between the acid group and ε-amino groups of lysine residues in the protein. Kellie et al.[1220] took advantage of the fact that steroid glucuronides already possess, on the glucuronic acid moiety, a suitable carboxylic acid group, and so themselves can be linked covalently to the lysine residues. Kellie has briefly reviewed the potentialities of the technique,[1221] which is increasingly employed.

A drawback is that the link between the steroid nucleus and the carrier protein should, for highest antigenic specificity, be as far from the functional groups of the steroid as possible. If the glucuronyl group of a conjugate is linked to the carrier, then higher specificity will result with estriol 3-glucuronide (D ring exposed) than with estriol 16α-glucuronide (A ring exposed); for estrone, estradiol, and estriol differ at their D ring, whereas all estrogens have a phenolic A ring; e.g., antiserum against estriol 3-glucuronide does not react with the 16α-glucuronide of estriol, nor with free estradiol or estrone, but exhibits weak cross-reactions between antisera for estrone, estradiol, and their glucuronides.[1222] The free steroid is often (as with the antiserum against estriol 3-glucuronide) antigenic to the same degree as the conjugate, but antiserum against estriol 16α-glucuronide exhibits only 2.2% cross-reactivity with estriol itself, and picograms of 16α-glucuronide could be measured in the presence of excess estriol and other estrogens in plasma and urine;[1223] cross-reactivity occurred with estriol 3-sulfate-16α-glucuronide, but this was present in only small amounts. (In such contexts the α refers to the configuration of the estriol hydroxyl group, not to the glucuronyl linkage which is always β).

These methods predict ovulation in nonpregnant women and monitor maternal and fetal health. A simple solid-phase radioimmunoassay of estriol 16α-glucuronide is described,[1224] the antiserum being coupled with agarose gel, considered superior to poly-acrylamide,[1225] as support.

Estriol 16α-glucuronide coupled through C-4 via an azobenzoyl derivative[1226] has been used as a hapten, but direct methods seem more convenient at present.

Drugs have long been measured by radioimmunoassay, but drug glucuronides are not yet regularly estimated in this manner. In one instance, using morphine 3-glucu-ronide linked to bovine serum albumin by a carbodiimide reaction,[1227] recognition of the glucuronyl moiety proved not very sharp, both glucuronide and free morphine being equally detectable, together with codeine. The 6-glucuronides of codeine and morphine exhibited less affinity. As morphine is excreted principally as 3-glucuronide, and codeine as 6-glucuronide, this procedure in urine could differentiate between abuse of the two types of drug.[1227]

2. Chromatography

It is not possible here even to list the many reports utilizing conventional simple chromatography of glucuronides, but some recent procedures with other types of chromatography may be mentioned.

Anionic liquid ion-exchangers, such as tetraheptylammonium chloride, were used[1228,1229] to separate steroid glucuronides. This compound, at 0.1 M in organic solvent, extracted over 99% of the 21-glucuronides of 11-deoxycorticosterone and cortisone; the more polar 3-glucuronide of β-cortol was only 43% extracted.[1228] The authors detail procedures[1229,1230] for transfer of less polar glucuronides back into an aqueous phase, and illustrate where the technique simply and rapidly concentrates conjugates from urines or eluates. Partition of glucuronides in this system appeared to be through ion-exchange rather than through hydrogen bonding. Corticosteroid glucuronides are quantitatively extracted into chloroform with tetrapentylammonium as counter ion;[1231] these ion-pair extraction methods have recently been reviewed.[1231a]

Glucuronides and sulfate esters have been separated by counter-current extraction, after removal of salts by ion exchangers.[1232] In high-pressure liquid chromatography the hydrophobic polystyrene-divinylbenzene-matrices are unsuitable anion exchangers for steroid glucuronides because of interactions with aglycon;[1233] cellulose matrices, if rigid enough for high pressure, were more promising, and ECTEOLA-cellulose satisfactory with constant control, separation even of the 16- and 17-glucuronides of estriol being possible. Bakke[1234] recently reviewed other examples of chromatographic extraction, concentration, and separation of glucuronides. He favors Porapak® Q as matrix.

Gas chromatography of unhydrolyzed glucuronides requires prior formation of derivatives, a drawback with labile glucuronide linkage or reactive aglycons. Gas chromatography of methyl and trimethylsilyl derivatives of O-glucuronides was demonstrated many years ago in Japan.[1235,1236] These derivatives[1237,1238] and others, such as methylacetyl,[1237,1239,1240] pertrimethylsilyl,[1241] permethyl,[1242] and methyl-trifluoroace-tyl[1243] derivatives, have been increasingly used for examination of urinary, biliary, and tissue glucuronides.[1234]

Chromatography of these derivatives is a useful accessory, but does not usually give enough information on the aglycon. A fuller procedure is usually necessary, such as the quantitative determination of four estriol conjugates in human pregnancy urine;[1244]

each conjugate was isolated intact by column chromatography on Sephadex® G-25 and LH-20, followed by enzymic hydrolysis, ether extraction, methylation, chromatography on alumina, and determination of the trimethylsilyl derivatives by gas chromatography, with an added check from the counting of labeled aglycon.[1244]

3. Mass Spectrometry

Identification and quantitation are not easily reconciled in one technique. Often, high-resolution ultraviolet liquid chromatography can be used to separate or measure the urinary conjugates without incurring inevitably destructive hydrolysis of the sample to liberate the aglycon. However, the conjugate must be identified and its molar absorptivity known, and as only microgram quantities are available and the excretory pattern is usually complex, mass spectrometry with or without gas chromatography is then required for identification. Molecular ion and fragmention pattern of higher molecular weight fragments of underivatized glucuronides are seldom seen in electron impact mass spectrometry because of the polar groups on glucuronic acid and on the fragments. Many glucuronides are degraded by heat before volatilization, destroying both uronic acid and aglycon. These points are discussed by Mrochek and Rainey[1241] who find mass spectrometry of trimethylsilyl derivatives of glucuronides useful to verify glucuronic acid and identify the aglycon. Quantitation by mass spectrometry is hazardous, liquid chromatography being preferable.[1241] Ester glucuronides need to be separated beforehand, for they are cleaved on derivatization, making it difficult to match up conjugating agent and aglycon. Gas chromatographic and mass spectrometric techniques were used in this way to identify estriol 3-, estradiol 3-, estriol 16α-, and estriol 17β-glucuronides using permethylation.[1245]

Mass spectrometry itself, widely used with steroid glucuronides,[1246,1247] is increasingly employed to characterize pesticides, drugs, and their metabolites;[1234,1248] e.g., glucuronides of cannabinol and metabolites,[1249] of diphenylhydantoin,[1250] and of hydroxyphenylalkylaminoethanols.[1251] In one typical mass spectrometric procedure for glucuronides,[1243] the drug was either added to perfusate of isolated liver or injected into the portal vein of the animal, the collected bile dried, permethylated with sodium dimethyl sulfoxide and methyl iodide, and the resulting products analyzed through both direct and gas chromatography inlets of the instrument. Glucuronides were identified by major ions at m/e 101 and 201, 1 to 5 $\mu\ell$ bile (some 0.1 to 1 μg glucuronide) being sufficient. The method detected ether and ester O-glucuronides, but was applicable to others; e.g., diglucuronides.[1243] A recent procedure is also described for mass spectrometry of steroid glucuronides.[1247] Bakke[1234] lists fragment ions and m/e values for detecting glucuronides in mass spectrometry.

C. Principles of Assay of Glucuronidation and of UDPGlucuronyltransferase

1. Introductory

The following sections consider very briefly various procedures utilized to study glucuronide formation in living systems — organism, tissue, or cell preparation. With intact-cell systems or animals, only the aglycon substrate, or its precursor, is presented; formation and channeling of UDPGlcUA to the transferase, as well as conjugation at the transferase, is undertaken by the system itself, which also usually releases formed glucuronide into the assay fluid. As a "no-UDPGlcUA" control is not possible, the product must be proved a glucuronide; and the assay should measure appearance of glucuronide and not disappearance of aglycon, certainly not disappearance of the compound originally administered (Section III.C.1). Isolated organs or cells may require added glucose to fuel UDPGlcUA synthesis if their glycogen is low, and always require oxygenation.

2. Use of the Intact Whole Organism

Pioneer studies on glucuronidation employed the whole organism, because of lack of knowledge of the organs involved, and because methods employing isolated tissues had not been developed. Data on the healthy whole organism are still essential, being the ultimate reference of all in in vitro studies, and not merely of those involving toxicity testing. Use of isotopes permits localization and assay of small amounts of aglycon or conjugate, thus removing a major objection to metabolic studies in the whole animal.

The metabolic fate of an administered compound can be investigated by identifying any glucuronides formed, and their aglycons in excretory fluids — urine, bile, feces. Sweat and tears do not seem to have been examined for glucuronides and saliva only recently, e.g., in antidoping precautions for race horses.[1252] A lipophilic membrane barrier could prevent excretion of most water-soluble organic molecules in saliva,[1252] and glucuronides may therefore be low there; none seems reported yet.

In urine, the classical source of glucuronidated metabolites, chemical hydrolysis is only likely with *N*-glucuronides in acid urines, but many aglycons (e.g., aminophenols) are chemically unstable on storage in solution even when conjugated. Enzymic hydrolysis is preventable by the addition of glucaro-1,4-lactone, the specific inhibitor of β-glucuronidase.

Bile is now recognized as a major route of excretion of glucuronides. Recent reviews[22,1095,1098] and Chapter 15 discuss factors necessary for biliary excretion. Storage of bile incurs chemical and enzymic hydrolysis, the former especially of bilirubin conjugates, where aglycon translocation within the conjugate molecule can occur (Chapter 7, Section III) and the latter through β-glucuronidase from damaged cells and bacteria. In feces or lower gut, bacteria can modify also the aglycon, making collection of little value.

Proportions of glucuronide excreted along these channels require assay in the whole animal. Biliary excretion studies involve fistulas or cannulae, and so an "abnormal" animal.

Comparison of routes of administration — oral, intravenous, intramuscular, intraperitoneal, by intubation, etc. — also requires the whole animal (Chapter 16, Section III.E).

The study of gross environmental change — temperature, circadian rhythms, birth, dietary restriction or dietary change, administration of hormones, or of inducing drugs — obviously needs the whole animal even if tissues themselves are subsequently examined, and this use shades off into straightforward toxicity testing. The use of "whole" animal preparations involving parabiosis, evisceration, etc., is usually as scientifically unrewarding as it is ethically undesirable.

A simple method of infusing hormones or xenobiotics into intact chick embryos in ovo may be mentioned here.[757] The compounds are placed in a home-brewer's fermentation lock and led by capillary attraction along a paper strip down to the chorioallantoic membrane of an embryonated egg, on to which the lock is taped; infusion for up to 4 days did not harm the embryo, and rates of flow are controllable by width and grade of paper.[757]

The welcome recent trend in glucuronidation studies towards awareness of in vivo complexity has brought back the once-discredited whole-organ techniques discussed below.

3. Perfusion and other Whole-Organ Techniques

Intact organs give information on glucuronidation in cell populations remaining organized as in the whole animal, but with greater controllability of environment, and experiments are shorter than with whole animal and longer (e.g., several hours) than with most in vitro techniques. Disadvantages are difficulty of oxygenation and need for skill.

The three intact organs or organ areas most used in glucuronida ion studies are liver, intestine, and lung. Perfused liver can be kept less unhealthy than formerly, and the elegant experiments of Bock and colleagues (Chapter 5) illustrate its value. Relation between glucuronidation and other reactions can be studied; "once through" and "cyclic" perfusions yield complementary results and are useful in comparing the three compartments of perfusing medium, bile, and intracellular environment.[1203]

Everted sacs or perfused loops of intestine provide information on its role in conjugation, deconjugation, and permeability to conjugates (Chapter 15); "normality" of intestinal cell membranes is suspect even in fresh preparations, and interpretation needs care.

Isolated perfused lung is used more frequently,[1121] with growing interest in its glucuronidation of possibly carcinogenic metabolites.

4. Use of Sliced Tissue

Sliced tissue, somewhat unfashionable, deserves its own rung in the ladder of techniques connecting the pure enzyme to whole animal. At optimal thickness — i.e., of the minimal proportion of damaged cells consistent with rapid diffusion of substrates and oxygen to the interior — cell-cell interaction and intracellular accumulation characteristics of organized tissue are available rapidly and simply; reproducibility and environmental control are little inferior to those of cell suspensions, provided the tissue is reasonably homogeneous (e.g., liver), or clearly zoned (e.g., kidney). Cold shock, osmotic shock, and overcrowding of slices must be avoided. Sliced liver and kidney have been frequently used for overall glucuronidation.[574] Lengths of rat or mouse intestine, treated as slices, give poor reproducibility, probably because they contract and exude mucus;[574] strips of intestinal mucosa are more active, are readily prepared from larger animals, but contain many damaged cells.

A useful procedure[286] is outlined in Section III.E.1 below; recent examples are where cell monolayers are used as "slices".[319,754]

5. Cell Suspensions

First used in glucuronidation by Henderson,[1006] they improve on slices in homogeneity and hence reproducibility; but their preparation involves enzymic digestion and activation of the transferase may occur (Chapter 15, Section II.A.). Examples of their use in glucuronidation as a link with Phase 1 systems cell suspensions are References 166, 278, and 1215. They glucuronidated naphthalene (as 1 naphthyl glucuronide) to the same degree as perfused liver.[278] If suspended in sterile nutrient media for several hours, they approach a cell suspension culture, and induction of transferase may be demonstrable.[1006,1122]

6. Culture Techniques

These techniques afford long-term study of regulation of transferase and associated enzymes, but require careful controls; specific examples are found in Chapter 15, Section II.

Their advantages are a long-term controllable environment, without the variable absorption, distribution, systemic toxicity, hormonal changes, or nutritional status of intact animal or tissue. They are particularly suited for induction studies. Induction can be demonstrated in perfused tissues, but, to be credible there, must be unmistakable increase within 6 hr, and whole organs are difficult to free of endogenous hormones in that time. In culture, the cells or tissues can be selected or combined artificially; e.g., adrenal cells can be cultured in contact with liver cells, the latter receiving products of the former. Advantages in rapid screening of drugs or carcinogens, especially for man, are evident.

However, the technique is essentially small-scale, for continuous culture of untransformed adult mammalian cells in a chemostat is not yet satisfactory. Also, several enzymes, e.g., glucose-6-phosphatase, fall progressively, indicating changes in the endoplasmic reticulum, and gradual shifts in metabolic pattern occur; selective enzyme loss through leakage may be considerable.[1253]

There are three main methods of culturing tissue: organ fragments, cell monolayers, and cell suspensions. Organ fragments used as explants are easily prepared, the tissue architecture and cell-cell interaction is preserved, there is no use of protease or collagenase to damage membranes, and no chance of transformation because cell division is minimal. However, difficulties ensue from irregular diffusion of gas or nutrients, central necrosis, and mechanically damaged periphery (the latter minimized in liver by snipping lobular edges, incurring only one injured surface). Inclusion of hormones can be minimized by preliminary perfusion of the organ. Monolayers and suspensions possess advantages of reproducible conditions per cell, washing away of extracellular hormones, availability of defined cell lines, and cellular division. Principal limitations include damage during preparation, especially with adult tissues, which require longer "softening-up", and potential transformation of the cell type: e.g., some hepatic epithelial cells retain many hepatic functions through several subcultures,[1124] but do not necessarily retain transferase activity. Suspensions could be developed to continuously monitor the appearance of glucuronides and sample for transferase and other enzymes.

Culture can be performed in vivo, e.g., in the anterior chamber of the rabbit eye, hamster cheek pouch, etc., but glucuronidation has only been studied on the chorioallantoic membrane of the egg.[711] Culture is more usually performed in dishes, flasks, on slides, or on inclined paper or membrane strips,[182,692] the latter holding promise for interaction of adjacent tissues.

The two main approaches for investigating glucuronidation have been (1) to reproduce the in vivo metabolism for drug screening and metabolite isolation; (2) to accentuate, controllably regulatory, responses less obvious in the intact animal and destroyed in homogenates (Chapter 15, Section II).

A very good critical review of general drug metabolism in cultured tissue exists.[1122]

7. Use of Homogenates and Microsomes

In these techniques, cofactors from broken cells are diluted or destroyed so that integrated metabolic pathways no longer exist, and cofactors are chosen to favor a particular pathway. For example, in the "classical" unfortified homogenate glucuronidation can only be studied by adding UDPGlcUA to a known concentration, and assayed on that basis. A 10% (or less) homogenate is recommended. The microsomal fraction can be washed free of most endogenous nucleotides but still contains UDPGlcUA pyrophosphatase and β-glucuronidase (Chapter 9); as its composition can vary with age (Chapter 10) and other factors, homogenates are better for screening.[180] Homogenization and microsome preparation can themselves activate the transferase (Chapter 5). The usual precautions, e.g., determination of linear reaction rate, initial, and maximum velocities, are needed with broken-cell preparations, the true V_{max} being especially important when the enzyme from different sources is being compared; e.g., from animals pretreated with different drugs or of different ages. Very lipid-soluble substrates may be sequestered in microsomal lipids,[263] and so rate of reaction does not depend on the concentration of substrate in the aqueous phase. Preincubation, suggested to overcome this problem,[263] brings about partial activation of the enzyme.

8. Use of Immobilized UDPGlucuronyltransferase

A column containing UDPglucuronyltransferase allows glucuronides to be obtained on a preparative scale free of protein; it can indicate if a drug gives glucuronides, or, when the fairly pure UDPglucuronyltransferase is bound, if a given compound is a substrate for it. One method[1254] entraps liver microsomes in polyvinylpyrrolidone gels by means of γ-irradiation; they contain the transferase, various Phase 1 activities, esterase, epoxide synthetase, and epoxide hydratase.[1254] Recovery was 20 to 40% of the activity in fresh microsomes, activity being stable for 2 months at 4° under nitrogen.[1254]

Rabbit-liver transferase was partially purified and after dialysis coupled to agarose by a cyanogen bromide method, at some 3 mg protein per mℓ packed gel; the time taken was 8 hr, and almost 100% of the activity in the coupling medium was immobilized. Half-life of the free enzyme at pH 8.0 and 4° was under 10 days, whereas when stored under the same condition immobilized on agarose, half-life was over 45 days.[349]

Solubilized and immobilized enzyme activities displayed similar pH optima and apparent K_m values for 4-nitrophenol, but different Lineweaver-Burk plots.[349]

The agarose-bound activity[1255] synthesized glucuronides of diethylstilbestrol, tetrahydrocannabinol, borneol, benzoate, bilirubin, meprobamate, and phenylthiol, products being characterized by chromatography and mass spectrometry. Drugs poorly soluble in water were added with ethanol, aqueous albumin, or propylene glycol, 20% aqueous ethanol inhibiting the transferase 50%; dioxane, dimethyl sulfoxide, and dimethylformamide were highly inhibitory.[349]

The transferase, considerably purified (to some 500-fold, including an activation step) was bound to copolymers of acrylamide, acrylic acid, or cyanogen bromide beads, along with cytochrome P-450 and NADPH-cytochrome c reductase[1256,1257] to give an extracorporeal detoxication cartridge through which blood could be passed. The transferase was up to 20-fold more stable than when free. UDPGlcUA and NADPH, when added, remained entrapped in the beads for about 1 hr these compounds could then be reloaded, but the technique is thus at present very expensive in prolonged use.

9. Some Recent Assay Procedures for UDPGlucuronyltransferase*

Selected popular assays are described and criticized in detail in Section III.

Benzo(a)pyrene metabolites — a method with tritium-labeled acceptor and C^{14} labeled donor substrate allows assay of their glucuronidation.[20]

Bilirubin — Lathe[461] has reviewed early work generally, and Heirwegh et al.[111] the assay itself; an exhaustive review by Heirwegh et al.,[460] describes various separations and analyses of diazopositive pigments, distinguishing between assaying bilirubin conjugation rates — where the azo colors are measured — and glycosyl transference — where the conjugated azopigments are isolated and quantitated. Much confusion has been caused by failure to realize that tetrapyrroles are split to dipyrrolic azopigments on diazotization, so that methods which diazotize before separation and assay cannot give information on relative amounts of mono- or diconjugates of bilirubin unless the conjugated and unconjugated dipyrrolic azopigments are subsequently separated and assayed. This important aspect is conveniently summarized.[678] Heirwegh et al. have also published a method for the direct separation of bilirubin conjugates.[473] A simplified needle-biopsy assay is described.[1258] One assay is detailed in Section III.C.

Morphine — Strickland et al.[1259] have described a colorimetric assay (see also Section III.C).

4-Nitrophenol — Two new radio-isotopic assays of increased sensitivity[1260,1261] and one directly measuring the appearance of the glucuronide[1262] have been published (see also Section III.C)

Tyramine — see Reference 422.

General — At least two methods have been reported for the assay of a wide range of solvent-extractable glucuronides using labeled UDPGlcUA. In these the glucuronide is separated either by solvents[133] or by column chromatoraphy on Amberlite®.[1264,1265] A useful review and modification[497] of previous assays needs reassessment in light of more recent work.

A continuous assay whereby the UDP produced is converted to UTP with phosphoenolpyruvate and pyruvate kinase and the resulting pyruvate then reduced to lactate with NADH and lactate dehydrogenase[402] is described in Section III.C.

10. Assays of UDPGlucose Dehydrogenase

The original assay method for UDPglucose dehydrogenase [517] has been only a little modified. A typical current assay is that employed by Gainey and Phelps[509] and a further modification, using stopped-flow techniques, is described by Gainey et al.[533]

Following the first histological demonstration of this enzyme,[1266,1267] another method, based on that of Stiller and Gorski,[1268] utilizes polyvinyl alcohol as gel former and phenazine methosulfate as an intermediary electron acceptor. It employs lowered NADH concentration and blocking by cyanide,[1269] and worked well in rat and rabbit liver and kidney, but not in hamster, where there seemed insufficient enzyme, or in mouse and guinea pig, where technical difficulties arose.[1269] The difficulties were probably those encountered[1270] in an unsuccessful attempt to apply the earlier method[1266] to chick embryo and mouse liver, using modifications[1271] limiting diffusion of this very soluble enzyme by an agar medium.

11. Assays of UDPGlucuronic Acid

There are no new methods of rapid routine assay; the most popular procedure is described in Section III.

12. Assays of Glucuronic Acid

Marsh[63] critically presented assays of glucuronic acid or uronic acid in glucuronides,[1272] total glucuronic acid in urine,[69,1273] plasma, and serum,[1274] by modifications of the naphthoresorcinol method; by the Fishman-Green[1274] hypoiodite method, which oxidizes the aldehydic group of free glucuronic acid without affecting conjugated glucuronic acid, and thus allows differential naphthoresorcinol-based estimation of total and conjugated (and hence free) acid; by the carbazole method for total uronic acid[1275] and, again after hypoiodite oxidation, for conjugated glucuronic acid.[1276] The carbazole reaction is more specific for uronic acids than is the naphthoresorcinol reaction and somewhat less liable to interference from other sugars. Ranges of these methods are some 10 to 80 μg, the accuracy some ±5%. One naphthoresorcinol procedure is a little more specific than usual,* and another modification has been reported.[1277]

A decarboxylation method[1278] quoted by Marsh[63] is more specific and accurate than these, but is tedious and requires much material (some 200 mg uronic acid).

In a different procedure, aldopentoses, aldohexoses, and hexuronic acids yield an orange color when added in aqueous solution to benzidine in acetic acid, measuring 50 to 600 μg hexuronic acid per milliliter;[1279] by a modification,[1280] a buffered solution allows direct spectrophotometric assay of glucuronic acid in the presence of the labile 'ester-type' glucuronides.[631]

A review[1281] of glucuronic and iduronic acid determination, chiefly in glycans, concludes that chromatographic methods are superior to colorimetric, and that gas-liquid chromatography (GLC) is fastest and most sensitive. Gas chromatography of these

---

* See also Chapter 7, Sections II. A, II. D.1, and II. D.2. and Reference 1219b.

epimeric acids as their volatile butaneboronates[1282,1283] has advantages over the acetates in shorter retention time and, over the trimethylsilylates, in good separation of simple alditols. Hexuronic acid in heparins is assayed by gas chromatography of their monomeric forms.[1284]

III. SELECTED PRACTICAL PROCEDURES

A. Introductory

The assays below are those commonly used in glucuronidation studies which have proved of value. Many require practical precautions not evident in the original publications. Tissue preparation was outlined in Section II above.

B. Direct Assay of UDPGlucuronic Acid in Tissues

1. General

Although chromatographic methods exist (Chapter 9, Section I.B), the following method is simple, rapid, and valuable for tissue comparisons and can be roughly quantitated. The protein from fresh tissue is precipitated by heat or denaturant, and the supernatant used directly, or after slight purification, as a UDPGlcUA source in a UDPglucuronyltransferase assay system. It is essential to remove the tissue rapidly after death and minimize exposure to undenatured protein. UDPGlcUA is thermostable at neutrality, but acid- and alkali-labile. In the first method below, the danger is enzymic destruction, in the second from acidic breakdown. Activators such as ATP and UDPGlcNAc that are present in tissues will raise apparent UDPGlcUA content of this crude extract by increasing the glucuronidation rate;[4] inhibitors can be detected by adding extract to a standard transferase determination. Controls for interference in the assay system itself should also be run: (1) in transferase assays employing diazotization (as for aminophenol or aminobenzoate), as many tissue extracts yield a weak background and (2) in assays measuring aglycon disappearance (as for 4-nitrophenol). PAPS in the extract may, especially if homogenate is present, remove aglycon by sulfate formation and so washed microsomes should be used. Sensitivity will be as high as permitted by the transferase assay employed.

2. Using Heat as Protein Denaturant[4,501]

Fresh tissue is rapidly scissored in ice to cubes of approximately 1 cm, weighed and dropped into boiling water (3 g in 6 m*l*). After 3 min at 100°, speed is no longer essential; the mixture is cooled, dispersed with a blender or a Teflon®-glass homogenizer, and centrifuged at low speed to sediment the coarse material, which is squeezed manually through clean gauze, the suspension is then centrifuged at 100,000 × g for 30 min to remove glycogen turbidity, and the clear supernatant used as the "UDPGlcUA" addition for a selected UDPglucuronyltransferase assay, e.g., 2-aminophenol,[500] or harmol or harmolol,[501] preferably with guinea pig liver microsomes (largely free of UDPGlcUA pyrophosphatase) as enzyme source. The final tissue:water ratio of the extract recommended[501] is 1:5 for guinea pig liver, 1:3 for guinea pig kidney, and 1:1 for other tissues; while 1:2 is more suitable for rat liver, where the nucleotide is low and pyrophosphatase active. Wong and Sourkes[501] added 0.6 m*l* of this extract to a final assay volume of 1 m*l*, containing 1 mg microsomal protein. Assay with harmalol or harmol is the more sensitive, Wong and Sourkes[501] claiming to detect 5 μg UDPglucuronic acid per gram tissue. Sensitivity is improved 25-fold[1285] with the radioactive assay for tyramine glucuronide.[422]

Bock et al.[180] report the following similar procedure, which they now prefer to the one described in Section III.B.3 below. Some 2 g liver was transferred to a preweighed homogenizer tube containing 5 m*l* hot distilled water, homogenized, and the mixture boiled for 3 min; after weighing and centriguation for 20 min at 100,000 × g, the supernatant was lyophilized, dissolved in 1 ml distilled water, and assayed as in Section III.B.3 below. Recovery of known amounts of UDPGlcUA was 65%.

3. Using Trichloroacetic Acid as Protein Denaturant

This procedure[117] appears more time-consuming and no more accurate or sensitive than the previous method (Section III.B.2 above).

Some 2 g tissue is rapidly removed to a preweighed homogenizer mortar containing 10 ml 0.3 M-glycine to 0.2 M-trichloroacetate buffer, pH 2.2, at 0°, and weighed by difference. After homogenization, (Teflon® pestle with a stainless steel shaft) and centrifugation (4°) at 100,000 × g for 10 min, 5 ml clear supernatant are removed and pH rapidly adjusted to 7.4. Six volumes of acetone at −20° are then added and after 30 min at −20° the precipitated UDPGlcUA is centrifuged down at 3000 × g for 5 min, washed with peroxide-free ether and recentrifuged; excess ether is removed *in vacuo.* The precipitate is dissolved in 1.0 ml 0.05 M-Tris-HCl buffer, pH 7.4; 0.5 ml of it was used[177] in a 1 ml total assay mixture containing 2 mg (solubilized) microsomal protein.

4. Using Acetone as UDPGlcUA Precipitant

Excess acetone at −20°C is added to supernatant from homogenate (see Reference 511a).

C. Practical Assay of UDPGlucuronyltransferase and Glucuronidation
1. General

Assay of UDPglucuronyltransferase measures glucuronidation due to added UDPGlcUA, i.e., is the difference between the plus and minus UDPGlcUA samples. Substrates disappearing, or conjugate appearing, in the absence of added UDPGlcUA should be minimal and, with dilute (<10% weight/volume) homogenates or washed microsomes, is usually <10% of that measured with UDPGlcUA present. When it exceeds 10%, or when glucuronidation need not be the only process occurring (as in all other tissue techniques), the conjugate must be identified as a β-D-glucuronide by complete hydrolysis with β-glucuronidase (preferably from rat preputial gland) at acid pH, and complete (i.e., >90%) inhibition of that hydrolysis by 100 mM-glucaro-1,4-lactone. Identification of liberated sugar as glucuronic acid by chromatographic behavior and formation of a lactone in acid chromatography solvents (galacturonic acid does not form a lactone) is also advised because β-glucuronidase possesses certain other hexuronidase activities.

Exploratory controls should check for the amount of substrate metabolized by enzymes other than UDPglucuronyltransferase; and for nonenzymic synthesis of glucuronide, always possible in the presence of, e.g., amino groups. Calibration curves are advised for all substrates.

Addition of activators, carrier protein, etc. have been discussed in previous chapters and are not usually included in descriptions of procedures below but should be added as desired. We find digitonin (1 to 3% wt/vol. in a cold-water stock homogenized suspension, added as 0.03 to 0.3% final concentration) suitable for all except estrogens, which form insoluble digitonides and for which we use Lubrol®[124-9] (200 to 400 µg/ml final concentration). We usually add MgCl₂ (final molarity 10 mM).

2. 2- or 4-Aminobenzoate as Substrate[4,329,497]

The method was originally described by Shirai and Ohkubo.[1286]

a. Principle

Aminobenzoates form ester-linked glucuroindes at the carboxyl group, leaving the amino group free for subsequent diazotization and assay. After incubation, unconjugated substrate is extracted with ether and polar glucuronide remains. A recent rapid procedure,[497] (1) diazotizes this conjugate and assays the chromogen. The earlier procedure, (2), recommended in exploratory work and essential when ± UDPGlcUA controls cannot be run, removes free substrate with ether and heats the conjugate in alkali to hydrolyze the ester link; the liberated aglycon is then extracted with ether, returned to alkaline aqueous solution, and assayed by diazotization. A shorter modification; (3), measures liberated aglycon directly after the alkaline hydrolysis.

b. Advantages

A method capable of assaying ester glucuronide synthesis by UDPglucuronyltransferase in broken-cell preparations (1), or by the glucuronidating system in intact cells, (2) and (3). Method (1) is rapid and the no-UDPGlcUA controls are virtually colorless.

c. Limitations

Sensitivity and reproducibility are not high in procedures (2) and (3) because of manipulation. As any alkali-labile link will be hydrolyzed, the presence in the conjugate of both D-glucuronic acid and a β-link must be established whenever whole cells are employed. With broken-cell preparations, an incubation of 30 min. to 1 hr is required.

d. Procedure

Substrate solution kept at $-20°$ is stable for weeks; storage under N_2 is not necessary.

1. **Rapid method.**[497] The reaction mixture contains 0.1 ml 0.25 M sodium phosphate, pH 7.6, 5 mM UDPglucuronic acid, 0.5 mM substrate, and enzyme (e.g., 2 mg microsomal protein) in a total of 0.8 ml.

   After incubation, samples of the mixture are added to 0.22 M trichloroacetic acid (1.5 volumes); after centrifugation, the supernatant is decanted into 0.3 ml 0.5 M-sodium phosphate, pH 3.9, extracted × 3 with an equal volume of ether and the ether layers discarded; 1.0 ml of 0.1 M-sodium phosphate, pH 1.0, is then added, followed by 0.1 ml each of 0.1% (wt/vol) NaNO$_2$, 1.0% ammonium sulfamate and 0.2% N-(1-naphthyl)ethylenediamine dihydrochloride, with precautions as when 2-aminophenol is substate (Section III.C.3 below). After 1 hr at room temperature color development is optimal, and optical density is determined at 550 to 555 nm.

   A modification (Burchell and Wishart, personal communication) conveniently utilizes many solutions prepared for assay of 2-aminophenol. The reaction mixture is as for 2-aminophenol, apart from the substrate itself, and precipitation is with 0.25 ml 0.5 M trichloracetic acid; 0.4 ml portions are then added to 0.3 ml 0.5 M phosphate, pH 3.9, and extracted × 3 with ether; to 0.5 ml samples is added 0.1 ml 2 M phosphate buffer at pH 2.25, followed by the diazotization reagents.

2. **Longer method.**[329,496] After incubation of 0.6 ml reaction mixture, add 1.4 ml water and 0.4 ml M trichloracetic acid. Centrifuge. To a 1.6 ml portion, add 0.28 ml of 0.5 M Na$_2$CO$_3$ and 0.12 ml water, checking that pH is 4; extract × 3 with 3 ml ether. To aqueous layer add 0.4 ml of M KOH, heat at 55°C for 1 hr. Bring to pH 4 with M HCl*; extract × 3 with 3 ml ether. Extract ethereal layer × 2 with 0.8 ml 0.1 M NaOH, and then add to the resulting aqueous layer 0.06 ml of 4 M HCl, checking that pH is below 3.0. Then add diazotization reagents as above.

3. A shorter version of 2 adds the diazotization reagents immediately after the first addition of M HCl (marked * above) to a pH below 3.0.

Paper chromatography of aminobenzoyl glucuronides has been described[329] and the method has been used with slices.[1082]

3. 2-Aminophenol as Substrate

The method was modified from an earlier procedure.[4,192,286,1287]

a. Principle

2-Aminophenol and its "ether" O-glucuronide are diazotized without being separated, but during colorimetric incubation selective conditions of pH and time[1287] ensure that only conjugated aglycon is assayed.

b. Advantages

The great advantage of this procedure is that it directly measures the glucuronide formed; it is rapid, reproducible, and reasonably specific. Sulfate gives no color, nor any conjugates (e.g., N-acetyl-2-aminophenol), with the amino group masked.

c. Limitations

Control of pH for color development is critical. Parallel use of authentic 2-aminophenyl glucuronide is essential until the technique is learned, and advisable for periodical checking thereafter. The rate or extent of color development may be slightly affected by many compounds added to the reaction mixture and calibration against a standard is recommended under new conditions (e.g., presence of detergents, etc.). The substrate readily oxidizes in neutral or alkaline solution to a yellow quinonoid compound, even under N_2. Ascorbate is therefore always added with it, but even then 2-aminophenol oxidizes on incubation in a recreation mixture if no tissue is present, the resulting quinone giving a pink color on subsequent acidification during assay. Such a "no-tissue" control is therefore not possible with 2-aminophenol. Concentrated tissue preparations (including boiled extracts) contain diazotizable substances; this small and controllable interference develops immediately on diazotization and so can be detected. All glycosides of 2-aminophenol (e.g., glucosides and xylosides) register similarly to the glucuronide, and must be checked for wherever ± UDPGlcUA controls are not possible.

d. Procedure

Commercial 2-aminophenol is toxic, gives a yellow solution, and must be purified by sublimation from a porcelain basin over a sand bath, a large round-bottomed flask filled with cold water serving as condenser. Two sublimations should produce a white or very pale pink condensate, preferable to the larger flakes because of readier solubility in water. The crystals should be stored as preweighed portions in stoppered vials, under N_2, at $-20°C$. When they are being dissolved in water, ascorbic acid must be present, heat should not be used, and vigorous shaking under N_2 is required. It is better if the stock solution is frozen before first use, ensuring the solution of even the flakiest crystal. The stock solution (5 mM 2-aminophenol, 10 mM ascorbic acid) is stable at least 5 months at $-20°$ if stored under N_2, and can be prepared for that period at one time, distributed among 2 to 3 ml vials. Accidental oxidation is readily seen by yellowing of the originally colorless or pale straw-colored thawed solution. Such oxidized solutions should not be used.

A typical reaction mixture contains 75 mM Tris-maleate buffer pH 7.4; 0.5 mM 2-aminophenol, 1 mM ascorbic acid, 2 to 5 mM UDPGlcUA, 200 to 400 μg microsomal protein, and water to 0.6 ml. Phosphate buffer is preferred by some[497] because of slight interference by Tris and similar amine-containing buffers in diazotization, but we have not found this serious. The pH optimum is subject to discussion and should be found experimentally (ca. pH 7.4). Incubation volumes down to 125 μl are readily processed. After incubation, add an equal volume of 1:1 (v/v) mixture of 2 M phosphate, pH 2.10, and 2 M trichloroacetate, pH 2.0; centrifuge, and take 0.8 ml supernatant. Alternatively, to 125 μl incubation volume, add an equal volume of 1 M trichloroacetic acid, centrifuge, and add 200 μl of the supernatant to 500 μl of 1.2 M phosphate (29.4 g orthophosphoric acid and 8.9 g KOH, diluted to 0.25l); this alternative allows more effective protein precipitation when large amounts of tissue have been used. The glucuronide is quite stable at acid pH in the cold, but if the solution is stored for long periods, darkness is advised as photochemical change of the conjugated aglycon occurs.

The diazotization reagents, 100 μl each of 0.2% $NaNO_2$, 1.0% ammonium sulfamate and 0.2% N-(1-naphthyl)ethylenediamine dihydrochloride (all w/v in water), are then added in that order, with extensive shaking and at least 3-min intervals between the additions; the nitrite requires time to react with the amino group (giving a straw yellow color) and the sulfamate requires time to destroy excess nitrite which otherwise will react with the dye to give a color. Failure to observe this simple precaution will result in erratic and spurious color development. Incubate at room temperature, for 1.5 hr in the dark. Read at 550 to 555 nm.

The pH for color development is critical and until experience has been gained a control of authentic 2-aminophenylglucuronide (Koch-Light Laboratories, Colnbrook, England) must be run at every determination, and is advised when kinetic experiments are being run from day to day. The pH in the color assay tubes must be 2.2 to 2.3, achieved by the above dilution of the phosphate buffer. Incorrect pH values are easily noted: at too low a pH, the controls containing 2-aminophenol only, are markedly purplish within an hour of incubation; in this event an immediate reading will minimize the error. At too high a pH, the initial clear straw-yellow of all tubes will either be replaced by, or rapidly become, a cloudy yellow-brown, with a pinkish tinge in "positive" tubes; in this case there is no way of saving the situation. A week or so of practice, however, ensures reproducible results with very useful assay, which is particularly applicable to sliced tissue and whole cells.

4. Bilirubin as Substrate

The method below (used in our laboratory by Dr. M. T. Campbell) is based on that developed by Heirwegh and his colleagues.[185,313] A detailed review of bilirubin assay methods has been published.[460] See limitations on assay methods in bile.[484] Wolkoff et al.[1284] describe the purification of the mono- and diglucuronides of bilirubin; Heirwegh (personal communication) warns against heavy metal ions complexing with and distorting azopigments during chromatographic separation.

a. Principle

Albumin-bound bilirubin is used as substrate, and digitonin is present to activate the enzyme. Diazotization splits the bilirubin molecule, free or conjugated, to two dipyrroles; the dipyrrole molecule containing the glycosyl residue preferentially develops color on diazotization at pH 2.7.

b. Advantages

Bilirubin is an endogenous substrate of great clinical importance, and this rapid assay gives reproducible results of sufficient sensitivity for routine clinical analysis, and for most research purposes. The same assay detects glucoside or xyloside conjugates if the appropriate sugar nucleotides are substituted for UDPGlcUA.

c. Limitations

1. The entire assay should be performed under dimmed light or, at the least, all solutions containing bilirubin should be shielded from direct sunlight or fluorescent light; photooxidation will introduce error.

2. The bilirubin:albumin ratio used gives a final molar ratio of 2:1 and maximum enzyme activity with albumin for several species (Campbell and Heirwegh, unpublished results). Increasing bilirubin concentration to 3:1, or even 4:1, does not markedly change activity but increasing albumin concentration progressively inhibits the enzyme, presumably by competition for binding. Omitting albumin completely lowers the enzyme activity 25 to 50%, possibly from precipitation of unbound bilirubin at the assay pH or from its readier oxidation. Effect on the albumin to bilirubin ratio of other untried additions to the assay system must be borne in mind.

3. Although the colorimetric procedure would measure both mono- and diconjugated bilirubin, the broken-cell in vitro reaction mixture forms virtually only monoconjugate. Man, rat, and several other species excrete bilirubin primarily as diconjugate, but as it is not yet certain if diglucuronide is formed sequentially, or how far monoglucuronidation is rate-limiting (Chapter 7, Section III.B), it is impossible to evaluate how far the in vitro assay underestimates the total bilirubin glucuronidation in these species.

4. Maximum reliable sensitivity of the colorimetric procedure described here is about 0.4 nmol conjugated bilirubin. Satisfactory for most studies, this limits studies of development or human hereditary transferase deficiency.

5. When enzyme activity is low and hemoglobin high (as in fetal liver), chromogenicity of hemoglobin in the diazotization procedure interferes significantly. When enzyme activity is low and tissue protein is high in the reaction mixture, sequestering of conjugate by the protein will prevent its complete assay colorimetrically.

d. Procedure

1. Preparation of bilirubin solution. Dissolve bilirubin (2.5 mg) in 250 $\mu\ell$ 0.1 M NaOH; add to 68 mg bovine serum albumin dissolved in 2.25 mℓ 40 mM Tris buffer at pH 7.4. This stock solution can be stored in the dark for 24 hr.

2. Incubation. The reaction mixture contains (as final concentrations) 50 mM Tris-HCl, pH 7.7, 7.5 mM MgCl$_2$, 0.34 mM bilirubin, 0.17 mM bovine serum albumin, 0.02% (w/v) digitonin suspended in 0.15 M KCl, 4 mM UDPGlcUA and 50 $\mu\ell$ of a 20% (w/v) homogenate or equivalent microsomes, in a total 250 $\mu\ell$.

   Preincubate for 5 min at 37° and start reaction by adding the UDPglucuronic acid. After 15 to 30 min shaking at 37° in the dark, add 1 mℓ glycine buffer (0.4 M HCl adjusted to pH 2.7 with solid glycine) and place on ice. Although tubes can be stored on ice in the dark for 2 hr with minimal loss in subsequent color development, it is best to perform the whole procedure at once.

3. Color development. The diazotization reagent is prepared according to Heirwegh et al.[486] Suspend 0.1 mℓ of ethyl anthranilate in 10 mℓ 0.15 M HCl, and add 0.3 mℓ of NaNO$_2$ (5 mg/mℓ water). Keep at room temperature for 5 min, then add 0.1 $\mu\ell$ ammonium sulfamate (10 mg/mℓ). Although Heirwegh et al.[486] recommend its use within 3 min, this solution is stable for several hours if kept cold in the dark. Loss of stability is indicated by a pinkish tinge.

   Add 0.5 mℓ of this solution to the reaction mixture from (2), mix well, and place in the dark at room temperature (20°) for 30 to 45 min, or in a 25° water bath for 30 min. Then add 0.1 mℓ 10% aqueous ascorbic acid, followed by 0.8

mℓ pentan-2-one. After vigorous shaking, chill, centrifuge, and read the (upper) organic phase at 546 nm. If high protein concentrations are present, the tubes should be frozen and thawed after addition of solvent. This assists sedimentation of the denatured protein and hence clarification of the organic phase following centrifugation.

5. Estradiol and Testosterone as Substrates

Early methods[155,355] assayed the conjugates themselves, but were time-consuming. Lucier[356] introduced a rapid, simple, radioactive assay, with the disadvantage of counting the disappearance of a small amount of substrate in the organic layer against a very high background. Rao et al.[1288] developed their early method into a convenient assay of the conjugate itself in the aqueous layer. The procedure below is a modification of this.[391]

a. Principle

Labeled steroid is incubated with UDPglucuronic acid. Unconjugated steroid is preferentially and wholly extracted into organic solvent, and labeled conjugate is counted in the aqueous layer.

b. Advantages

Steroids are important endogenous substrates, and the method is rapid and simple, with low background.

c. Limitations

It is assumed that the aglycon of the conjugate is in fact the substrate placed in the reaction mixture. The original designers of the method checked this for rat liver, but with new species and tissues, confirmatory chromatography would seem desirable.

d. Procedure

A 3 mℓ tube contains 75 mM Trismaleate buffer, pH 7.4, 0.8 mM estradiol or testosterone, 0.04 μCi 4^{14}-C steroid (ca. 100,000 cpm, dissolved in 10 $\mu\ell$ ethanol; propylene glycol[1288] is not necessary), 50 $\mu\ell$ of 10% (w/v) homogenate or equivalent microsomes, and 4 mM-UDPGlcUA in a total reaction volume of 250 $\mu\ell$. A preincubation of 5 min before addition of UDPGlcUA to remove the ethanol was originally performed, but is not necessary.

After 5 to 10 min incubation at 37°, add 2 mℓ dichloromethane. Boiling[1288] is not necessary; shake, spin, add 125 $\mu\ell$ of aqueous layer to 4 mℓ Aquasol® and count.

6. Harmol and Harmolol as Substrates

The method was first developed by Wong and Sourkes.[407,501,739,1199]

a. Principle

Both the substrate and its conjugates are fluorescent and are readily separable chromatographically. Quantitation is ensured by the elution and fluorimetric assay of the areas located by chromatography.

b. Advantages

A complete picture is readily obtained of the unchanged substrate and its conjugates. As conjugates other than glucuronide (e.g., glucoside, sulfate) can also be readily detected at the same time, this method is useful for comparing conjugatory pathways,[1199] especially as under some circumstances,[1199] harmol and harmolol are preferentially sulfated and glucuronidated, respectively. The assay is extremely sensitive.[739]

c. Limitations

The essential chromatographic step necessarily involves extra manipulation, with loss of time and increased sources of error. However, with preformed plastic-backed thin-layer chromatography (TLC) plates, the assays of a dozen or so reaction tubes can readily be performed in an afternoon. Harmalol is less good as a substrate than harmol; it tends to oxidize to harmol (the oxidation is chromatographically detectable, however) and gives less discrete chromatographic spots, free or conjugated.[501]

d. Procedure

Types of reaction mixtures contain (1). 0.5 M Tris buffer, pH 7.8, 0.4 mM harmol or harmalol, 2 mM UDPGlcUA (the 0.4 mM quoted[501] is too low) and 1 mg microsomal protein in 1 ml; (2).[1199] 0.75 M Tris buffer, pH 7.8, with 5 mM-MgCl$_2$, 0.15 mM harmalol or harmol, 1.5 mM UDPGlcUA, and 200 μg (activated) microsomal protein, all per ml of reaction mixture; (3).[739] 0.1 M Tris buffer pH 7.8 with 10 mM MgCl$_2$, 0.2 mM harmol, 1.25 mM UDPGlcUA and tissue in a total 0.25 ml.

After incubation, for procedures (1) and (2) add 0.2 ml and 0.11 ml, respectively, of 5% (w/v) ZnSO$_4$ solution; for (3), heat at 100° for 3 min. After centrifugation, portions of the supernatant are added to t.l.c. plates.

Procedure 1 — Add 15 μl to cellulose-coated (0.25 mm) plates and develop in 0.1 M HC1 at room temperature for 30 to 60 min. R$_f$ values are harmol 0.11, harmalol 0.21, harmol glucuronide 0.43, harmalol glucuronide, 0.55. Spots are located by UV light, removed while the medium is still moist, and eluted by overnight suspension in 3 ml 0.1 M HCl. After centrifugation, fluoescence is read at 420 nm (activation 320 nm) for harmol glucuronide, and at 505 nm (activation 370 nm) for harmalol glucuronide. Aglycons differed fluorimetrically from their glucuronides by less than 5 nm and could be measured similarly.[501]

Procedure 2 — Add 15 to 30 μl supernatant to plates of Kieselgel® 60F 254, develop 1.5 hr with chloroform:methanol:isopropanol:concentrated ammonia solution (90:10:95:5) at room temperature in the dark. R$_f$ values were harmol, 0.78; harmol glucuronide, 0; harmol sulfate, 0.38; harmalol, 0.42; harmolol glucuronide, 0; harmalol sulfate, 0.18. After localization in UV light (350 nm) spots are eluted in 3 ml 0.1 M HCl in the dark overnight. Mulder and Hagedoorn[1199] found elution of these compounds, in the above order, to be 82, 90. 87, 41, and 90%, respectively; recovery of harmalol sulfate was not determined, and the low value for harmalol was because of loss of fluorescence during chromatography in their solvent system, recoveries of fresh conjugate being 80%. These authors also found that conjugates of harmol had a fluorescence ratio to free harmol of 1.0:0.8; conjugates of harmalol had a ratio to free harmolol of 1.0:1.7.

Procedure 3 — Add 5 μl supernatant to plates of silica gel G (Merck), 250 μm thickness; develop for 8 cm in *n*-butanol:glacial acetic acid: water (4:1:5). R$_f$ values were 0.6 for harmol and 0.2 for its glucuronide. Sherer[739] scanned the plates directly, with 247 nm as activating, and 420 nm as emitting wavelength; he measured reasonably accurately down to 10 pmol of harmol, with further possible extensions of sensitivity.

We have found (Dutton, unpublished work) assays of harmol glucuronidation speeded by the use of plastic Eastman-Kodak Chromagram® strips precoated with silica gel. Wong[409] describes a microassay modification of the method, suitable for 10 mg of tissue, but Sherer's modification (3)[739] appears more suited for small-scale work.

7. 4-Methylumbelliferone as Substrate[4,923,1072,1289]

a. Principle

The substrate fluorescence is wholly quenched on conjugation. Unconjugated substrate is extracted with chloroform. The conjugate is hydrolyzed with β-glucuronidase, alkali is added, and the free aglycon is estimated fluorimetrically.

b. Advantages

Extremely sensitive. Aitio[1070] quotes 10 pmol of glucuronide detected.

c. Limitations

Unsuitable for large amounts of tissue, where quenching can be serious. All additions require checking for quenching. Trace positives require very careful control.

d. Procedure

Substrate solution stable at 2° for several weeks in dark. To 100 μl of 0.1 M phosphate pH 7.0, 1.6 mM UDPGlcUA and 0.1 mM-4-methylumbelliferone, add 25 μl tissue preparation.[1070] After 30 min, add 5 ml chloroform and 5 ml 0.05 M acetate, pH 5.0. Extract ×3 with 5 ml chloroform, add 0.5 ml aqueous phase to 1.5 ml 0.01 M acetate (Ph 5.0) containing 100 units β-glucuronidase (0.1 mg Type II from bovine liver, Sigma®), or a rat-preputial gland preparation. In controls, add 10 mM glucaro-1,4-lactone to specifically inhibit the β-glucuronidase. Incubate 30 min at 37°, add 1 mll of 1.6 M glycine, pH 10.3; measure fluorescence at 370-nm activation and 450-nm emission.

A fluorescence twice that in the "no-UDPGlcUA" blanks is the sensitivity limit.

8. Morphine as Substrate

The method is that described by del Villar et al.[365]

a. Principle

Labeled morphine is incubated with enzyme and UDPGlcUA. The conjugate is extracted into an aqueous layer and counted.

b. Advantages

Morphine is a substrate of pharmacological interest. The method is fairly rapid.

c. Limitations

We have not found this assay so convenient as similar ones for 1-naphthol and the steroids, owing to significant readings in the "no-UDPGlcUA" blanks. Use of detergents as activators may increase this drawback by making selective extraction of morphine and its polar metabolites more difficult. The concentration of UDPGlcUA used by del Villar et al.[365] incurs great expense.

d. Procedure

Reaction mixture contains 75 mM Tris-maleate, pH 7.4 (or 50 mm Tris-HCl, pH 8.0[365]), 2.4 mM (1.5 mM[365]) morphine sulfate, 0.2 μCi per ml reaction mixture of (^{14}C-N-methyl) morphine hydrochloride, UDPglucuronic acid 4 mM (20 mM[365]) and 50 μl homogenate (10% w/v) in a total 500 μl (or 8 mg microsomal protein in a total 2 ml[365]) reaction mixture. After 10 min at 37°, to each ml of reaction mixture, add 1 ml trichloroacetic acid (10% w/v), and centrifuge. Add 1 ml supernatant to 45-ml centrifuge tubes containing 0.2 ml 1 M NaOH, 0.3 ml water, 2.0 ml KH$_2$PO$_4$ (40% w/v) and 0.5 ml morphine sulfate (0.05% w/v) as carrier. Add 15 ml of ethylene chloride:amyl alcohol (7:3 v/v), and shake for 30 min. Re-extract the aqueous phase. Add 0.5 ml aqueous phase to 6 ml water and 10 ml Aquasol® and count. Recovery of morphine glucuronide is quoted as 100% by del Villar et al.,[365] and identification of morphine and its 3-glucuronide can be established by chromatography and hydrolysis by β-glucuronidase against the authentic conjugate.[365]

9. 1-Naphthol as Substrate

The method is based on that of Lucier et al.,[133] but counts the conjugate itself in the aqueous layer.[179]

a. Principle

Labeled 1-naphthol is conjugated, unconjugated aglycon removed into chloroform and conjugate measured by liquid scintillation in the aqueous phase.

b. Advantages

A very rapid reproducible assay.

c. Limitations

Only suitable for rapid assay when UDPGlcUA is employed with broken-cell preparations and other conjugatory pathways are operating at negligible level.

d. Procedure

Various modifications can be made. We have used the following: an incubation tube contains 0.5 mM 1-naphthol, dissolved in 0.3% (aqueous v/v) dimethylsulfoxide, 0.04 μCi[^{14}C] 1-napththol (ca. 100,000 cpm.), 4 mM UDPGlcUA, 75 mM Tris-maleate pH 7.4, and 25 μl of a 10% (w/v) homogenate, or equivalent microsomes, in a total 125 μl reaction mixture. Incubate at 37° for 5 min. Add 375 μl of 1:1 mixture of 0.50 M trichloracetic acid and 0.80 M glycine, pH 2.2. Add 2 ml chloroform, shake, spin, take 125 ml of the aqueous phase, add to 4 ml Aquasol® and count; extraction of conjugate is virtually complete. Dimethylsulfoxide appears at this concentration not to affect the reaction.

10. 4-Nitrophenol as Substrate

The method was originally devised by Isselbacher and Axelrod.[227,228,498,570,601]

a. Principle

The substrate is colored, its conjugates colorless. A simple difference therefore measures conjugate synthesis. The color is usually measured at highly alkaline pH, after the reaction.

b. Advantages

The method being very rapid, is suitable for kinetic studies, especially with a serial assay.[497] With broken-cell preparations, incubations of 5 to 10 mm are adequate.

c. Limitations

As in all "difference" assays, accurate technique is essential. The molar extinction coefficient varies with the pH value and the buffered or alkali solutions added for colorimetry require constant checking. 4-Nitrophenol displays various kinetic anomalies[633,1290] possibly from protein-binding; a substrate concentration:enzyme activity curve is desirable with each species and tissue studied. The assay is not recommended at a pH below 7.0[497] because of the liberation then of a nonprecipitable yellow substance from microsomes in the presence of 4-nitrophenol. Other sources of error have been noted,[497,1291] and the upper limit of substrate concentration is advised as 0.6 mM over pH range 7.0 to 8.0[497]

d. Procedure

A typical reaction mixture contains: phosphate (50 mM), pH 7.1,[497] or Tris-maleate (75 mm), pH 7.4;[192] 0.2 (0.6 mM^{497}) 4-nitrophenol; 1.0 mM (5 mM^{497}) UDPGlcUA;

and 0.5 to 1 mg microsomal protein; in a total 0.5 m*l*. After incubation, various procedures remove protein and develop the color. (1).[227] To 0.3 m*l* reaction mixture add 2 m*l* ethanol, centrifuge, and add an aliquot to 0.1 M NaOH. (2).[570] To 0.30 m*l* reaction mixture add 1.2 m*l* 0.1 M trichloracetic acid, centrifuge, and add to 0.03 m*l* 10 M KOH; another version[497] adds 0.1 m*l* reaction mixture to 2 m*l* 0.1 M trichloracetic acid, centrifuges, and decants into 0.05 m*l* 10 M KOH. (3).[395,601] To 0.3 m*l* reaction mixture, add 0.5 m*l* 0.2 M trichloroacetic acid, centrifuge; add 0.5 m*l* supernatant to 1.0 m*l* buffer pH 9.0 (containing 0.27 M glycine, 0.17 M NaHCO$_3$, and 0.12 M NaCl). Extinctions are read at 400 nm. Molar extinction coefficient at pH 9.0 and above 1S 18×10^{-3} cm^2.

A chromatographic modification[1072] and a radioisotopic method of greater sensitivity exist.[1260]

11. 4-Nitrothiophenol as Substrate

The method is a modification[395] of that used for β-glucuronidase.[601]

a. Principle

As for 4-nitrophenol.

b. Advantages

A simple assay for an *S*-glucuronide. The others (see Chapter 7, Section II.A) are longer and less easy to quantitate.[395]

c. Limitations

As for 4-nitrophenol. Also, 4 nitrothiophenol is readily oxidized and must be stored under N$_2$.

d. Procedure

As for (3). of Section III.C.10 above. Assay should be under N$_2$. Extinctions are read at 410 nm.

The conjugate is hydrolyzable by β-glucuronidase and the liberated aglycon is readily distinguished from 4-nitrophenol by the disappearance of the yellow color on addition of H$_2$O$_2$ to its alkaline solution.

12. Substrate Assay by the Mulder and Van Doorn Method

This method was recently published for a variety of phenols[402] and has been modified (Dr. B. Burchell, personal communication) to cater for a wide range of substrates with improved control.

a. Principle

The reactions involved are[402]

$$\text{R.OH} + \text{UDP-glucuronic acid} \rightarrow \text{R.O.glucuronide} + \text{UDP} \quad (1)$$

$$\text{UDP} + \text{phosphoenopyruvate} \rightarrow \text{UTP} + \text{pyruvate} \quad (2)$$

$$\text{Pyruvate} + \text{NADH} + \text{H}^+ \rightarrow \text{lactate} + \text{NAD}^+ \quad (3)$$

(Enzymes concerned are (1) UDPglucuronyltransferase, (2) pyruvate kinase, and (3) lactate dehydrogenase). The conversion of NADH to NAD$^+$ is followed spectrophotometrically.

b. Advantages

The assay is followed continuously, and the effect of additions noted immediately; reaction course can be inspected with coupled pen recorder.

Glucuronidation can be measured of any substrate which does not interfere with the coupling Reactions 2 and 3. Use of a dual beam recording spectrophotometer (see below) enables highly colored or turbid reaction mixtures to be used.

c. Limitations

The original unmodified procedure[402] was limited to clear, virtually colorless, reaction mixtures; controls for the effect of new substrates on Reactions 2 and 3 are essential, but were difficult to perform before the modification reported below. All constituents must be of highest purity. (Additional severe limitations have recently been reported: see Section IV. Additional Notes at the end of the chapter.)

d. Procedure

Components of the assay mixture:[402] 75 mM Tris-HCl buffer, pH 7.3; 5 mM MgCl$_2$; 0.02 mM phosphoenolpyruvate, monopotassium salt; 0.2 mM NADH; 25 μg pyruvate kinase (200 units/mg assayed at 25°, phosphoenolpyruvate as substrate); 2.5 μg lactate dehydrogenase (250 units/mg assayed at 25°, pyruvate as substrate); 1.5 mM UDPGlcUA; usually 0.3 m M acceptor substrate, dissolved in ethanol-water (1:1 by vol) to a final concentration of 5% by volume ethanol in the incubation medium; and some 150 μg microsomal protein per ml. With aminophenols, ascorbic acid (1.1 mM final) was also present, being added together with the substrates. Incubation was in a cuvette of 10 mm light path, or with certain substrates (such as 4-nitrophenol which absorbs markedly at 340 nm), of 5-mm light path. Extinction at 340 nm was measured continuously against water for some 6 min at 29°. Controls originally employed were "no-UDPglucuronic acid" and "no phosphoenolpyruvate".

In the modification, the procedure is facilitated and extended by use of the double-beam recording spectrophotometer of Haddock and Garland,[1292] which accurately measures optical density changes in highly colored or turbid solutions, so that the above procedure can be satisfactorily used for many other substrates than the original simple phenols; e.g., morphine, 1-naphthol, 4-aminobenzoate, and even bilirubin. Any color or turbidity changes are constantly controlled with the two curvettes, but the activity is only measured in one. The many controls essential with this complex coupled assay are now conveniently performed: e.g., with UDPGlcUA and aglycon but without phosphoenolpyruvate or NADH. Interference from nucleoside diphosphatase is minimized by addition of 100 mM-sodium azide.

D. Reverse Reaction[118,497]

1. Principle

The products of the forward reaction, UDP and a glucuronide (those of 4-nitrophenol, 1-naphthol, phenolphthalein and possibly estrone, are known to be suitable, while that of 2-aminophenol is not suitable), are incubated with the enzyme in the presence of glucaro-1,4-lactone to diminish hydrolysis of the glucuronides (though this is minimal).

2. Advantages

For investigative work; because the reverse reaction with 4-nitrophenol as aglycon measures appearance of product, low rates can be measured more accurately than with the forward reaction.[497]

3. Limitations

The rate is rapid in rat and beef liver microsomes, which break down UDPGlcUA by a side reaction. The rate soon declines in guinea pig liver microsomes which break down UDP rapidly.[497]

4. Procedure

UDP (5 mM), glucaro-1,4-lactone (5 mM), 4-nitrophenyl glucuronide (10 mM), buffer (sodium phosphate, 50 mM or Tris-maleate 100 mM), pH 7.1, are equilibrated at 37°, and then 1 mg microsomal protein is added per 0.5 mℓ reaction volume. Immediately, and 2, 4 and 6 min later, remove 0.1 mℓ aliquots into 2.0 mℓ 0.1 M trichloroacetic acid. Centrifuge, add supernatant to 0.5 mℓ 10 M KOH, mix and read at 400 nm.

E. Assay of Glucuronidation

1. Tissue Slices

The principal advantages, and limitations of the technique have been discussed in Section II.B.4 above.

a. Procedure

i. Apparatus

The following simple rapid procedure is excellent for cutting slices of adult liver or kidney with reproducible glucuronidating ability (± 10% variation on wet weight or protein basis over triplicate assays).

To the bottom of a 100 mℓ glass beaker is cemented a ground glass square 5 × 5 cm. A microscope slide is ground on one side. A thin commercial safety razor blade is washed in acetone to remove packaging wax.

ii. Tissue

With or without previous perfusion by Ringer solution, liver is freshly excised and immediately placed in Ringer solution at 30°. Osmotic shock must be avoided; we have not noted serious effects from thermal shock but we now keep the tissue at room temperature or slightly above; originally, the hollow beaker served for ice, which may still be useful with gut (see below).

iii. Slicing

A 5 mm cube of liver is cut out, blotted on filter paper and placed at the corner of the ground glass square nearest the operator's hand (e.g., lower right-hand corner if operator is right-handed). The microscope slide is laid, ground-glass downwards, on the tissue and very gently pressed as the razor is worked gently in through the tissue exactly as if cutting bread. The thickness of the slice should be such that any lettering on the blade can be read through both slide and slice. The blade is carried right through the tissue, the slide removed, and the slice floated off the blade into fresh Ringer. If the thickness is correct, the slice (when from unperfused liver) should be pink and should undulate as it sinks; too thick slices are stiff, too thin slices are almost colorless and clearly frayed. Blade, slide, and ground glass square should be washed and blotted before the next slice is cut. With practice, 20 or 30 slices can be obtained from one rat liver within 10 to 20 min; of these at least 75% should be usable. No more than three of such 5 × 5 mm slices should be placed in each 2 mℓ of medium, contained in a 25 mℓ flask. Larger flasks are used proportionately as scale is increased.
able. No more than three of such 5 × 5 mm slices should be placed in each 2 mℓ of medium, contained in a 25 mℓ flask. Larger flasks are used proportionately as scale is increased.

Certain tissues require different treatment. Skin (epidermis) is best peeled from a guinea pig or rabbit ear and the strips squared into slices. Gastrointestinal mucosa from guinea pig and larger animals are scraped by a blunt edge in sheets from opened up sections of the tract which have been stretched across the ground glass square; the sheets are then shaped into slices. Tract from smaller animals can be used entirely, and the sections opened up and cut to not more than 1 cm^2; however, much contraction and mucous secretion ensues and circulation of gas and medium about the tissue is impaired. Operations with gut are best conducted at 0°, because of possible damage by proteases when stripping the mucosa.

iv. Incubation

The incubation mixture is Krebs' bicarbonate Ringer, with $MgSO_4$ replaced by $MgCl_2$. With 2-aminophenol as substrate, a final concentration of 0.23 mM, with 1 mM ascorbic acid, has been used.[286] Incubation is under O_2:CO_2 (95:5 v/v) at 37° with shaking to ensure movement of slices through medium. Linear glucuronidation has been observed for as long as 2 hr, but fragmentation may ensue after 1 hr of vigorous shaking.

v. Assay

The slices and any fragments are removed and blotted. They may be dried at 110° and weighed; or homogenized, and protein content then measured by the usual methods. The amount of conjugate remaining in the slice is readily ascertained by homogenization of the washed slice in cold trichloroacetic acid and subsequent assay.

The conjugate in the medium is assayed, if of 2-aminophenol, by the usual method, protein being first precipitated by trichloroacetic acid. With "difference" assays, the conjugate should be identified as a β-glucuronide by specific enzymic hydrolysis.

IV. ADDITIONAL NOTES

Section II.B.2: High-performance liquid chromatography (h.p.l.c.) is now the method of choice for microassay of a variety of glucuronides, as in developmental studies on biopsy samples of human liver with the previously refractory substrate bilirubin;[1292a] it can also provide for small-scale preparation of glucuronide.[1292b] Other instructive articles on the application of h.p.l.c. to glucuonidation are References 1292c-e.

Section II.B.9: Fluorescent methods measuring transferase activity to 3-hydroxybenzo(a)pyrene have been reported [1292f, 1292g] picogram amounts of glucuronide being detected.[1292g]

Section II.B.10: A small-scale assay for UDPglucose dehydrogenase employed labeled glucose in UDPglucose and utilized 10$\mu\ell$ of a 10% (w/v) homogenate in 60$\mu\ell$ incubation mixture;[695] it hydrolysed the nucleotides enzymically and chromatographed them in ethanol:water (1:1), glucuronic acid remaining at the origin and glucose moving with the solvent front.[695]

Section III.C.12.c: This assay has been reported "useless" with rabbit liver,[170a] where the rapid breakdown of UDP causes gross overestimation of activity.

References
____

REFERENCES

1. **Dutton, G. J.**, Preface, in *Glucuronic Acid*, Dutton, G. J., Ed., Academic Press, New York, 1966, vii.
2. **Williams, R. T.**, Introduction: pathways of drug metabolism, *Handb. Exp. Pharmacol.*, 28 (2), 226, 1971.
3. **Smith, J. N.**, Comparative detoxication of invertebrates, in *Drug Metabolism from Microbe to Man*, Parke, D. V. and Smith, R. L., Eds., Taylor & Francis, London, 1977, 219.
4. **Dutton, G. J., Ed.**, The biosynthesis of glucuronides, in *Glucuronic Acid*, Academic Press, New York, 1966, 186.
5. **Williams, R. T.**, *Detoxication Mechanisms*, 2nd ed., Chapman & Hall, London, 1959, chap. 21.
6. **Robinson, D., Smith, J. N., and Williams, R. T.**, Studies in Detoxication. 52. The apparent dissociation constants of some glucuronides, mercapturic acids and related compounds, *Biochem. J.*, 55, 151, 1953.
7. **Smith, J. N. and Williams, R. T.**, The metabolism of phenacetin in the rabbit, *Biochem. J.*, 44, 239, 1949.
8. **Smith, R. L. and Williams, R. T.**, Implications of the conjugation of drugs and other exogenous compounds, in *Glucuronic Acid*, Dutton, G. J., Ed., Academic Press, New York, 1966, 457.
9. **Mason, H. S., North, J. C. and Vanneste, M.**, Microsomal mixed-function oxidations: the metabolism of xenobiotics, *Fed. Proc. Fed. Am. Soc. Exp. Biol.*, 24, 1172, 1965.
10. **Albert, A.**, *Selective Toxicity*, 5th ed., Chapman & Hall, London, 1973, chapter 15.
11. **Dutton, G. J.**, Detoxication in normal and damaged liver, in *Alcoholic Cirrhosis and other Toxic Hepatopathias*, Nordiska Bokhandelns Förlag, Stockholm, 1970, 174.
12. **Brodie, B. B. and Maickel, R. P.**, Comparative biochemistry of drug metabolism, in *Proc. 1st. Int. Pharmacology Meeting*, Vol. 6, Brodie, B. B. and Erdös, E. G., Eds., Pergamon Press, Oxford, 1962, 299.
13. **Mazur, A. and Shorr, E.**, The isolation of stilbestrol monoglucuronide from the urine of rabbits, *J. Biol. Chem.*, 144, 283, 1942.
14. **Dodgson, K. S., Garton, G. A., Stubbs, A. L., and Williams, R. T.**, Studies in detoxication. 15. On the glucuronides of stilboestrol, hexoestrol, and dineoestrol, *Biochem. J.*, 42, 357, 1948.
15. **Wilder Smith, A. E. and Williams, P. C.**, Urinary elimination of synthetic oestrogens and stilboestrol glucuronide in animals, *Biochem. J.*, 42, 253, 1948.
16. **Fevery, J., Campbell, M. T., Blanckaert, N., and Heirwegh, K. P. M.**, Conjugation of bilirubin IXα, in *Liver and Bile (Falk Symposium 23)*, Bianchi, L., Gerok, W., and Sickinger, K., Eds., MTP Press, Lancaster, England, 1977, 139.
17. **Bonnett, R., Davies, J. E., and Hursthouse, M. B.**, Structure of bilirubin, *Nature (London)*, 262, 326, 1976.
18. **Durston, W. E. and Ames, B. N.**, A simple method for the detection of mutagens in urine: studies with the carcinogen 2-acetylaminofluorene, *Proc. Natl. Acad. Sci. U.S.A.*, 71, 737, 1974.
19. **Commoner, B., Vithayathil, A. J., and Henry, J. I.**, Detection of metabolic carcinogen intermediates in urine of carcinogen-fed rats by means of bacterial mutagenesis, *Nature (London)*, 249, 850, 1974.
20. **Nemoto, N. and Gelboin, H. V.**, Enzymatic conjugation of benzo(α)pyrene oxides, phenols and dihydrodiols with UDP-glucuronic acid, *Biochem. Pharmacol.*, 25, 1221, 1976.
21. **Toivonen, L., Hänninen, O., and Hartiala, K.**, Drug metabolism in canine duodenum, *Ann. Med. Exp. Biol. Fenn.*, 51, 8, 1973.
22. **Smith, R. L.**, *The Excretory Function of Bile*, Chapman & Hall, London, 1973.
23. **Brewster, D., Jones, R. S., and Parke, D. V.**, The metabolism of cyclohexane-carboxylate in the rat, *Biochem. J.*, 164, 595, 1977.
24. **Levy, G. and Procknal, J. A.**, Drug biotransformation reactions in man. I. Mutual inhibition in glucuronide formation of salicylic acid and salicylamide in man, *J. Pharm. Sci.*, 57, 1330, 1968.
25. **Levy, G.**, Saturation of glucuronide formation in man and its clinical implications, *Chem. Biol. Interact.*, 3, 291, 1971.
26. **Levy, G., Tsuchiya, T., and Amsel, L. P.**, Limited capacity for salicyl glucuronide formation and its effect on the kinetics of salicylate elimination in man, *Clin. Pharmacol. Ther.*, 13, 258, 1972.
27. **Smith, J. N.**, The comparative metabolism of xenobiotics, *Adv. Comp. Physiol. Biochem.*, 3, 173, 1968.
28. **Baulieu, E. E., Corpéchot, C., Dray, F., Emilliozi, R., Lebeau, M. C., Mauvais-Jarvis, P., and Robel, P.**, An adrenal secreted 'androgen': dehydroisoandrosterone sulfate. Its metabolism and a tentative generalisation on the metabolism of other steroid conjugates in man, *Recent Prog. Horm. Res.*, 21, 411, 1965.

29. **Lebeau, M. C. and Baulieu, E. E.,** On the significance of the metabolism of steroid hormone conjugates, in *Metabolic Conjugation and Metabolic Hydrolysis,* Vol. 3, Fishman, W. H., Ed., Academic Press, New York, 1973, 151.

30. **Einarsson, K., Gustafsson, J. A., Ihre, T., and Ingelman-Sundberg, M.,** Specific metabolic pathways of steroid sulfates in human liver microsomes, *J. Clin. Endocrinol. Metab.,* 43, 56, 1976.

31. **Siiteri, P. K. and Lieberman, S.,** *In vivo* studies with radioactive steroid conjugates. I. The fate of randomly tritiated androsterone glucuronoside in humans, *Biochemistry,* 2, 1171, 1963.

32. **Underwood, R. H. and Tait, J. F.,** Purification, partial characterization and metabolism of an acid labile conjugate of aldosterone, *J. Clin. Endocrinol. Metab.,* 24, 1110, 1964.

33. **Kirdani, R. Y., Slaunwhite, Jr., W. R., and Sandberg, A. A.,** Studies on phenolic steroids in human subjects. XIV. Studies on the fate of injected 6,7,-H^3-oestriol-3-C^{14}-glucosiduronate, *J. Steroid Biochem.,* 1, 265, 1970.

34. **Kappas, A. and Granick, S.,** Steroid induction of porphyrin systhesis in liver cell cell culture. II. The effect of heme, uridine diphosphate glucuronic acid and inhibitors of nucleic acid and protein systhesis on the induction process, *J. Biol. Chem.,* 243, 346, 1968.

35. **Robel, P., Emiliozzi, R., and Baulieu, E. E.,** Studies on testosterone metabolism. V. Testosterone-^{3}H 17-glucuronide-^{14}C to 5β-androstane-3α, 17β-diol-^{3}H 17-glucuronide-^{14}C, the 'direct' 5β-metabolism of testosterone glucuronide, *J. Clin. Endocrinol. Metab.,* 27, 1290, 1967.

36. **Hobkirk, R. and Nilsen, M.,** Metabolism of estrone-3-glucosiduronate in the human female, *Steroids,* 15, 649, 1970.

37. **Robel, P., Emiliozzi, R., and Baulieu, E. E.,** Metabolism of dehydroisoandresterone glucuronide, *J. Biol. Chem.,* 241, 5879, 1966.

38. **Hobkirk, R., Green, R. N., Nilsen, M., and Jennings, B. A.,** Direct conversion of 17β-estradiol 3-glucosiduronate and 17β-estradiol 3-sulfate to their 17-keto forms by human kidney homogenates, *Can. J. Biochem.,* 52, 15, 1974.

39. **Matsui, M., Fukuo, A., Kimura, K., and Okada, M.,** Comparative studies on the metabolic hydrogenation of the ring A in testosterone and its conjugates by male rat liver *in vitro, Chem. Pharm. Bull.,* 20, 1913, 1972.

40. **Fukushima, D. K., Matsui M., Bradlow, H. L., and Hellman, L.,** Metabolism of testosterone *N*-acetylglucosaminide, *Steroids,* 19, 385, 1972.

41. **Williamson, D. G.,** Evidence for rabbit liver phenolic steroid 17α-dehydrogenase which is more active on estradiol 17α,3α-glucuronide than on free steroid, *Proc. Can. Fed. Biol. Soc.,* 14, 97, 1971.

42. **Hasnain, S. and Williamson, D. G.,** Purification of multiple forms of the soluble 17α-hydroxy steroid dehydrogenase of rabbit liver, *Biochem. J.,* 147, 457, 1975.

42a. **Hasnain, S. and Williamson, D. G.,** Properties of the multiple forms of the soluable 17α-hydroxy steroid dehydrogenase rabbit liver, *Biochem. J.,* 161, 269, 1977.

43. **Rao, G. S., Rao, M. L., Haueter, G., and Breuer, H.,** Steroid glucuronyltransferases. V. Formation and hydrolysis of oestrogen glucuronides by the liver, kidney and intestine of the pig, *Hoppe Seylers Z. Physiol. Chem.,* 355, 881, 1974.

44. **Honma, S. and Nambara, T.,** Evidence for O-methylation of catechol estrogen 2-glucuronoside with retention of conjugate in the rat, *Chem. Pharm. Bull.,* 22, 1952, 1974.

45. **Honma, S. and Nambara, T.,** Studies on steroid conjugates. XIV. Participation of glucuronidation in selective O-methylation of catechol estrogen in the rat, *Chem. Pharm. Bull.,* 23, 747, 1975.

46. **Ekstrom, B., Vorys, N., Neri, A., Taylor, J. N., and Wieland, R. G.,** Free and conjugated $C_{19}O_2$-steroids from spermatic vein blood, *Am. J. Med. Sci.,* 255, 202, 1968.

47. **Cottle, W. H. and Veress, A. T.,** Absorption of glucuronide conjugate of triiodothyronine, *Endocrinology,* 88, 522, 1971.

48. **Schulz, R. and Goldstein, A.,** Inactivity of narcotic glucuronides as analgesics and on guinea pig ileum, *J. Pharmacol. Exp. Ther.,* 183, 404, 1972.

49. **Yoshimura, H., Îda, S., Oguri, K., and Tsukamoto, H.,** Biochemical basis for analgesic activity of morphine-6-glucuronide. I. Penetration of morphine-6-glucuronide in the brain of rats, *Biochem. Pharmacol.,* 22, 1423, 1973.

50. **Miller, J. A. and Miller, E. C.,** The metabolic activation of carcinogenic aromatic amines and amides, *Prog. Exp. Tumor Res.,* 11, 273, 1969.

51. **Kadlubar, F. F., Miller, J. A., and Miller, E. C.,** Hepatic metabolism of *N*-hydroxy-*N*-methyl-4-aminobenzene and other *N*-hydroxyarylamines to reactive sulfuric acid esters, *Cancer Res.,* 36, 2350, 1976.

52. **Irving, C. C.,** Conjugates of *N*-hydroxy compounds, in *Metabolic Conjugation and Metabolic Hydrolysis,* Vol. 1, Fishman, W. H., Ed., Academic Press, New York, 1970, 53.

53. **Irving, C. C.,** Metabolic activation of *N*-hydroxy compounds by conjugation, *Xenobiotica,* 1, 387, 1971.

54. **Maher, V. M. and Reuter, M. A.,** Mutations and loss of transforming activity of DNA caused by the *O*-glucuronide conjugate of the carcinogen, *N*-hydroxy-2-aminofluorene, *Mutat. Res.,* 21, 63, 1973.

55. **Cardona, R. A. and King, C. M.,** Activation of the *O*-glucuronide of the carcinogen *N*-hydroxy-*N*-2-fluorenylacetamide by enzymatic deacetylation *in vitro*: formation of fluorenylamine-tRNA adducts, *Biochem. Pharmacol.,* 25, 1051, 1976.

56. **Irving, C. C.,** Influence of the aryl group on the reaction of glucuronides of *N*-arylacethydroxamic acids with polynucleotides, *Cancer Res.,* 37, 524, 1977.

57. **Mulder, G. J., Hinson, J. A., and Gillette, J. R.,** Generation of reactive metabolites of *N*-hydroxyphenacetin by glucuronidation and sulfation, *Biochem. Pharmacol.,* 26, 189, 1977.

58. **Mulder, G. J., Hinson, J. A., and Gillette, J. R.,** Conversion of the N-O-glucuronide and N-O-sulfate conjugates of *N*-hydroxyphenacetin to reactive intermediates, *Biochem. Pharmacol.,* 27, 1641, 1978.

59. **Bicker, U.,** Application of β-D-glucuronides and glucose together suggests a new direction for cancer chemotherapy, *Nature (London),* 252, 726, 1974.

60. **Allen, M. J., Boyland, E., Dukes, C. E., Horning, E. S., and Watson, J. G.,** Cancer of the urinary bladder induced in mice with metabolites of aromatic amines and tryptophan, *Br. J. Cancer,* 11, 212, 1957.

60a. **Mulder, G. J.,** Detoxification or toxification? Modification of the toxicity of foreign compounds by conjugation in the liver, *TIBS,* 4, 86, 1979.

60b. **Ashurst, S. W., Mehta, R., and Cohen, G. M.,** Importance of conjugation reactions in determining the qualitative nature of polycyclic aromatic hydrocarbon-DNA interactions, *Med. Biol.* 57, 313, 1979.

61. **Arcos, M. and Lieberman, S.,** 5-Pregnene-3β,20α-diol-3-sulfate-20-(2′-acetamido-2′-deoxy-α-D-glucoside) and 5-pregnene-3β,20α-diol-3,20-disulfate. Two novel urinary conjugates, *Biochemistry,* 6, 2032, 1967.

62. **Matsui, M. and Fukushima, D. K.,** On the configuration of naturally-occurring steroid *N*-acetylglucosaminides, *Biochemistry,* 8, 2997, 1969.

63. **Marsh, C. A.,** Chemistry of D-glucuronic acid and its glycosides, in *Glucuronic Acid,* Dutton, G. J., Ed., Academic Press, New York, 1966, 3.

63a. **Keglević, D.,** Glycosiduronic acids and related compounds, *Adv. Carbohydr. Chem. Biochem.,* 36, 57, 1979.

64. **Goodwin, B. L.,** *Handbook of Intermediary Metabolism of Aromatic Compounds,* Chapman & Hall, London, 1976.

65. **Porteous, J. W. and Williams, R. T.,** Studies in detoxication. 19. The metabolism of benzene. I. (a) The determination of phenol in urine with 2:6-dichloroquinonechloroimide. (b) The excretion of phenol, glucuronic acid and ethereal sulphate by rabbits receiving benzene and phenol. (c) Observations on the determination of catechol, quinol and muconic acid in urine, *Biochem. J.,* 44, 46, 1949.

66. **Katzman, P. A., Straw, R. F., Buehler, H. J., and Doisy, E. A.,** Hydrolysis of conjugated estrogens, *Recent Prog. Horm. Res.,* 9, 45, 1954.

67. **Wakabayishi, M., Wotiz, H. H., and Fishman, W. H.,** Action of β-glucuronidase on a steroid-enol-β-glucosiduronic acid, *Biochim. Biophys. Acta,* 48, 198, 1961.

68. **Matsui, M., Kawase, K., and Okada, M.,** Preparation and stability in solution of androstenedione 3-enol glucosiduronate, *Chem. Pharm. Bull.,* 22, 2530, 1974.

69. **Mead, J. A. R., Smith, J. N., and Williams, R. T.,** Studies in detoxication. 71. The metabolism of hydroxycoumarins, *Biochem. J.,* 68, 61, 1958.

70. **Schachter, D.,** The chemical estimation of acyl glucuronides and its application to studies on the metabolism of benzoate and salicylate in man, *J. Clin. Invest.,* 36, 297, 1957.

71. **Kadlubar, F. F., Miller, J. A., and Miller, E. C.,** Hepatic microsomal *N*-glucuronidation and nucleic and binding of *N*-hydroxy arylamines in relation to urinary bladder carcinogenesis, *Cancer Res.,* 37, 805, 1977.

72. **Walker, S. R. and Williams, R. T.,** Metabolites of sulphadimethoxine in the urine of various species, *Biochem. J.,* 115, 61, 1969.

73. **Diekmann, H. and Garbe, A.,** Species differences of the metabolism of the psychoactive compound EMD 16 923 and the detection of a new type of glucuronide in man, *Naunyn Schmiedebergs Arch. Pharmacol.,* 287, R77, 1975.

74. **Wakabayashi M.,** β-Glucuronidases in metabolic hydrolysis, in *Metabolic Conjugation and Metabolic Hydrolysis,* Vol. 2, Fishman, W. H., Ed., Academic Press, New York, 1970, 519.

75. **Bridges, J. W., Kibby, M. R., and Williams, R. T.,** The structure of the glucuronide of sulphadimethoxine formed in man, *Biochem. J.,* 96, 829, 1965.

76. **Haynes, L. J.,** Naturally occurring *C*-glycosyl compounds, *Adv. Carbohydr. Chem.,* 20, 357, 1965.

77. **Richter, W. J., Alt, K. O., Dieterle, W., Faigle, J. W., Kriemler, H.-P., Mory, H., and Winkler, T.**, C-Glucuronides, a novel type of drug metabolites, *Helv. Chim. Acta,* 58, 2512, 1975.

78. **Dieterle, W., Faigle, J. W., Früh, F., Mory, H., Theobald, W., Alt, K. O., and Richter, W. J.**, Metabolism of phenylbutazone in man, *Arzneim. Forsch.,* 26, 572, 1976.

79. **Dieterle, W., Faigle, J. W., Mory, Richter, W. J., and Theobald, W.**, Biotransformation and pharmacokinetics of sulfinpyrazone (Anturan®) in man, *Eur. J. Clin. Pharmacol.,* 9, 135, 1975.

80. **Jayle, M. F. and Pasqualini, J. R.**, Implication of conjugation of endogenous compounds — steroids and thyroxine, in *Glucuronic Acid,* Dutton, G. J., Ed., Academic Press, New York, 1966, 507.

81. **Artz, N. E. and Osman, E. M.**, *Biochemistry of Glucuronic Acid,* Academic Press, New York, 1950.

82. **Bray, H. G.**, D-glucuronic acid in metabolism, *Adv. Carbohydr. Chem.,* 8, 251, 1953.

83. **Teague, R. S.**, The conjugates of D-glucuronic acid of animal origin, *Adv. Carbohydr. Chem.,* 9, 185, 1954.

84. **Hemingway, A., Pryde, J., and Williams, R. T.**, XX. The biochemistry and physiology of glucuronic acid. V. The site and mechanism of the formation of conjugated glucuronic acid, *Biochem. J.,* 28, 136, 1934.

85. **Lipschitz, W. L. and Bueding, E.**, Mechanism of the biological formation of conjugated glucuronic acids, *J. Biol. Chem.,* 129, 333, 1939.

86. **Crépy, O.**, Étude *in vitro* de la glucuroconjugaison hépatique des oestrogènes naturels (oestriol, oestrone, alpha-oestradiol), *Arch. Sci. Physiol.,* 1, 427, 1947.

87. **Storey, I. D. E.**, The synthesis of glucuronides by tissue slices, *Biochem. J.,* 47, 212, 1950.

88. **Florkin, M., Crismer, R., Duchateau, G., and Houet, R.**, Sur les glucuronides et la β-glucuronidase, *Enzymologia,* 10, 220, 1942.

89. **Karunairatnam, M. C. and Levvy, G. A.**, The inhibition of β-glucuronidase by saccharic acid and the role of the enzyme in glucuronide synthesis, *Biochem. J.,* 44, 599, 1949.

90. **Dutton, G. J. and Storey, I. D. E.**, Uridine compounds in glucuronic acid metabolism. 1. The formation of glucuronides in liver suspensions, *Biochem. J.,* 57, 275, 1954.

91. **Dutton, G. J. and Storey, I. D. E.**, Glucuronide synthesis in liver homogenates, *Biochem. J.,* 48, XXIX, 1951.

92. **Dutton, G. J. and Storey, I. D. E.**, The isolation of a compound of uridine diphosphate and glucuronic acid from liver, *Biochem. J.,* 53, XXXVII, 1953.

93. **Storey, I. D. E. and Dutton, G. J.**, Uridine compounds in glucuronic acid metabolism. 2. The isolation and structure of 'uridine-diphosphate-glucuronic acid', *Biochem. J.,* 59, 279, 1955.

94. **Caputto, R. F., Leloir, L. F., Cardini, C. E., and Paladini, A. C.**, Isolation of the coenzyme of the galactose phosphate-glucose phosphate transformation, *J. Biol. Chem.,* 184, 333, 1950.

95. **Mills, G. T., Lochhead, A. C. and Smith, E. E. B.**, Uridine pyrophosphoglycosyl compounds and the formation of glucuronides by isolated enzyme systems, *Biochim. Biophys. Acta,* 27, 103, 1958.

96. **Goldberg, N. D.**, The Effects of Nucleotide Substrates containing 5-Fluorouracil and 6-Azauracil on Certain Kinetic Parameters of UPDG Dehydrogenase and UDP-Glucuronyltransferase, Ph.D. thesis, University of Wisconsin, Madison, 1963.

97. **Honjo, M., Furukawa, Y., Imai, K., Moriyama, H., and Tanaka, K.**, Synthesis of uridine diphosphate-glucuronic acid, *Chem. Pharm. Bull.,* 10, 225, 1962.

98. **Honjo, M., Furukawa, Y., and Kanai, Y.**, Synthesis of pseudouridine 5'-phosphate glucuronic acid, *Biochim. Biophys. Acta,* 91, 525, 1964.

99. **Dutton, G. J.**, Comparison of glucuronide synthesis in developing mammalian and avian liver, *Ann. N.Y. Acad. Sci.,* 111, 259, 1963.

100. **Nilsson, R., Petterson, E., and Dallner, G.**, Permeability of microsomal membranes isolated from rat liver, *J. Cell Biol.,* 56, 762, 1973.

101. **Winsnes, A.**, Kinetic properties of different forms of hepatic UDPglucuronyltransferase, *Biochim. Biophys. Acta,* 284, 394, 1972.

102. **Berry, C. and Hallinan, T.**, Summary of a novel, three-component regulatory model for uridine diphosphate glucuronyltransferase, *Biochem. Soc. Trans.,* 4, 652, 1976.

103. **Dallner, G., Behrens, N. H., Parodi, A. J., and Leloir, L. F.**, Subcellular distribution of dolichol phosphate, *FEBS Lett.,* 24, 315, 1972.

104. **Behrens, N. F., Parodi, A. J., Leloir, L. F., and Krisman, C. R.**, The role of dolichol monophosphate in sugar transfer, *Arch. Biochem. Biophys.,* 143, 375, 1971.

105. **Winses, A.**, Regulation of Hepatic UDP-Glucuronyltransferase, thesis, University of Oslo, Norway, 1973.

106. **Puhakainen, E. and Hänninen, O.**, Lipid acceptor in UDPglucuronic acid metabolism in rat liver microsomes, *FEBS Lett.,* 39, 144, 1974.

107. **Stern, N. and Tietz, A.**, Biosynthesis of glucuronyl diglyceride by a cell-free system obtained from a moderately halophilic-halotolerant bacterium, *FEBS Lett.,* 19, 217, 1971.

108. **Shaw, J. M. and Pieringer, R. A.**, Glucuronosyl diacylglycerol of *Pseudomonas diminuta* ATCC 11568, *J. Biol. Chem.*, 252, 4391, 1977.

109. **Turco, S. J. and Heath, E. C.**, Formation of an oligosaccharide-lipid containing GlcUA and GlcNAC in SV_{40}-transformed human lung fibroblasts, *Fed. Proc. Fed. Am. Soc. Exp. Biol.*, 35, 1374, (Abstr. 104), 1976.

110. **Hopwood, J. J. and Dorfman, A.** Isolation of lipid glucuronic-acid and *N*-acetylglucosamine derivatives from a rat fibrosarcoma, *Biochem. Biophys. Res. Commun.*, 75, 472, 1977.

111. **Heirwegh, K. P. M., Meuwissen, J. A. T. P., and Fevery, J.**, Critique of the assay and significance of bilirubin conjugation, *Adv. Clin. Chem.*, 16, 239, 1973.

112. **Heath, E. C. and Dingell, J. V.**, The interaction of foreign chemical compounds with the glucuronidation of estrogens in vitro, *Drug Metab. Dispos.*, 2, 556, 1974.

113. **Williamson, D. G., Polakova, A., and Layne, D. S.**, Estrogen glucosides and galactosides: formation by rabbit liver microsomes in vitro, *Biochem. Biophys. Res. Commun.*, 42, 1057, 1971.

114. **Dutton, G. J.**, The specificity of uridine diphosphate glucuronyltransferase — a practical problem, *Report of the 10th Anniversary Symposium on Glucuronic Acid*, Tokyo Biochemical Research Foundation, Tokyo, 1965, 37.

115. **Gregory, J. D. and Lipmann, F.**, The transfer of sulfate among phenolic compounds with 3',5'-diphosphoadenosine as coenzyme, *J. Biol. Chem.*, 229, 1081, 1957.

116. **Vessey, D. A. and Zakim, D.**, Regulation of microsomal enzymes by phospholipids. II. Activation of hepatic uridine diphosphate-glucuronyltransferase, *J. Biol. Chem.*, 246, 4649, 1971.

117. **Vessey, D. A. and Zakim, D.**, Regulation of microsomal enzymes by phospholipids. V. Kinetic studies of hepatic uridine diphosphate-glucuronyltransferase, *J. Biol. Chem.*, 247, 3023, 1972.

118. **Berry, C. and Hallinan, T.**, 'Coupled transglucuronidation': a new tool for studying the latency of UDP-glucuronyl transferase, *FEBS Lett.*, 42, 73, 1974.

119. **Fishman, W. H. and Green, S.**, Glucosiduronic acid synthesis by β-glucuronidase in a transfer reaction, *J. Am. Chem. Soc.*, 78, 880, 1956.

120. **Fishman, W. H. and Green, S.**, Enzymic catalysis of glucuronyl transfer, *J. Biol. Chem.*, 225, 435, 1957.

121. **Fishman, W. H.**, *Biochemistry of Glucuronic Acid; Lectures in Japan*, Faculty of Pharmaceutical Science, University of Tokyo, 1959.

122. **Yoshida, K. , Kato, K., and Tsukamoto, H.**, Metabolism of drugs. XLIV. Glucuronyl transfer reaction catalysed by β-glucuronidase by the use of ester glucuronides as substrates, *Chem. Pharm. Bull.*, 12, 656, 1964.

123. **Tsukamoto, H., Kato, K., Yoshida, K., and Tatsumi, K.**, Metabolism of drugs. XLIII. Comparison of glucuronyl transfer activity between β-glucuronidase and uridine diphosphate transglucuronylase, *Chem. Pharm. Bull.*, 12, 734, 1964.

124. **Hawley, D. E. and Marsh, C. A.**, Glucuronyltransferase activity of purified mammalian β-glucuronidase, *Proc. Aust. Biochem. Soc.*, 2, 7, 1969.

125. **Jansen, P. L. M., Chowdhury, J. R., Fischberg, E. B., and Arias, I. M.**, Enzymatic conversion of bilirubin monoglucuronide to diglucuronide by rat liver plasma membranes, *J. Biol. Chem.*, 252, 2710, 1977.

126. **Bridges, J. W. and Williams, R. T.**, *N*-Glucuronide formation *in vivo* and *in vitro*, *Biochem. J.*, 83, 27P, 1962.

127. **Tamura, S., Tomizawa, S., Tsutsumi, S., Tsugoro, S. M., Kizu, K., Ito, H , Nakai, K., Kita, S., and Masuda, M.**, Metabolism of glucuronic acid in fatigue due to physical exercise, *Jpn. J. Pharmacol.*, 16, 138, 1966.

127a. **Neimann, R. and Buddecke, E.**, Acceptor-specific glucuronyl transfer catalysed by β-glucuronidase, *Biochim. Biophys. Acta*, 567, 196, 1979.

128. **Aitio, A. and Vainio, H.**, UDPglucuronosyltransferase and mixed function oxidase activity in microsomes prepared by differential centrifugation and calcium aggregation, *Acta Pharmacol. Toxicol.*, 39, 554, 1976.

129. **Amar-Costesec, A., Beaufay, H., Wibo, M., Thinès-Sempoux, D., Feytmans, E., Robbi, M., and Berthet, J.**, Analytical study of microsomes and isolated subcellular membranes from rat liver. II. Preparation and composition of the microsomal fraction, *J. Cell Biol.*, 61, 201, 1974.

130. **Gram, T. E., Hansen, A. R., and Fouts, J. R.**, The submicrosomal distribution of hepatic uridine diphosphate glucuronyltransferases in the rabbit, *Biochem. J.*, 106, 587, 1968.

131. **Eriksson, L. C.**, Studies on the biogenesis of endoplasmic reticulum in the liver cell, *Acta Pathol. Microbiol. Scand., Sect. A*, Suppl. 239, 1973.

132. **Bergman, A. and Dallner, G.**, Properties of a rat liver smooth microsomal subfraction not aggregated by Mg^{++}, *Life Sci.*, 18, 1083, 1976.

133. **Lucier, G. W., McDaniel, O. S., and Matthews, H. B.,** Microsomal rat liver UDPglucuronyltransferase: effects of piperonyl butoxide and other factors on enzyme activity, *Arch. Biochem. Biophys.,* 145, 520, 1971.

134. **von Bahr, C. E., Hietanen, E., and Glaumann, H.,** Oxidation and glucuronidation of certain drugs in various subcellular fractions of rat liver: binding of desmethylimipramine and hexobarbital to cytochrome P-450 and oxidation and glucuronidation of desmethylimipramine, aminopyrine, *p*-nitrophenol and 1-naphthol, *Acta Pharmacol. Toxicol.,* 31, 107, 1972.

135. **Nyqvist, S. E. and Morré, D. J.,** Distribution of UDP-glucuronyl transferase among cell fractions of rat liver, *J. Cell. Physiol.,* 78, 9, 1971.

135a. **Thaler, M. M., Erickson, R. P., and Pelger, A.,** Gentically determined abnormalities of microsomal enzymes in liver of mutant newborn mice, *Biochem. Biophys. Res. Commun.,* 72, 1244, 1976.

136. **Halac, E., Dipiazza, M., and Detwiler, P.,** The formation of bilirubin mono- and diglucuronide by rat liver microsomal preparations, *Biochim. Biophys. Acta,* 279, 544, 1972.

137. **Gregory, D. H. and Strickland, R. D.,** Solubilization and characterization of hepatic bilirubin UDPglucuronyltransferase, *Biochim. Biophys. Acta,* 327, 36, 1973.

138. **Rao, G. S., Haueter, G., Rao, M. L., and Breuer, H.,** Steroid glucuronyltransferases of rat liver. Properties of estrone and testosterone glucuronyltransferases and the effect of ovariectomy, castration and administration of steroids on the enzymes, *Biochem. J.,* 162, 545, 1977.

139. **Mulder, G. J.,** The effect of phenobarbital on the submicrosomal distribution of uridine diphosphate glucuronyltransferase from rat liver, *Biochem. J.,* 117, 319, 1970.

140. **Rane, A., von Bahr, C., Orrenius, S., and Sjöqvist, F.,** Drug metabolism in the human fetus, in *Fetal Pharmacology,* Boréus, L., Ed., Raven Press, New York, 1973, 287.

141. **Ackermann, E., Rane, A., and Ericsson, J. L. E.,** The liver microsomal mono-oxygenase system in the human fetus: distribution in different centrifugal fractions, *Clin. Pharmacol. Ther.,* 13, 652, 1972.

142. **Chatterjee, J. B., Price, Z. H., and McKee, R. W.,** Biosynthesis of L-ascorbic acid in different subcellular fractions of prenatal and postnatal rat liver, *Nature (London),* 207, 1168, 1965.

143. **Banjo, A. O. and Nemeth, A. M.,** Proliferation of endoplasmic reticulum with its enzyme, UDPglucuronyltransferase, in chick embryo liver during culture. Effects of phenobarbital, *J. Cell Biol.,* 70, 319, 1976.

144. **DePierre, J. W. and Dallner, G.,** Structural aspects of the membrane of the endoplasmic reticulum, *Biochim. Biophys. Acta,* 415, 411, 1975.

145. **Beaufay, H., Amar-Costesec, A., Feytmans, E., Thinès-Sempoux, D., Wibo, M., Robbi, M., and Berthet, J.,** Analytical study of microsomes and isolated subcellular membranes from rat liver. I. Biochemical methods, *J. Cell Biol.,* 61, 188, 1974.

145a. **Wishart, G. J., Campbell, M. T., and Dutton, G. J.,** Functional heterogeneity of UDP-glucuronyltransferase, in *Conjugation Reactions in Drug Biotransformation,* Aitio, A., Ed., Elsevier, Amsterdam, 1978, 179.

146. **Fry, D. J. and Wishart, G. J.,** Apparent induction by phenobarbital of uridine diphosphate glucuronyltransferase activity in nuclear envelopes of embryonic chick liver, *Biochem. Soc. Trans.,* 4, 265, 1976.

147. **Wishart, G. J. and Fry, D. J.,** Uridine diphosphate glucuronyltransferase activity in nuclei and nuclear envelopes of rat liver and its apparent induction by phenobarbital, *Biochem. Soc. Trans.,* 5, 705, 1977.

148. **Fry, D. J.,** Isolation of nuclear envelope from whole tissue, in *Subnuclear Components,* Birnie, G. D., Ed., Butterworths, London, 1976, 59.

149. **Gorski, J. P.,** Rat Liver UDP-Glucuronosyltransferase: Phospholipid Dependence, Purification and Biochemical Characterization, Ph.D. thesis, University of Wisconsin, Madison, 1975.

150. **Gorski, J. P. and Kasper, C. B.,** Purification and properties of microsomal UDP-glucuronosyltransferase from rat liver, *J. Biol. Chem.,* 252, 1336, 1977.

151. **Kashnig, D. M. and Kasper, C. B.,** Isolation, morphology and composition of the nuclear membrane from rat liver, *J. Biol. Chem.,* 244, 3786, 1969.

152. **Grube, E., Götze, W., Rao, G. S., Rao, M. L., and Breuer, H.,** Steroidglucuronyltransferasen. I. Morphologische Untersuchungen und Bestimmung der Östradiol-17β-3-Glucuronyltransferase in subzellulären Fraktionen des Schweinedünndarmes, *Hoppe Seylers Z. Physiol. Chem.,* 352, 1215, 1971.

153. **Schumacher, R., Rao, G. S., Rao, M. L., and Breuer, H.,** Steroidglucuronyltransferases. III. Oestradiol 17-β 3-glucuronyltransferase of the mitochondria of pig intestine, *Hoppe Seylers Z. Physiol. Chem.,* 353, 1784, 1972.

154. **Rao, G. S., Schumacher, R., Rao, M. L., and Breuer, H.,** Steroidglucuronyltransferases. IV. Localization of the oestradiol-17β 3-glucuronyltransferase in the outer membrane of the mitochondria of pig intestine, *Hoppe Seylers Z. Physiol. Chem.,* 353, 1789, 1972.

155. **Rao, G. S., Rao, M. L., and Breuer, H.,** Partial purification and kinetics of oestriol 16α-glucuron-yltransferase from the cytosol of human liver, *Biochem. J.,* 118, 625, 1970.

155a. **Mäntylä, E., Suolinna, E.-M., Ahotuka, M., and Aitio, A.,** UDP-glucuronsyltransferase activity of nonparachymal hepatic cells in the rat, in *Conjugation Reactions in Drug Biotransformation,* Aitio, A., Ed., Elsevier, Amsterdam, 1978, 49.

156. **Nilsson, O. S. and Dallner, G.,** Enzyme and phospholipid asymmetry in liver microsomal membranes, *J. Cell Biol.,* 72, 568, 1977.

157. **Hänninen, O. and Alanen, K.,** The competitive inhibition of p-nitrophenyl β-D-glucopyranosiduronic acid synthesis by aliphatic alcohols *in vitro, Biochem. Pharmacol.,* 15, 1465, 1966.

158. **Vainio, H.,** Activation and inactivation of membrane-bound UDP-glucuronyltransferase by organic solvents *in vitro, Acta Pharmacol. Toxicol.,* 34, 152, 1974.

159. **Illing, H. P. A. and Benford, D.,** Observations on the accessibility of acceptor substrate to the active center of UDP-glucuronosyltransferase *in vitro, Biochem. Biophys. Acta,* 429, 768, 1976.

160. **Hänninen, O. and Puukka, R.,** The effect of digitonin on UDP-glucuronyltransferase in microsomal membranes, *Suom. Kemistil., B.,* 43, 451, 1970.

161. **Graham, A. B., Pechey, D. T., Wood, G. C., and Woodcock, B. G.,** The role of microsomal membrane in control of uridine-diphosphate glucuronyltransferase activity *in vitro* and *in vivo, Biochem. Soc. Trans.,* 2, 1167, 1974.

162. **Wilkinson, J. and Hallinan, T.,** Trypsin-susceptibility of UDP-glucuronyltransferase, *FEBS Lett.,* 75, 138, 1977.

163. **Zakim, D. and Vessey, D. A.,** The effects of lipid-protein interactions on the kinetic parameters of microsomal UDP-glucuronyltransferase, in *The Enzymes of Biological Membranes,* Vol. 2, Marto-nosi, A., Ed., Plenum Press, New York, 1976, 443.

163a. **Wishart, G. J. and Fry, D. J.,** Activation characteristics of UDP-glucuronosyltransferase in nuclear and microsomal membranes of rat liver, in *Conjugation Reactions in Drug Biotransformation,* Aitio, A., Ed., Elsevier, Amsterdam, 1978, 498.

163b. **Gessner, T. and Drasner, J.,** Nuclear UDP-glucuronyltransferase activity, *Pharmacologist,* 19, 191, 1977.

164. **Berry, C., Stellon, A., and Hallinan, T.,** Guinea pig microsomal UDP-glucuronyltransferase: com-partmented or phospholipid-constrained?, *Biochim. Biophys. Acta,* 403, 335, 1975.

165. **von Bahr, C. and Borgå, O.,** Uptake, metabolism and excretion of desmethylimipramine in the iso-lated perfused rat liver, *Acta Pharmacol. Toxicol.,* 29, 359, 1971.

166. **Moldéus, P., Vadi, H., and Berggren, M.,** Oxidative and conjugative metabolism of p-nitroanisole and p-nitrophenol in isolated rat liver cells, *Acta Pharmacol. Toxicol.,* 39, 17, 1976.

167. **Vainio, H.,** Drug hydroxylation and glucuronidation in liver microsomes of phenobarbital-treated rats, *Xenobiotica,* 3, 715, 1973.

167a. **Hallinan, T.,** Comparison of compartmented or of conformational phospholipid-constraint models for the intramembraneous arrangement of UDP-glucuronyltransferase, in *Conjugation Reactions in Drug Biotransformation,* Aitio, A., Ed., Elsevier, Amsterdam, 1978, 257.

167b. **Vessey, D. A. and Zakim, D.,** Are glucuronidation reactions compartmented?, in *Conjugation Re-actions in Drug Biotransformation,* Aitio, A., Ed., Elsevier, Amsterdam, 1978, 247.

168. **Vainio, H.,** Linkage of microsomal drug oxidation and glucuronidation, *Proc. 6th Int. Congr. Phar-macol.,* 6, 53, 1975.

169. **Vessey, D. A. and Zakim, D.,** Membrane fluidity and the regulation of membrane-bound enzymes, *Horizons Biochem. Biophys.,* 1, 138, 1974.

170. **Zakim, D. and Vessey, D. A.,** Membrane-bound estrone as substrate for microsomal UDP glucuron-yltransferase, *J. Biol. Chem.,* 252, 7534, 1977.

170a. **Finch, S. A. E., Slater, T., and Stier, A.,** Nucleotide metabolism by microsomal UDP-glucuronyl-transferase and nucleoside diphosphatase as determined by ^{31}P nuclear-magnetic resonance spectro-scopy, *Biochem. J.,* 177, 925, 1979.

170b. **Hallinan, T., Pohl, K. R. E., and de Brito, R.,** Studies on the inhibition of hepatic microsomal glucuronidation by uridine nucleotides or adenosine triphosphate, *Med. Biol.,* 57, 269, 1979.

171. **Dutton, G. J. and Burchell, B.,** Activation and induction of glucuronidating enzymes during devel-opment, *Biochem. Soc. Trans.,* 2, 1176, 1974.

172. **Lueders, K. K. and Kuff, E. L.,** Spontaneous and detergent activation of a glucuronyltransferase *in vitro, Arch. Biochem. Biophys.,* 120, 198, 1967.

173. **Winsnes, A. and Dutton, G. J.,** Comparison between o-aminophenol glucuronidation in liver slices and homogenates from control and phenobarbital-treated Wistar and Gunn rats, *Biochem. Pharma-col.,* 22, 1765, 1973.

174. **Hamada, N. and Gessner, T.,** Effect of 3-methylcholanthrene (3MC) on glucuronide synthesis in perfused rat liver, *Proc. Am. Assoc. Cancer Res.,* 14, 62, 1973.

175. **Hamada, N. and Gessner, T.**, Effect of 3-methylcholanthrene pretreatment on glucuronidation and sulfation in perfused rat liver, *Drug Metab. Dispos.*, 3, 407, 1975.
176. **Bock, K. W. and Fröhling, W.**, UDP-glucuronyltransferase activity in isolated perfused rat liver, *Naunyn Schmiedebergs Arch. Pharmakol.*, 277, 103, 1973.
177. **Bock, K. W. and White, I. N. H.**, UDP-glucuronyltransferase in perfused rat liver and in microsomes: influence of phenobarbital and 3-methylcholanthrene, *Eur. J. Biochem.*, 46, 451, 1974.
178. **Bock, K. W.**, Oxidation of barbiturates and the glucuronidation of 1-naphthol in perfused rat liver and in microsomes, *Naunyn Schmiedebergs Arch. Pharmakol.*, 283, 319, 1974.
179. **Otani, G., Abou-el-Makarem, M. M., and Bock, K. W.**, UDP-glucuronyltransferase in perfused rat liver and in microsomes. III. Effects of galactosamine and carbon tetrachloride on the glucuronidation of 1-naphthol and bilirubin, *Biochem. Pharmacol.*, 25, 1293, 1976.
180. **Bock, K. W., Huber, E., and Schlote, W.**, UDP-glucuronyltransferase in perfused rat liver and in microsomes. Effects of CCl_4 injury, *Naunyn Schmiedebergs Arch. Pharmakol.*, 296, 199, 1977.
181. **Abou-el-Makarem, M. M. and Bock, K. W.**, UDP-glucuronyltransferase in perfused rat liver and microsomes. Glucuronidation of bilirubin, *Eur. J. Biochem.*, 62, 411, 1976.
181a. **Orrenius, S., Andersson, B., Jernström, B., and Moldéus, P.**, Isolated hepatocytes as an experimental tool in the study of drug conjugation reactions, in *Conjugation Reactions in Drug Biotransformation*, Aitio, A., Elsevier, Amsterdam, 1978, 273.
182. **Dutton, G. J., Wishart, G. J., Leakey, J. E. A., and Goheer, M. A.**, Conjugation with glucuronic acid and other sugars, in *Drug Metabolism — from Microbe to Man*, Parke, D. V. and Smith, R. L., Eds., Taylor & Francis, London, 1976, 71.
183. **Burchell, B., Leakey, J., and Dutton, G. J.**, Relationship between activation of 'detoxicating' enzymes in stored broken-cell preparations and in autolysing liver, *Enzyme*, 20, 156, 1975.
184. **Halac, E. and Reff, A.**, Studies on bilirubin UDP-glucuronyltransferase, *Biochim. Biophys. Acta*, 139, 328, 1967.
185. **Heirwegh, K. P. M., Van der Vijver, M., and Fevery, J.**, Assay and properties of digitonin-activated bilirubin uridine diphosphate glucuronyltransferase from rat liver, *Biochem. J.*, 129, 605, 1972.
186. **Winsnes, A.**, Inhibition of hepatic UDPglucuronyltransferase by nucleotides, *Biochim. Biophys. Acta*, 289, 88, 1972.
187. **Zakim, D. and Vessey, D. A.**, Membrane dependence of uridine diphosphate glucuronyltransferase, *Biochem. Soc. Trans.*, 2, 1165, 1974.
188. **Keppler, D. O. R., Rudigier, J. F. M., Bischoff, E., and Decker, K. F. A.**, The trapping of uridine phosphates by D-galactosamine, D-glucosamine, and 2-deoxy-D-galactose — a study on the mechanism of galactosamine hepatitis, *Eur. J. Biochem.*, 17, 246, 1970.
189. **Keppler, D., Rudigier, J., and Decker, K.**, Enzymic determination of uracil nucleotides in tissues, *Anal. Biochem.*, 28, 105, 1970.
190. **Vessey, D. A., Goldenberg, J., and Zakim, D.**, Kinetic properties of microsomal UDP-glucuronyltransferase. Evidence for cooperative kinetics and activation by UDP-*N*-acetylglucosamine, *Biochim. Biophys. Acta*, 309, 58, 1973.
191. **Zakim, D., Goldenberg, J., and Vessey, D. A.**, Influence of membrane lipids on the regulatory properties of UDP-glucuronyltransferase, *Eur. J. Biochem.*, 38, 59, 1973.
192. **Winses, A.**, Studies on the activation *in vitro* of glucuronyltransferase, *Biochim. Biophys. Acta*, 191, 279, 1969.
193. **Burchell, B., Dutton, G. J., and Winsnes, A.**, Comparison of culture-induced, phenobarbital-induced and naturally-developing UDP-glucuronyltransferase, *Enzyme*, 17, 146, 1974.
194. **Henderson, P. Th. and Dewaide, J. H.**, Metabolism of drugs in isolated rat hepatocytes, *Biochem. Pharmacol.*, 18, 2087, 1969.
195. **Henderson, P. Th.**, Activation *in vitro* of rat hepatic UDP-glucuronyltransferase by ultrasound, *Life Sci.*, 9, 511, 1970.
196. **Burchell, B.**, Observations of uridine diphosphate glucuronyltransferase activity towards oestriol and xenobiotics in developing and cultured tissues from mouse and man, *Biochem. Soc. Trans.*, 1, 1212, 1973.
197. **Burchell, B.**, UDP-glucuronyltransferase activity towards oestriol in fresh and cultured foetal tissues from man and other species, *J. Steroid Biochem.*, 5, 261, 1974.
198. **Zakim, D. and Vessey, D. A.**, The effect of a temperature-induced change within membrane lipids on the regulatory properties of microsomal uridine diphosphate glucuronyltransferase, *J. Biol. Chem.*, 250, 342, 1975.
199. **Eletr, S., Zakim, D., and Vessey, D. A.**, A spin-label study of the role of phospholipids in the regulation of membrane-bound microsomal enzymes, *J. Mol. Biol.*, 78, 351, 1973.
200. **Pechey, D. T., Graham, A. B., and Wood, G. C.**, The phospholipid dependence of uridine diphosphate glucuronyltransferase. Temperature-dependence of microsomal enzyme activity and thermotropic changes in membrane structure, *Biochem. J.*, 175, 115, 1978.

201. **Duppel, W. and Ullrich, V.**, Membrane effects on drug monooxygenation activity in hepatic microsomes, *Biochim. Biophys. Acta*, 426, 399, 1976.
202. **Norling, A. and Hänninen, O.**, Activation energy of glucuronide biosynthesis, *Biochim. Biophys. Acta*, 350, 183, 1974.
203. **Dutton, G. J. and Burchell, B.**, Newer aspects of glucuronidation, *Prog. Drug. Metab.*, 2, 1, 1977.
204. **Chien, K. R. and Farber, J. L.**, Microsomal membrane dysfunction in ischemic rat liver cells, *Arch. Biochem. Biophys.*, 180, 191, 1977.
205. **Marzella, L. and Glaumann, H.**, Effect of storage and in vitro ischemia on the ultrastructure of microsomal membranes and on microsomal enzymes, *Virchows Arch. B.*, 22, 1, 1976.
206. **Illing, H. P. A. and Dutton, G. J.**, Some properties of the UDP-glucuronyltransferase activity synthesizing thio-β-D-glucuronides, *Biochem. J.*, 131, 139, 1973.
207. **White, A. E.**, The distribution of glucuronyl transferase in cell membranes, in *Bilirubin Metabolism*, Bouchier, I. A. T. and Billing, B. H., Eds., Blackwell Scientific, Oxford, 1967, 183.
208. **Graham, A. B. and Wood, G. C.**, Factors affecting the response of microsomal UDP-glucuronyltransferase to membrane perturbants, *Biochim. Biophys. Acta*, 311, 45, 1973.
209. **Vainio, H. and Hietanen, E.**, Induction deficiency of the microsomal UDPglucuronosyltransferase by 3-methylcholanthrene in Gunn rats, *Biochim. Biophys. Acta*, 362, 92, 1974.
210. **Graham, A. B. and Wood, G. C.**, Studies of the activation of UDP-glucuronyltransferase, *Biochim. Biophys. Acta*, 276, 392, 1972.
211. **Vainio, H. and Hänninen, O.**, Effect of surfactants on drug metabolism in hepatic microsomes, *Suom. Kemistil. B.*, 45, 56, 1972.
212. **Winsnes, A.**, Age and sex dependent variability of the activation characteristics of UDP-glucuronyltransferase in vitro, *Biochem. Pharmacol.*, 20, 1249, 1971.
213. **Jacobson, M. M., Levin, W., and Conney, A. H.**, Studies on bilirubin and steroid glucuronidation by rat liver microsomes, *Biochem. Pharmacol.*, 24, 655, 1975.
214. **Zakim, D. and Vessey, D. A.**, Regulation of microsomal enzymes by phospholipids. IX. Production of uniquely modified forms of microsomal UDP-glucuronyltransferase by treatment with phospholipase A and detergents, *Biochim. Biophys. Acta*, 410, 61, 1975.
215. **Somani, S. M. and Anderson, J. H.**, Microsomal glucuronidation of a quaternary ammonium compound, 3 hydroxyphenyltrimethylammonium, *Drug Metab. Dispos.*, 5, 15, 1977.
216. **Aitio, A.**, Effect of chrysene and carbon tetrachloride administration on rat hepatic microsomal mono-oxygenase and UDPglucuronosyltransferase activity, *FEBS Lett.*, 42, 46, 1974.
217. **Vainio, H.**, Action of chaotropic agents on drug-metabolizing enzymes in hepatic microsomes, *Biochim. Biophys. Acta*, 307, 152, 1973.
218. **Hatefi, Y. and Hanstein, W. G.**, Solubilization of particulate proteins and nonelectrolytes by chaotropic agents, *Proc. Natl. Acad. Sci. U.S.A.*, 62, 1129, 1969.
219. **Notten, W. R. F. and Henderson, P. Th.**, Action of n-alkanes on drug-metabolizing enzymes from guinea-pig liver, *Biochem. Pharmacol.*, 24, 1093, 1975.
220. **Notten, W. R. F. and Henderson, P. Th.**, The influence of n-hexane treatment on the glucuronic acid pathway and activity of some drug-metabolizing enzymes in guinea pig, *Biochem. Pharmacol.*, 24, 127, 1975.
221. **Järvisalo, J., Kipiö, J., Elovaara, E., and Vainio, H.**, Deleterious effects of subacute carbon disulphide exposure on mouse liver, *Biochem. Pharmacol.*, 26, 1521, 1977.
222. **Järvisalo, J., Savolainen, H., Elovaara, E., and Vainio, H.**, The in vivo toxicity of CS_2 to liver microsomes: binding of labeled CS_2 and changes of the microsomal enzyme activities, *Acta Pharmacol. Toxicol.*, 40, 329, 1977.
222a. **Lalani, E. M. A. and Burchell, B.**, Stimulation of detective Gunn-rat liver uridine diphosphate glucuronyltransferase activity in vitro by alkyl ketones, *Biochem. J.*, 177, 993, 1979.
223. **Pogell, B. M. and Leloir, L. F.**, Nucleotide activation of liver microsomal glucuronidation, *J. Biol. Chem.*, 236, 293, 1961.
224. **Zakim, D., Goldenberg, J., and Vessey, D. A.**, Kinetic properties of microsomal UDP-glucuronyltransferase. Regulation by metal ions, *Biochim. Biophys. Res. Commun.*, 52, 453, 1973.
225. **Wong, K. P. and Lau, Y. K.**, Assay of UDPglucuronic acid pyrophosphatase and its relation to transglucuronidation, *Biochim. Biophys. Acta*, 220, 61, 1970.
226. **Halac, E. and Bonevardi, E.**, Solubilization and activation of liver UDPglucuronyltransferase by EDTA, *Biochim. Biophys. Acta*, 67, 498, 1963.
227. **Isselbacher, K. J., Chrabas, M. F., and Quinn, R. C.**, The solubilisation and partial purification of a glucuronyl transferase from rabbit liver microsomes, *J. Biol. Chem.*, 237, 3033, 1962.
228. **Storey, I. D. E.**, The inhibition of uridine diphosphate transglucuronylase activity of mouse liver, *Biochem. J.*, 95, 201, 1965.
229. **Winsnes, A.**, The effects of sulfhydryl reacting agents on hepatic UDP-glucuronyltransferase *in vitro*, *Biochim. Biophys. Acta*, 242, 549, 1971.

230. **Zakim, D. and Vessey, D. A.**, Regulation of microsomal enzymes by phospholipids. III. The role of −SH groups in the regulation of microsomal UDP-glucuronyltransferase, *Arch. Biochem. Biophys.*, 148, 97, 1972.

231. **Fuchs, M., Rao, G. S., Rao, M. L., and Breuer, H.**, Studies on the properties of an enzyme forming the glucuronide of pregnanediol and the pattern of development of steroid glucuronyltransferases in rat liver, *J. Steroid Biochem.*, 7, 235, 1977.

232. **Rao, G. S., Rao, M. L., and Breuer, H.**, Investigations on the kinetic properties of estrone glucuronyltransferase from pig kidney, *Biochim. Biophys. Acta*, 452, 89, 1976.

233. **Stevenson, I. H. and Greenwood, D. T.**, Inhibition of hexobarbital metabolism by diethylnitrosamine, *Biochem. Pharmacol.*, 17, 842, 1968.

234. **Stevenson, I. H., Greenwood, D. T. and McEwen, J.**, Hepatic UDP-glucuronyltransferase in Wistar and Gunn rats — *in vitro* activation by diethylnitrosamine, *Biochem. Biophys. Res. Commun.*, 32, 866, 1968.

235. **Zakim, D., Goldenberg, J., and Vessey, D. A.**, Regulation of microsomal enzymes by phospholipids. VI. Abnormal enzyme-lipid interaction in liver microsomes from Gunn rats, *Biochim. Biophys. Acta*, 297, 497, 1973.

236. **Nakata, D., Zakim, D., and Vessey, D. A.**, Defective function of a microsomal UDP-glucuronyltransferase in Gunn rats, *Proc. Natl. Acad. Sci. U.S.A.*, 73, 289, 1976.

237. **Mowat, A. P. and Arias, I. M.**, Observations of the effect of diethylnitrosamine on glucuronide formation, *Biochim. Biophys. Acta*, 212, 175, 1970.

238. **Nakata, D., Zakim, D. A., and Vessey, D. A.**, Modification of protein-lipid interactions in the Gunn rat by treatment of microsomal UDP-glucuronyltransferase with diethylnitrosamine, *Biochem. Pharmacol.*, 24, 1823, 1975.

239. **Burchell, B.**, Studies on the purification of rat liver uridine diphosphate glucuronyltransferase, *Biochem. J.*, 161, 543, 1977.

240. **Burchell, B.**, Purification of UDP-glucuronyltransferase from untreated rat liver, *FEBS Lett.*, 78, 101, 1977.

241. **Burchell, B.**, Substrate specificity and properties of uridine diphosphate glucuronyltransferase purified to apparent homogeneity from phenobarbital-treated rat liver, *Biochem. J.*, 173, 749, 1978.

242. **Burchell, B. and Hallinan, T.**, Phospholipid content and activity of pure uridine diphosphate glucuronyltransferase from rat liver, *Biochem. J.*, 171, 821, 1978.

243. **Puukka R., Laitinen, M., Vainio, H., and Hänninen, O.**, Hepatic UDPglucuronosyltransferase. Partial purification after 3-methylcholanthrene pretreatment of the rat, *Int. J. Biochem.*, 6, 267, 1975.

244. **Marniemi, J.**, On the Control of Glucuronide Biosynthesis in Hepatic Microsomes and Different Tissues of the Rat, Ph.D. thesis, University of Turku, Turku, Finland, 1974.

245. **Hietanen, E. and Vainio, H.**, Effect of administration route of DDT on acute toxicity and on drug biotransformation in various rodents, *Arch. Environ. Contam. Toxicol.*, 4, 201, 1976.

246. **Hallinan, T.**, Lipid effects on enzyme activities: phospholipid dependence and phospholipid constraint, *Biochem. Soc. Trans.*, 2, 817, 1974.

247. **Fleischer, S., Brierley, G., Klowen, H., and Slautterback, D. B.**, Studies on the electron transfer system. XLVII. The role of phospholipids in electron transfer, *J. Biol. Chem.*, 237, 3264, 1962.

248. **Berry, C.**, Effect of Membrane Environment upon the Activity of Guinea Pig Liver Microsomal UDP-Glucuronyltransferase and Nucleoside Diphosphatase, Ph.D. thesis, University of London, 1975.

249. **Tomlinson, G. A. and Yaffe, S. J.**, The formation of bilirubin and *p*-nitrophenyl glucuronides by rabbit liver, *Biochem. J.*, 99, 507, 1966.

250. **Graham, A. B. and Wood, G. C.**, The phospholipid-dependence of UDP-glucuronyltransferase, *Biochem. Biophys. Res. Commun.*, 37, 567, 1969.

251. **Attwood, D., Wood, G. C., and Graham, A. B.** The phospholipid-dependence of uridine diphosphate glucuronyltransferase. Reactivation of the phospholipase A-inactivated enzyme by phospholipids and detergents, *Biochem. J.*, 123, 875, 1971.

252. **Hänninen, O. and Puukka, R.**, Activation of microsomal UDP glucuronyltransferase by phospholipases, *Chem. Biol. Interact.*, 3, 282, 1971.

253. **Vessey, D. A. and Zakim, D.**, Regulation of microsomal enzymes by phospholipids. IV. Species differences in the properties of microsomal UDP-glucuronyltransferase, *Biochim. Biophys. Acta*, 268, 61, 1972.

254. **Berry, C., Allistone, J., and Hallinan, T.**, Phospholipid dependence of UDP-glucuronyltransferase, *Biochim. Biophys. Acta*, 507, 198, 1978.

255. **Graham, A. B. and Wood, G. C.**, On the activation of microsomal UDPglucuronyltransferase by phospholipase A, *Biochim. Biophys. Acta*, 370, 431, 1974.

256. **Graham, A. B., Pechey, D. T., Toogood, K. C., Thomas, S. B., and Wood, G. C.**, The phospholipid dependence of uridine diphosphate glucuronyltransferase, *Biochem. J.*, 163, 117, 1977.

257. **Jansen, P. L. M. and Arias, I. M.**, Delipidation and reactivation of UDPglucuronyltransferase from rat liver, *Biochim. Biophys. Acta*, 391, 28, 1975.

258. **Rao, G. S., Rao, M. L., and Breuer, H.**, Studies on a solubilized oestriol 16α-glucuronyltransferase from human liver microsomes, *J. Steroid Biochem.*, 3, 1, 1972.

259. **Berry, C., Caldecourt, M., and Hallinan, T.**, Effect on uridine diphosphate glucuronyltransferase activity of depleting and restoring phospholipids to guinea pig liver microsomal preparations, *Biochem. J.*, 154, 783, 1976.

260. **Allistone, J., Caldecourt, M., Berry, C., and Hallinan, T.**, Reactivation of guinea pig liver UDP-glucuronyltransferase with monoacyl phosphatidylcholine after extensive phospholipid-depletion at 5° using pure *B. cereus* phospholipase C, *Biochem. Soc. Trans.*, 5, 297, 1977.

261. **Erickson, R. H., Zakim, D., and Vessey, D. A.**, Preparation and properties of a phospholipid-free form of microsomal UDP-glucuronyltransferase, *Biochemistry*, 17, 3706, 1978.

261a. **Gorski, J. P. and Kasper, C. B.**, UDP-Glucuronyltransferase: phospholipid dependence and properties of the reconstituted apoenzyme, *Biochemistry*, 17, 4600, 1978.

262. **Bock, K. W., von Clausbruch, U. C., Josting, D., and Ottenwälder, H.**, Separation and partial purification of two differentially-inducible UDP-glucuronyltransferases from rat liver, *Biochem. Pharmacol.*, 26, 1097, 1977.

263. **Zakim, D. and Vessey, D. A.**, The properties of uridine diphosphate glucuronyltransferase(s) which catalyse the synthesis of steroid glucuronides in microsomal fractions from guinea-pig liver, *Biochem. J.*, 157, 667, 1976.

264. **Labow, R. S., Williamson, D. G., and Layne, D. S.**, Effect of proteases and of crude phospholipases on steroid glycosyltransferases from rabbit liver, *Biochemistry*, 12, 1548, 1973.

265. **Puukka, R. and Hanninen, O.**, Effect of fatty acids on UDPglucuronyltransferase activity in the rat liver microsomes, *Int. J. Biochem.*, 3, 357, 1972.

266. **Hargreaves, T.**, Breast milk jaundice, *Br. Med. J.*, 3, 647, 1970.

267. **Hargreaves, T. and Piper, R. F.**, Breast milk jaundice. Effect of inhibitory breast milk and 3α,20β-pregnanediol on glucuronyl transferase, *Arch. Dis. Child.*, 46, 195, 1971.

268. **Hargreaves, T.**, Effect of fatty acids on bilirubin conjugation, *Arch. Dis. Child.*, 48, 446, 1973.

269. **Di Augustine, R. P. and Fouts, J. R.**, The effects of unsaturated fatty acids on hepatic microsomal drug metabolism and cytochrome P-450, *Biochem. J.*, 115, 547, 1969.

270. **Levillain, P., Odièvre, M., Luzeau, R., and Lemonnier, A.**, Possibilities d'inhibition de la bilirubine en fonction de la teneur en acides gras libres du lait maternel, *Biochim. Biophys. Acta*, 264, 538, 1972.

271. **Tam, B. K. and McCay, P. B.**, Reduced triphosphopyridine nucleotide oxidase-catalysed alterations of membrane phospholipids. III. Transient formation of phospholipid peroxides, *J. Biol. Chem.*, 245, 2295, 1970.

272. **Hogberg, J., Bergstrand, A., and Jakobsson, S. V.** Lipid peroxidation of rat-liver microsomes. Its effect on the microsomal membrane and some membrane-bound microsomal enzymes, *Eur. J. Biochem.*, 37, 51, 1973.

273. **Adlard, B. P. F. and Lathe, G. H.**, The effect of steroids and nucleotides on solubilized bilirubin uridine diphosphate-glucuronyltransferase, *Biochem. J.*, 119, 437, 1970.

274. **Black, M., Billing, B. H., and Heirwegh, K. P. M.**, Determination of bilirubin UDP-glucuronyl transferase activity in needle-biopsy specimens of human liver, *Clin. Chim. Acta*, 29, 27, 1970.

275. **Strebel, L. and Odell, G. B.**, Bilirubin uridine diphospho-glucuronyltransferase in rat liver microsomes: genetic variation and maturation, *Pediatr. Res.*, 5, 548, 1971.

276. **Vessey, D. A., Goldenberg, J., and Zakim, D.**, Differentiation of homologous forms of hepatic microsomal UDP-glucuronyltransferase. II. Characterization of the bilirubin conjugating form, *Biochim. Biophys. Acta*, 309, 75, 1973.

277. **Smith, H. G., Smith, W. R. D., Jepson, J. B., and Sorensen, K.**, The metabolism and excretion of indolylacrylic acid in the rat, *Biochem. Pharmacol.*, 19, 1689, 1970.

278. **Bock, K. W., van Ackeren, G., Lorch, F., and Birke, F. W.**, Metabolism of naphthalene to naphthalene dihydrodiol glucuronide in isolated hepatocytes and in liver microsomes, *Biochem. Pharmacol.*, 25, 2351, 1976.

279. **Wilkinson, J. and Hallinan, T.**, Stimulation of steroid glucuronidation by uridine diphosphate N-acetylglucosamine and by microsomal disruption, *Biochem. J.*, 168, 125, 1977.

280. **Zakim, D. and Vessey, D. A.**, Regulation of microsomal UDP-glucuronyltransferase — mechanism of activation by UDP-N-acetylglucosamine, *Biochem. Pharmacol.*, 26, 129, 1977.

281. **Zakim, D., Goldenberg, J., and Vessey, D. A.**, Effects of metals on the properties of hepatic microsomal uridine diphosphate glucuronyltransferase, *Biochemistry*, 12, 4068, 1973.

281a. **Berry, C. S.**, Critical evaluation of UDP-N-acetylglucosamine and product glucuronides as allosteric effectors of UDP-glucuronyltransferase, in *Conjugation Reactions in Drug Biotransformation*, Aitio, A., Ed., Elsevier, Amsterdam, 1978, 233.

282. **Schöllhammer, I., Poll, D. S., and Bickel, M. H.,** Liver microsomal β-glucuronidase and UDP-glucuronyltransferase, *Enzyme,* 20, 269, 1975.

283. **Marniemi, J. and Vainio, H.,** Action of mild alkali on UDPglucuronosyltransferase of hepatic microsomes, *Acta Pharmacol. Toxicol.,* 38, 393, 1976.

284. **Howland, R. D., and Bohm, R.,** Possible multiple binding sites for *o*-aminophenol uridine diphosphate glucuronyltransferase, *Biochem. J.,* 163, 125, 1977.

285. **Howland, R. D. and Burkhalter, A.,** Some effects of 3-methylcholanthrene on uridine diphosphate glucuronyltransferase in the rat and guinea pig, *Biochem. Pharmacol.,* 20, 1463, 1971.

286. **Dutton, G. J. and Storey, I. D. E.,** Glucuronide-forming enzymes, *Methods Enzymol.,* 5, 159, 1962.

287. **Mulder, G. J.,** Heterogeneity of uridine diphosphate glucuronyltransferase from rat liver, *Biochem. J.,* 125, 9, 1971.

288. **Howland, R. D., Burkhalter, A. J., Trevor, A. J., Hageman, S., and Shirachi, D. Y.,** Properties of Lubrol-extracted uridine diphosphate glucuronyltransferase, *Biochem. J.,* 125, 991, 1971.

289. **Potrepka, R. F. and Spratt, J. L.,** A study on the enzymatic mechanism of guinea pig hepatic-microsomal bilirubin glucuronyl transferase, *Eur. J. Biochem.,* 29, 433, 1972.

290. **Zakim, D. and Vessy, D. A.,** Regulation of microsomal UDP-glucuronyltransferase by metal ions. Differential effects of Mn^{++} on forward and reverse reactions, *Eur. J. Biochem.,* 64, 459, 1976.

291. **Lage, G. L. and Spratt, J. L.,** Bilirubin conjugation by hepatic microsomes from adult male guinea pigs, *Arch. Biochem. Biophys.,* 126, 175, 1968.

292. **Storey, I. D. E.,** Some differences in the conjugation of *o*-aminophenol and *p*-nitrophenol by the uridine diphosphate transglucuronylase of mouse-liver homogenates, *Biochem. J.,* 95, 209, 1965.

293. **Sanchez, E., and Tephly, T. R.,** Activation of hepatic microsomal glucuronyl transferase by bilirubin, *Life Sci.,* 13, 1483, 1973.

294. **del Villar, E., Sanchez, E., Autor, A. P., and Tephly, T. R.,** Morphine metabolism. III. Solubilization and separation of morphine and *p*-nitrophenol UDP-glucuronyltransferases, *Mol. Pharmacol.,* 11, 236, 1975.

295. **Yeary, R. A., Wise, K. J., and Davis, D. R.,** Effects of bilirubin on glucuronidation of *p*-nitrophenol and morphine, *Res. Commun. Chem. Pathol. Pharmacol.,* 16, 463, 1977.

296. **Hargreaves, T. and Lathe, G. H.,** Inhibitory aspects of bile secretion, *Nature (London),* 200, 1172, 1963.

297. **Hsia, D. Y.-Y., Riabov, S., and Dowben, R. M.,** Inhibition of glucuronosyl transfer by steroid hormones, *Arch. Biochem. Biophys.,* 103, 181, 1963.

298. **Hänninen, O. and Marniemi, J.,** Effect of glucuronides on glucuronide biosynthesis, *Eur. J. Biochem.,* 18, 282, 1971.

299. **Marniemi, J. and Hänninen, O.,** Action of drug conjugates on glucuronide biosynthesis, *Int. J. Biochem.,* 4, 95, 1973.

300. **Mulder, G. J. and Pilon, A. H. E.,** UDP-glucuronyltransferase and phenolsulfotransferase from rat liver *in vivo* and *in vitro, Biochem. Pharmacol.,* 24, 517, 1975.

301. **Mulder, G. J.,** On non-specific inhibition of rat liver microsomal UDP-glucuronyltransferase by some drugs, *Biochem. Pharmacol.,* 23, 1283, 1974.

302. **Vessey, D. A. and Zakim, D.,** Stimulation of microsomal uridine diphosphate glucuronyltransferase by glucuronic acid derivatives, *Biochem. J.,* 139, 243, 1974.

302a. **Norling, A., Moldéus, P., Andersson, B., and Hänninen, O.,** Passage of glucuronides and sulphates into isolated hepatocytes and action on conjugation reactions, in *Conjugation Reactions in Drug Biotransformation,* Aitio, A., Ed., Elsevier, Amsterdam, 1978, 303.

302b. **Norling, A., Andersson, B., Berggren, M., and Moldéus, P.,** Uptake of glucuronides into isolated hepatocytes and their effects on glucuronide and sulphate conjugation, *Acta Pharmacol. Toxicol.,* 43, 311, 1978.

303. **Götze, W., Grube, E., Rao, G. S., Rao, M. L., and Breuer, H.,** Steroidglucuronyltransferasen. II. Solubilisierung, Anreigherung, und kinetische Eigenschaften einer Östradiol-17β-3-Glucuronyltransferase aus der Mikrosomen-Fraktion des Schweinedunndarmes, *Hoppe Seylers Z. Physiol. Chem.,* 352, 1223, 1971.

304. **Ogawa, H., Sawada, M., and Kawada, M.,** Purification and properties of a pyrophosphatase from rat liver microsomes capable of catalysing the hydrolysis of UDP-glucuronic acid, *J. Biochem. (Tokyo),* 59, 126, 1966.

305. **Hänninen, O. and Marniemi, J.,** Inhibition of glucuronide synthesis by physiological metabolites in liver slices, *FEBS Lett.,* 6, 177, 1970.

306. **Marniemi, J., Nurminen, J., and Hänninen, O.,** Studies on the action of D-galactose and D-galactosamine on the glucuronide synthesis in the rat tissues, *Int. J. Biochem.,* 4, 227, 1973.

307. **Royer, M. and Ortiz, G.,** Bilirubin output and clearance in isolated dog liver perfusion, *Acta Hepato. Gastroenterol.,* 18, 312, 1971.

308. **Bachorik, P. S. and Dietrich, L. S.,** The purification and properties of detergent-solubilized rat liver nucleotide pyrophosphatase, *J. Biol. Chem.,* 247, 5071, 1972.

309. **Schröter, W.,** Mechanismen der Stoffwechselkontrolle in der Neugeborenenperiode, in *Stoffwechsel des Neugeborenen,* Joppich, G. and Wolf, H., Eds., Hippokrates Verlag, Stuttgart, 1970, 254.

310. **Schröter,** Die Activitat der UDP-Glucuronsäurepyrophosphatase in der Leber wahrend der Entwicklung, *Z. Kinderheilk.,* 108, 93, 1970.

311. **Hänninen, O.,** On the metabolic regulation in the glucuronic acid pathway in the rat tissues, *Ann. Acad. Sci. Fenn. Ser. A2,* 142(96 pp) 1968.

312. **Halac, E. and Sicignano, C.,** Re-evaluation of the influence of age, sex, pregnancy and phenobarbital on the activity of UDP-glucuronyltransferase in rat liver, *J. Lab. Clin. Med.,* 73, 677, 1969.

313. **Van Roy, F. P. and Heirwegh, K. P. M.,** Determination of bilirubin glucuronide and assay of glucuronyltransferase with bilirubin as acceptor, *Biochem. J.,* 107, 507, 1968.

314. **Winsnes, A.,** Variable effect of phenobarbital treatment of mice on hepatic UDP-glucuronyltransferase activity when judged by slightly different enzyme-assay techniques, *Biochem. Pharmacol.,* 20, 1853, 1971.

315. **Mowat, A. P. and Arias, I. M.,** Partial purification of hepatic UDP-glucuronyltransferase. Studies of some of its properties, *Biochim. Biophys. Acta,* 212, 65, 1970.

316. **Fevery, J., Leroy, P., and Heirwegh, K. P. M.,** Enzymic transfer of glucose and xylose from uridine diphosphate glucose and uridine diphosphate xylose to bilirubin by untreated and digitonin-activated preparations from rat liver, *Biochem. J.,* 129, 619, 1972.

317. **Gabaldón, M. and Lacomba, T.,** Glucuronyltransferase activity towards diethylstilbestrol in rat and hamster: effect of activators, *Eur. J. Cancer,* 8, 275, 1972.

318. **Miller, K. W., Heath, E. C., Easton, K. H., and Dingell, J. V.,** Effect of Triton X-100 on the conjugation of tetrahydrocortisone *in vitro, Biochem. Pharmacol.,* 22, 2319, 1973.

319. **Wishart, G. J., Goheer, M. A., Leakey, J. E. A., and Dutton, G. J.,** Precocious development of UDP-glucuronyltransferase activity during organ culture of foetal rat liver in the presence of glucocorticoids, *Biochem. J.,* 166, 249, 1977.

320. **Miettinen, T. A. and Leskinen, E.,** Enzymic levels of glucuronic acid metabolism in the liver, kidney and intestine of normal and fasted rats, *Biochem. Pharmacol.,* 12, 565, 1963.

321. **Wong, K. P.,** Bilirubin glucuronyltransferase. Specific assay and kinetic studies, *Biochem. J.,* 125, 27, 1971.

322. **Greenwood, D. T. and Stevenson, I. H.,** The stimulation of glucuronide formation *in vitro* and *in vivo* by a carcinogen, diethylnitrosamine, *Biochem. J.,* 96, 37P, 1965.

323. **Weatherill, P. J. and Burchell, B.,** Reactivation of a pure defective UDP-glucuronyltransferase from homozygous Gunn rat liver, *FEBS Lett.,* 87, 207, 1978.

324. **Vainio, H. and Aitio, A.,** Enhancement of microsomal drug hydroxylation and glucuronidation in rat liver by phenobarbital and 3-methylcholanthrene in combination, *Acta Pharmacol. Toxicol.,* 34, 130, 1974.

325. **Vainio, H. and Hänninen, O.,** A comparative study on drug hydroxylation and glucuronidation in liver microsomes of phenobarbital and 3-methylcholanthrene treated rats, *Acta Pharmacol. Toxicol.,* 35, 65, 1974.

326. **Aitio, A.,** Effect of various membrane perturbing agents on the UDP-glucuronosyltransferase activity in different rat tissues, *Int. J. Biochem.,* 5, 617, 1974.

327. **Lucier, G., McDaniel, O., Brubaker, P., and Klein, R.,** Effects of methylmercury hydroxide on rat liver microsomal enzymes, *Chem. Biol. Interact.,* 4, 265, 1971-2.

327a **Wishart, G. J. and Fry, D. J.,** Evidence from rat liver nuclear preparations that latency of microsomal UDP-glucuronosyltransferase is associated with vesiculation, *Biochem. J.,* 186, 687, 1980.

327b. **Burchell, B.,** Identification and purification of multiple forms of UDP-glucuronosyltransferase, *Rev. Biochem. Toxicol.,* 3, 1980, in press.

327c. **Tukey, R. H., Billings, R. E., Autor, A. P., and Tephly, T. R.,** Phospholipid-dependence of oestrone UDP-glucuronyltransferase and *p*-nitrophenyl UDP-glucuronyltransferase, *Biochem. J.,* 179, 59, 1979.

327d. **Bentley, D. N., Wood, G. C., and Graham, A. B.,** Effects of lipid peroxidation on the activity of microsomal UDPglucuronosyltransferase, *Med. Biol.,* 57, 274, 1979.

327e. **Stier, A., Kuhnle, W., Bosterling, B., and Finch, S. A. E.,** The integrated system of drug metabolism and its perturbation on induction, in *The Induction of Drug Metabolism,* Estabrook, R. W. and Lindenlaub, E., Eds., Schattauer Verlag, Stuttgart, 1978, 225.

328. **Razin, S.,** Reconstitution of biological membranes, *Biochim. Biophys. Acta,* 265, 241, 1972.

329. **Dutton, G. J.,** Uridine diphosphate glucuronic acid as glucuronyl donor in the synthesis of 'ester', aliphatic and steroid glucuronides, *Biochem. J.,* 64, 693, 1956.

330. **Vollrath, W., Rao, G. S., Rao, M. L., and Breuer, H.,** New results on oestrogen glucuronyltransferase, *J. Steroid Biochem.,* 5, 877, 1974.

331. **Labow, R. S., Williamson, D. G., and Layne, D. S.,** Properties of some glycosyl transferases from rabbit tissues, *Biochemistry,* 10, 2553, 1971.

332. **Hänninen, O., Laitinen, M., and Puhakainen, E.,** Uridine diphosphate glucuronic acid: its relation to the hydroxylation system and its possible purification, *Biochem. Soc. Trans.,* 2, 1180, 1974.

333. **Hänninen, O., Lang, M., Koivusaari, U., and Ollikainen, T.,** UDPglucuronosyltransferase: substrate specificity and reactivation after partial separation from other membrane components, in *Biological Reactive Intermediates,* Jollow, D. J., Kocsis, J. J., Snyder, R., and Vainio, H., Eds., Plenum Press, New York, 1977, 239.

334. **Lucier, G. W., McDaniel, O. S., and Hook, G. E. R.,** Nature of the enhancement of hepatic uridine diphosphate glucuronyltransferase activity by 2,3,7,8-tetradichlorodibenzo-p-dioxin in rats, *Biochem. Pharmacol.,* 24, 325, 1975.

335. **del Villar, E., Sanchez, E., and Tephly, T. R.,** Morphine metabolism. V. Isolation of separate glucuronyltransferase activities for morphine and p-nitrophenol from rabbit liver microsomes, *Drug Metab. Dispos.,* 5, 273, 1977.

336. **Bock, K. W., Fröhling, W., Remmer, H., and Rexer, B.,** Effects of phenobarbital and 3-methylcholanthrene on substrate specificity of rat liver microsomal UDPglucuronyltransferase, *Biochim. Biophys. Acta,* 327, 43, 1973.

337. **Bock, K. W., Kittel, J., and Josting, D.,** Purification of rat liver UDP-glucuronyltransferase: separation of two enzyme forms with different substrate specificity and differential inducibility, in *Conjugation Reactions in Drug Biotransformation,* Aitio, A., Ed., Elsevier, Amsterdam, 1978, 357.

337a. **Yuasa, A.,** Purification and properties of uridine diphosphate glucuronyl transferase from rabbit liver microsomes, *J. College Dairying (Japan),* 7 (Suppl.), 103, 1977.

338. **Mulder, G. J.,** Heterogeneity of hepatic microsomal uridine diphosphate glucuronyltransferase: a critical evaluation, *Biochem. Soc. Trans.,* 2, 1172, 1974.

339. **Dutton, G. J.,** Glucuronide-forming enzymes, *Handb. Exp. Pharmacol.,* 28(2), 378, 1971.

340. **Banerjee, R. K. and Roy, A. B.,** Kinetic studies of the phenol sulphotransferase reaction, *Biochim. Biophys. Acta,* 151, 573, 1968.

341. **Roy, A. B.,** Sulphate conjugation enzymes, *Handb. Exp. Pharmacol.,* 28(2), 536, 1971.

342. **Wishart, G. J. and Dutton, G. J.,** Release by phenobarbital of the repression of UDP-glucuronyltransferase activity *in ovo, Biochem. Pharmacol.,* 24, 451, 1975.

343. **Katchalski, E., Silman, I., and Goldman, R.,** Effect of the microenvironment on the mode of action of immobilized enzymes, *Adv. Enzymol.,* 34, 445, 1971.

344. **Zaborsky, O. R.,** *Immobilized Enzymes,* CRC Press, Cleveland, 1973.

345. **Engasser, J. M. and Horvath, C.,** Inhibition of bound enzymes. I. Antienergistic interaction of chemical and diffusional inhibition, *Biochemistry,* 13, 3845, 1974.

346. **Engasser, J. M. and Horvath, C.,** Inhibition of bound enzymes. II. Characterization of product inhibition and accumulation, *Biochemistry,* 13, 3849, 1974.

347. **Engasser, J. M. and Horvath, C.,** Inhibition of bound enzymes. III. Diffusion-enhanced regulating effect with substrate inhibition, *Biochemistry,* 13, 3855, 1974.

348. **Marniemi, J. and Laitinen, M.,** Regulatory properties of partially-purified UDP-glucuronyltransferase (p-nitrophenol), *Int. J. Biochem.,* 6, 345, 1975.

349. **Parikh, I., MacGlashan, D. W., and Fenselau, C.,** Immobilized glucuronosyltransferase for the synthesis of conjugates, *J. Med. Chem.,* 19, 296, 1976.

350. **Arias, I. M.,** Ethereal and N-linked glucuronide formation by normal and Gunn rats *in vitro* and *in vivo, Biochem. Biophys. Res. Commun.,* 6, 81, 1961.

351. **Adamson, R. H., Bridges, J. W., Kibby, M. R., Walker, S. R., and Williams, R. T.,** The fate of sulphadimethoxine in primates compared with other species, *Biochem. J.,* 118, 41, 1970.

352. **Hammar, C. H. and Gempp-Friedrich, W.,** Glucuronyltransferasen des Meerschweinchens bei experimenteller Hepatopathie infolge Langzeitbehandlung mit Thioacetamid, *Res. Exp. Med.,* 160, 112, 1973.

353. **Dahm, K. and Breuer, H.,** Reingung und Charakterisierung einer löslichen Uridin Diphosphat-Glucuronat:17β-Hydroxysteroid-Glucuronyl Transferase beim Menschen, *Biochim. Biophys. Acta,* 128, 306, 1966.

354. **Dahm, K., Lindlau, M., and Breuer, H.,** Biogenese von Östriol-3-monoglucuronid, *Acta Endocrinol.,* 56, 403, 1967.

355. **Rao, G. S., Rao, M. L., and Breuer, H.,** Studies on a testosterone glucuronyltransferase from the cytosol fraction of human liver, *Biochem. J.,* 119, 635, 1970.

356. **Lucier, G. W.,** Microsomal glucuronidation of selected steroids using a rapid radiometric assay, *J. Steroid Biochem.,* 5, 681, 1974.

357. **Breuer, H. and Wessendorf, D.,** Enzymatische Bildung von Östradiol-(17β)-glucuroniden in der Mikrosomen-Fraktion der Kaninchenleber, *Hoppe Seylers Z. Physiol. Chem.,* 345, 1, 1966.

358. **Kirdani, R. Y., Slaunwhite, W. R., Jr., and Sandberg, A. A.,** Studies on phenolic steroids in human subjects. XI. The biosynthesis and proof of structure of estriol-3,16-diglucosiduronate, *Steroids,* 12, 171, 1968.

359. **Slaunwhite, W. R., Jr., Lichtman, M. A., and Sandberg, A. A.,** Studies on phenolic steroids in human subjects. VI. Biosynthesis of estriolglucosiduronic acid-16-^{14}C by human liver, *J. Clin. Endocrinol.,* 24, 638, 1964.

360. **Sandberg, A. A. and Slaunwhite, W. R., Jr.,** Studies on phenolic steroids in human subjects. VIII. Metabolic fate of estriol and its glucuronide, *J. Clin. Invest.,* 44, 694, 1965.

361. **Schriefers, H., Ghraf, R., and Pohl, F.,** Biosynthese von Steroidglucuroniden durch Rattenlebermikrosomen. Zur Frage des Verhaltens der UDP-Glucuronyl-Transferase-Activität unter Nahrungsentzug und in Alloxandiabetes, *Hoppe Seylers Z. Physiol. Chem.,* 344, 25, 1966.

362. **Drucker, W. D.,** Glucuronic acid conjugation of tetrahydrocortisone and *p*-nitrophenol in the homozygous Gunn rat, *Proc. Soc. Exp. Biol.,* 129, 308, 1968.

363. **Jansen, P. L. M.,** The enzyme-catalyzed formation of bilirubin diglucuronide by a solubilized preparation from cat liver microsomes, *Biochim. Biophys. Acta,* 338, 170, 1974.

364. **Sanchez, E. and Tephly, T. R.,** Morphine metabolism. I. Evidence for separate enzymes in the glucuronidation of morphine and *p*-nitrophenol by rat hepatic microsomes, *Drug Metab. Disp.,* 2, 247, 1974.

365. **Del Villar, E., Sanchez, E., and Tephly, T. R.,** Morphine metabolism. II. Studies on morphine glucuronyltransferase activity in intestinal microsomes of rats, *Drug Metab. Disp.,* 2, 370, 1974.

366. **Temple, A. R., Clement, M. S., and Done, A. K.,** Studies of glucuronidation. IV. Evidence of different processes for *o*-aminophenol and *p*-nitrophenol, *Proc. Soc. Exp. Biol.,* 128, 307, 1968.

367. **Zakim, D., Goldenberg, J., and Vessey, D. A.,** Differentiation of homologous forms of UDP-glucuronyltransferase. I. Evidence for the glucuronidation of *o*-aminophenol and *p*-nitrophenol by separate enzymes, *Biochim. Biophys. Acta,* 309, 67, 1973.

368. **Gram, T. E., Litterst, C. L., and Mimnaugh, E. G.,** Enzymatic conjugation of foreign chemical compounds by rabbit lung and liver, *Drug Metab. Disp.,* 2, 254, 1974.

369. **Dutton, G. J.,** Developmental studies on some enzymes associated with detoxication, *Enzyme,* 15, 304, 1973.

370. **Dutton, G. J.,** Variations in glucuronide formation by perinatal liver, *Biochem. Pharmacol.,* 15, 947, 1966.

371. **Millburn, P.,** Factors affecting glucuronidation *in vivo, Biochem. Soc. Trans.,* 2, 1182, 1974.

372. **Abrams, L. S. and Elliott, H. W.,** Morphine metabolism *in vivo* and *in vitro* by homozygous Gunn rats, *J. Pharmacol. Exp. Therap.,* 189, 285, 1974.

373. **Capel, I. D., Millburn, P., and Williams, R. T.,** The conjugation of 1- and 2-naphthols and other phenols in the cat and pig, *Xenobiotica,* 4, 601, 1974.

374. **Batt, A.-M., Ziegler, J. M., and Siest, G.,** Competitive inhibition of glucuronidation by *p*-hydroxyphenyl hydantoin, *Biochem. Pharmacol.,* 24, 152, 1975.

375. **Lacomba, T., Antonio, P., Gabaldón, M., and Barbera, F.,** Effect of activation on the formation of diethylstilbestrol diglucuronide by hamster liver homogenates, *Rev. Esp. Fisiol.,* 30, 53, 1974.

376. **Gartner, L. M. and Arias, I. M.,** Hormonal control of hepatic bilirubin transport and conjugation, *Am. J. Physiol.,* 222, 1091, 1972.

377. **Burchell, B. and Dutton, G. J.,** Delayed induction by phenobarbital of UDP-glucuronyltransferase activity towards bilirubin in fetal liver, *Biol. Neonate,* 26, 122, 1975.

378. **Van Leusden, H. A. J. M., Bakkeren, J. A. J. M., Zilliken, F., and Stolte, L. A. M.,** *p*-Nitrophenylglucuronide formation by homozygous adult Gunn rats, *Biochem. Biophys. Res. Commun.,* 7, 67, 1962.

379. **Frei, J., Birchmeier, H. and Schmid, E.,** Multiplicity and specificity of UDP-glucuronyltransferase, *Enzymol. Biol. Clin.,* 11, 385, 1970.

380. **Hakim, J., Feldmann, G., Troube, H., Boucherot, J., and Boivin, P.,** Étude comparatif des activités bilirubine et paranitrophénol-glucuronyl transférasiques. I. Effets de l'intoxication alcoolique subaiguë chez le rat normal et partiellement hépatectomisé, *Pathol. Biol.,* 18, 627, 1970.

381. **Hakim, J., Feldmann, G., Troube, H., Boucherot, J., and Boivin, P.,** Étude comparatif des activités bilirubine et paranitrophénol-glucuronyl transférasiques. II. Effets de l'intoxication alcoolique chronique chez l'homme, *Pathol. Biol.,* 20, 277, 1972.

382. **Okalicsanyi, L., Frei, J., Magnenat, P., and Naccarado, R.,** Multiplicity and specificity of UDPglucuronyl transferase. III. UDP-glucuronyl transferase and β-glucuronidase activities assayed with different substrates in inherited and acquired liver diseases, *Enzyme,* 12, 658, 1971.

383. **Jansen, P. L. M. and Henderson, P. Th.,** Influence of phenobarbital treatment on 4-nitrophenol conjugation in Wistar rat, Gunn rat and cat, *Biochem. Pharmacol.,* 21, 2457, 1972.

384. **Mulder, G. J.,** Bilirubin and the heterogeneity of microsomal uridine diphosphate glucuronyltransferase from rat liver, *Biochim. Biophys. Acta,* 289, 284, 1972.

385. **Rugstad, H. E. and Dybing, E.**, Glucuronidation in cultures of human skin epithelial cells, *Eur. J. Clin. Invest.*, 5, 133, 1975.

386. **Javitt, N. B.**, Ethereal and acyl glucuronide formation in the homozygous Gunn rat, *Am. J. Physiol.*, 211, 424, 1966.

387. **Tukey, R. H., Billings, R. E., and Tephly, T. R.**, Separation of estrone UDP-glucuronyltransferase and p-nitrophenyl UDP-glucuronyltransferase activities, *Biochem. J.*, 171, 659, 1978.

388. **Lucier, G., Sonawane, B. R., McDaniel, O. J., and Hook, G. E. R.**, Postnatal stimulation of hepatic microsomal enzymes following administration of TCDD to pregnant rats, *Chem. Biol. Interact.*, 11, 15, 1975.

389. **Lucier, G. W. and McDaniel, O. S.**, Steroid and non-steroid UDP glucuronyltransferase: glucuronidation of synthetic estrogens as steroids, *J. Steroid Biochem.*, 8, 867, 1977.

390. **Wishart, G. J., Mossman, S., Donald, A., and Dutton, G. J.**, Differential stimulation of foetal rat liver UDP-glucuronosyltransferase activity towards certain substrates after glucocorticoid treatment in culture and *in utero*, and during natural development, *Biochem. Soc. Trans.*, 5, 721, 1977.

391. **Wishart, G. J.**, Functional heterogeneity of UDP-glucuronosyltransferase as indicated by its differential development and inducibility by glucocorticoids. Demonstration of two groups within the enzyme's activity to twelve substrates, *Biochem. J.*, 174, 485, 1978.

391a. **Billings, R. E., Tephly, T. R., and Tukey, R. H.**, The separation and purification of estrone and p-nitrophenol UDP-glucuronyltransferase activities, in *Conjugation Reactions in Drug Biotransformation*, Aitio, A., Ed., Elsevier, Amsterdam, 1978, 365.

392. **Sedlak, J., Demey, L., and van Couwenberge, H.**, Salicylate sodique et glucuronyl transférase dans les homogénats de foie de rat, *Arch. Int. P-armacodyn.*, 200, 389, 1972.

392a. **Smith, R. L. and Caldwell, J.**, Drug metabolism in nonhuman primates, in *Drug Metabolism from Microbe to Man*, Parke, D. V. and Smith, R. L., Eds., Taylor & Francis, London, 1977, 331.

392b. **Axelrod, J., Inscoe, J., and Tomkins, G. M.**, Enzymatic synthesis of N-glucosyluronic acid conjugates, *J. Biol. Chem.*, 232, 835, 1958.

392c. **Bock, K. W., v. Clausbruch, U. C., Kaufmann, R., Lilienblum, W., Desch, F., Pfeil, H., and Platt, K. L.**, Functional heterogeneity of UDP-glucuronyltransferase in rat tissues, *Biochem. Pharamacol.*, 29, 495, 1980.

393. **Puhakainen, E., Lang, M., Ilvonen, A., and Hänninen, O.**, Biosynthetic production of ^{14}C-labelled reference glucuronides, *Acta Chem. Scand.*, B30, 685, 1976.

394. **Kindel, P. K. and Watson, R. R.**, Synthesis, characterization and properties of uridine 5-(α-D-apio-D-furanosyl pyrophosphate), *Biochem. J.*, 133, 227, 1973.

395. **Dutton, G. J. and Illing, H. P. A.**, Mechanism of biosynthesis of thio-β-D-glucuronides and thio-β-D-glucosides, *Biochem. J.*, 129, 539, 1972.

396. **Lindsay, R. H., Vaughn, A. W., Kelly, K., and Aboul-Enein, H. Y.**, Site of glucuronide conjugation to the antithyroid drug 6-n-propyl-2-thiouracil, *Biochem. Pharmacol.*, 26, 833, 1977.

397. **Lindsay, R. H., Cash, A. G., Vaughn, A. W., and Hill, J. B.**, Glucuronide conjugation of 6-n-propyl-2-thiouracil and other antithyroid drugs by guinea pig liver microsomes *in vitro*, *Biochem. Pharmacol.*, 26, 617, 1977.

398. **Bauer, J. E., Gerber, N., Lynn, R. K., Smith, R. G., and Thompson, R. H.**, New N-glucuronide metabolite of carbamezapine, *Experientia*, 32, 1032, 1976.

399. **Bickel, M. H., Minder, R., and di Francesco, C.**, Formation of N-glucuronide of desmethylimipramine in the dog, *Experientia*, 29, 960, 1973.

400. **Hammar, C. -G. and Holmstedt, B.**, Mass Fragmentography. Identification of chlorpromazine and its metabolites in human blood by a new method, *Analyt. Biochem.*, 25, 532, 1968.

401. **Lin, T.-S., and Kolattukudy, P. E.**, Glucuronyl glycine, a novel N-terminus in a glycoprotein, *Biochem. Biophys. Res. Commun.*, 75, 87, 1977.

402. **Mulder, G. J. and van Doorn, A. B. D.**, A rapid NAD$^+$-linked assay for microsomal uridine diphosphate glucuronyl transferase of rat liver and some observations on substrate specificity of the enzyme, *Biochem. J.*, 151, 131, 1975.

403. **Rance, M. J. and Shillingford, J. S.**, The role of the gut in the metabolism of strong analgesics, *Biochem. Pharmacol.*, 25, 735, 1976.

404. **Rance, M. J. and Shillingford, J. S.**, The metabolism of phenolic opiates by rat intestine, *Xenobiotics*, 7, 529, 1977.

405. **Batt, A. M., Ziegler, J. M., and Siest, G.**, Rate of glucuronoconjugation of phenolic substrates and metabolites, *Abstr. 6th Int. Congr. Pharmacol.*, 294, 124, 1975.

406. **Wong, K. P. and Sourkes, T. L.**, Glucuronidation of 3-O-methylnoradrenaline, harmalol and some related compounds, *Biochem. J.*, 110, 99, 1968.

407. **Wong, K. P.**, Microassay of hepatic UDP-glucuronyltransferase, *Clin. Chim. Acta*, 26, 119, 1969.

408. **Dingell, J. V., Caldwell, J., Moffat, J. R., Smith, R. L., and Williams, R. T.**, Studies on the conjugation of 4'-hydroxyamphetamine *in vitro*, *Biochem. Soc. Trans.*, 2, 306, 1974.

409. **von Bahr, C. and Bertilsson, L.,** Hydroxylation and subsequent glucuronide conjugation of desmethylimipramine in rat liver microsomes, *Xenobiotica,* 1, 205, 1971.
410. **Manzo, L., Berte, F., and de Bernardi, M.,** Oxazepam glucuronidation "in vitro" in some maternal and fetal tissues of the rat. Effects of pretreatment with oxazepam and phenobarbital, *Boll. Chim. Farm.,* 108, 19, 1969.
411. **Labow, R. S. and Layne, D. S.,** The formation of glucosides of isoflavones and of some other phenols by rabbit liver microsomal fractions, *Biochem. J.,* 128, 491, 1972.
412. **Sullivan, L. J.,** 5,6-Dihydro-5,6-dihydroxycarbaryl glucuronide as a significant metabolite of carbaryl in the rat, *J. Agric. Food Chem.,* 20, 980, 1972.
413. **Roerig, S., Fujimoto, J. M., and Wang, R. J. H.,** Isolation of hydromorphone and dihydromorphine glucuronides from urine of the rabbit after hydromorphone administration, *Proc. Soc. Exp. Biol. Med.,* 143, 230, 1973.
414. **Kabacoff, B. L., Fairchild, C. M., and Burnett, C.,** Pyridinethione glucuronide as a metabolite of sodium pyridinethione, *Food Cosmet. Toxicol.,* 9, 519, 1971.
415. **Thompson, R. M., Gerber, N., Seibert, R. A., and Desideri, D. M.,** Identification of 2-methyl-1,4-naphthohydroquinone monoglucuronide as a metabolite of 2-methyl-1,4-naphthoquinone (menadione) in rat bile, *Res. Commun. Chem. Pathol. Pharmacol.,* 4, 543, 1972.
416. **Harvey, D. J., Martin, B. R., and Paton, W. D. M.,** Identification of the glucuronides of cannabidiol and hydroxycannabidiols in mouse liver, *Biochem. Pharmacol.,* 25, 2217, 1976.
417. **Thompson, R. M., Gerber, N., and Seibert, R. A.,** Metabolism of methocarbamol (robaxin) in the isolated perfused rat liver and identification of glucuronides, *Xenobiotica,* 5, 145, 1975.
418. **Lynn, R. K., Smith, R. G., Leger, R. M., Deinzer, M. L., Griffin, D., and Gerber, N.,** Identification of glucuronide metabolites of benzomorphan narcotic analgesic drugs in bile from the isolated perfused rat liver by gas chromatography and mass spectrometry, *Drug Metab. Dispos.,* 5, 47, 1977.
419. **Axelrod, J., Senoh, S., and Witkop, B.,** O-Methylation of catechol amines *in vivo, J. Biol. Chem.,* 233, 697, 1958.
420. **Tacker, M., McIsaac, W. M., and Creaven, P. J.,** Metabolism of tyramine-1-^{14}C by the rat, *Biochem. Pharmacol.,* 19, 2763, 1970.
421. **Tacker, M., Creaven, P. J., and McIsaac, W. M.,** Preliminary observations on the metabolism of [1-^{14}C] tyramine in man, *J. Pharm. Pharmacol.,* 24, 247, 1972.
422. **Wong, K. P.,** The biosynthesis of tyramine glucuronide by liver microsomes, *Biochem. J.,* 164, 529, 1977.
423. **Inwang, E. E., Madubuile, P. U., and Mosnain, A. D.,** Evidence for the excretion of 2-phenylethylamine glucuronide in human urine, *Experientia,* 29, 1080, 1973.
424. **Landsberg, L., Berardino, M. B., and Silva, P.,** Metabolism of ^{3}H-L-dopa by the rat gut *in vivo* — evidence for glucuronide conjugation, *Biochem. Pharmacol.,* 24, 1167, 1975.
424a. **Crooks, P. A., Breakefield, X. O., Suleus, C. H., Castiglione, C. M., and Coward, J. K.,** Extensive conjugation of dopamine (3,4-dihydroxyphenethylamine) metabolites in cultured human skin fibroblasts and rat hepatoma cells, *Biochem. J.,* 176, 187, 1978.
425. **Clark, W. G., Drell, W. and Pogrund, R. S.,** Purification and properties of L-epinephrine glucuronide, *Arzneim. Forsch.,* 24, 1769, 1974.
426. **Airaksinen, M., Miettinen, T. A., and Huttunen, J. K.,** Glucuronidation of 5-hydroxyindole derivatives *in vitro, Biochem. Pharmacol.,* 14, 1019, 1965.
426a. **Leakey, J. E. A.,** An improved assay technique for uridine diphosphate glucuronosyltransferase activity towards 5-hydroxytryptamine and some properties of the enzyme, *Biochem. J.,* 175, 1119, 1979.
427. **Watanabe, M., Ohkubo, K., and Tamura, W.,** Studies on carcinogenic tryptophan metabolites. I. Enzymatic formation and hydrolysis of glucuronide of 3-hydroxyanthranilic acid, *Biochem. Pharmacol.,* 21, 1337, 1972.
428. **Back, P., Spaczynski, K., and Gerok, W.,** Bile salt glucuronides in urine, *Hoppe Seylers Z. Physiol. Chem.,* 355, 749, 1974.
429. **Back, P.,** Bile acid glucuronides. II. Isolation and identification of a chenodeoxycholic acid glucuronide from human plasma in intrahepatic cholestasis, *Hoppe Seylers Z. Physiol. Chem.,* 357, 213, 1976.
430. **Back, P. and Bowen, D. V.,** Bile acid glucuronides. III. Chemical Synthesis and characterization of glucuronic acid coupled mono-, di-, and trihydroxy bile acids, *Hoppe Seylers Z. Physiol. Chem.,* 357, 219, 1976.
431. **Back, P.,** Discussion in *Liver and Bile (Falk Symposium 23),* Bianchi, L., Gerok, W., and Sickinger, K., Eds., MTP Press, Lancaster, England, 1977, 397.
432. **Fröhling, W. and Stiehl, A.,** Bile salt glucuronides — identification and quantitative analysis in the urine of patients with cholestasis, *Eur. J. Clin. Invest.,* 6, 67, 1976.
433. **Fröhling, W., Stiehl, A., Ast, E., Czygan, P., and Kommerall, B.,** Induction and activation of rat liver microsomal bile salt glucuronyltransferase, *Biochim. Biophys. Acta,* 44, 525, 1976.

434. **Lippel, B. and Olson, J. A.,** Biosynthesis of β-glucuronides of retinol and of retinoic acid in vivo and in vitro, *J. Lipid Res.*, 9, 168, 1968.

435. **Schriefers, H., Ghraf, R., and Lehnen, B.,** Glucuronidierung von Hydroxy- Testosteron- und Hydroxy-Androstendion-Verbindungen durch die mikrosomale UDP-Glucuronyltransferase der Rattenleber, *Z. Naturforsch.*, 27, 49, 1972.

436. **Layne, D. S.,** New metabolic conjugates of steroids, in *Metabolic Conjugation and Metabolic Hydrolysis*, Vol. 1, Fishman, W. H., Ed., Academic Press, New York, 1970, 21.

437. **Harkness, R. A., Davidson, D. W. and Strong, J. A.,** The metabolism of small and of large amounts of progesterone in man, *Acta Endocrinol.*, 60, 221, 1969.

438. **Rhamy, R. K.,** The identification of androstenedione and androstenedione 3-enol glucuronide in the blood of a woman with an interstitial cell ovarian tumor after 4-^{14}C-testosterone, *Steroids*, 22, 719, 1973.

439. **Weeks, C. M., Rohren, D. C., and Duax, W. L.,** The molecular structure of an aldosterone 18-glucopyranosiduronate, *J. Steroid Biochem.*, 7, 545, 1976.

440. **Shaw, J. M. and Pieringer, R. A.,** Biosynthesis of glucuronosyl diglyceride by particulate fractions of *Pseudomonas diminuta*, *Biochem. Biophys. Res. Commun.*, 46, 1201, 1972.

441. **Stern, N. and Tietz, A.,** Glycolipids of a halotolerant moderately halophilic bacterium. II. Biosynthesis of glucuronosyldiglyceride in cell free particles, *Biochim. Biophys. Acta*, 296, 136, 1973.

442. **Collins, D. C., Jirku, H., and Layne, D. S.,** Steroid *N*-acetylglucosaminyl transferase. Localization and some properties of the enzyme in rabbit tissues, *J. Biol. Chem.*, 243, 2928, 1968.

443. **Felger, C. B. and Katzman, P. A.,** Mono- and diglucuronides of estriol in human pregnancy urine, *Fed. Proc. Fed. Am. Soc. Exp. Biol.*, 20, 199, 1961.

444. **De la Torre, B.,** Conjugation of 16-epiestriol by pregnant women, *Acta Endocrinol.*, 75, 159, 1974.

445. **Rothstein, M. and Greenberg, D. M.,** Studies on the metabolism of xanthurenic acid-4-C^{14}, *Arch. Biochem. Biophys.*, 68, 206, 1957.

446. **Marsh, C. A. and Levvy, G. A.,** Glucuronide metabolism in plants. 3. Triterpene glucuronides, *Biochem. J.*, 63, 9, 1956.

447. **Fischer, L. J., Millburn, P., Smith, R. L., and Williams, R. T.,** The fate of [^{14}C]stilboestrol in the rat, *Biochem. J.*, 100, 69 P, 1966.

448. **Fischer, L. J. and Millburn, P.,** Stilboestrol transport and glucuronide formation in everted sacs of rat intestine, *J. Pharm. Exp. Ther.*, 175, 267, 1970.

449. **Quamme, G. A., Layne, D. S., and Williamson, D. G.,** Species differences in the formation *in vitro* of estrogen conjugates, *Comp. Biochem. Physiol.*, 39B, 25, 1971.

450. **Compernolle, F., Van Hees, G. P., Blanckaert, N., and Heirwegh, K. P. M.,** Glucuronic acid conjugates of bilirubin-IXα in normal bile compared with post-obstructive bile. Transformation of the 1-*O*-acylglucuronide into 2-, 3-, and 4-*O*-acylglucuronides, *Biochem. J.*, 171, 185, 1978.

451. **Gwilliam, C., Paquet, A., Williamson, D. G., and Layne, D. S.,** Isolation from rabbit urine of the 3-glucuronide-17-*N*-acetylglucosaminide of 17α-estradiol, *Can. J. Biochem.*, 52, 452, 1974.

452. **Williamson, D. G., Collins, D. C., and Layne, D. S.,** Steroid estrogen glycosides. Formation of glucosides and galactosides by human liver and kidney, *J. Biol. Chem.*, 247, 3286, 1972.

453. **Labow, R. S., Williamson, D. G., and Layne, D. S.,** The transfer of glucose to steroids by sheep liver microsomes, *Can. J. Biochem.*, 52, 203, 1974.

454. **Labow, R. S. and Layne, D. S.,** A comparison of glucoside formation by liver preparations from rabbit and mouse, *Biochem. J.*, 142, 75, 1974.

455. **Sandberg, A. A., Slaunwhite, W. R., Jr., and Kirdani, R. Y.,** Metabolic conjugation and hydrolysis of estrogens and progesterone in the enterohepatic circulation, in *Metabolic Conjugation and Metabolic Hydrolysis*, Vol. 2, Fishman, W. H., Ed., Academic Press, New York, 1970, 123.

456. **Barford, P. A., Olavesen, A. H., Curtis, C. G., and Powell, G. M.,** Metabolic fates of diethylstilboestrol sulphates in the rat, *Biochem. J.*, 164, 423, 1977.

457. **Wortmann, W., Cooper, D. Y., and Touchstone, J. C.,** Sulfoglucuronides from incubation of ^{3}H-estradiol-17β with rat liver microsomes, *Steroids*, 20, 321, 1972.

458. **Javitt, N. B.,** Biochemical probes for the study of binding and conjugation of glutathione 5-transferases, in *Glutathione: Metabolism and Function*, Arias, I. M. and Jakoby, W. B., Eds., Raven Press, New York, 1976, 309.

459. **Hykes, P., Jirsu, M., Jirsová, V., and Heringová, A.,** The metabolism of phenoltetrachlorphthalein and its analogues in Wistar and Gunn rats, *Acta Hepato Gastroenterol.*, 19, 417, 1972.

460. **Heirwegh, K. P. M., Compernolle, F., Desmet, V., Fevery, J., Meuwissen, J. A. T. P., Van Roy, F. P., and De Groote, J.,** Recent advances in separation and analysis of diazo-positive bile pigments, *Methods Biochem. Anal.*, 22, 205, 1974.

461. **Lathe, G. H.,** The degradation of haem by mammals and its excretion as conjugated bilirubin, *Essays Biochem.*, 8, 107, 1972.

462. Fevery, J., De Groote, J., and Heirwegh, K. P. M., Properties of bilirubin UDP-glycosyltransferases, *Front. Gastrointest. Res.*, 2, 243, 1976.

463. Schmid, R. and McDonagh, A. F., The enzymatic formation of bilirubin, *Ann. N.Y. Acad. Sci.*, 244, 533, 1975.

464. Heirwegh, K. P. M., Blanckaert, N., Compernolle, F., Fevery, J., and Zaman, Z., Detection and properties of the non-α-isomers of bilirubin IX, *Biochem. Soc. Trans.*, 5, 316, 1977.

465. Blanckaert, N., Heirwegh, K. P. M., and Zaman, Z., Comparison of the biliary excretion of the four isomers of bilirubin-IX in Wistar and homozygous Gunn rats, *Biochem. J.*, 164, 229, 1977.

466. Schmid, R. and Lester, R., Implication of conjugation of endogenous compounds — bilirubin, in *Glucuronic Acid*, Dutton, G. J., Ed., Academic Press, New York, 1966, 493.

467. Billing, B. H., Cole, P. G., and Lathe, G. H., The excretion of bilirubin as a diglucuronide giving the direct van den Bergh reaction, *Biochem. J.*, 65, 774, 1957.

468. Schmid, R., The identification of direct-acting bilirubin as bilirubin glucuronide, *J. Biol. Chem.*, 229, 881, 1957.

469. Talafant, E., Properties and composition of bile pigment giving a direct diazo reaction, *Nature, (London)*, 178, 312, 1956.

470. Gregory, C. H., Studies of conjugated bilirubin. 'Pigment I' a complex of conjugated and free bilirubin, *J. Lab. Clin. Med.*, 61, 917, 1963.

471. Weber, A. P., Schalm, L., and Witmans, J., Bilirubin monoglucuronide (pigment 1): a complex, *Acta Med. Scand.*, 173, 19, 1963.

472. Jansen, P. L. M., The isomerisation of bilirubin monoglucuronide, *Clin. Chim. Acta*, 49, 233, 1973.

473. Heirwegh, K. P. M., Fevery, J., Michiels, R., Van Hees, G. P., and Compernolle, F., Separation by thin-layer chromatography and structure elucidation of bilirubin conjugates isolated from dog bile, *Biochem. J.*, 145, 185, 1975.

474. McDonagh, A. F., The photochemistry and photometabolism of bilirubin, in *Phototherapy in the Newborn: an Overview*, Odell, G. B., Ed., National Academy of Science, Washington, 1974, 56.

475. Fevery, J., van der Vijver, M., Michiels, R., and Heirweigh, K. P. M., Comparison in different species of biliary bilirubin IXα conjugates with the activities of hepatic and renal bilirubin-IXα uridine diphosphate glycosyltransferases, *Biochem. J.*, 164, 737, 1977.

476. Fevery, J., Van Damme, B., Michiels, R., De Groote, J., and Heirwegh, K. P. M., Bilirubin conjugates in bile of man and rat in the normal state and in liver disease, *J. Clin. Invest.*, 51, 2482, 1972.

477. Gordon, E. R., Goresky, C. A., Chan, T.-H., and Perlin, A. S., The isolation and characterization of bilirubin diglucuronide, the major bilirubin conjugate in dog and human bile, *Biochem. J.*, 155, 477, 1976.

478. Gordon, E. R., Chan, T.-H., Samodai, K., and Goresky, C. A. The isolation and further characterization of the bilirubin tetrapyrroles in bile-containing human duodenal juice and dog gallbladder bile, *Biochem. J.*, 167, 1, 1977.

479. Fevery, J., Leroy, P., Van der Vijver, M., and Heirwegh, K. P. M., Structures of bilirubin conjugates synthesized *in vitro* from bilirubin and uridine diphosphate glucuronic acid, uridine diphosphate glucose or uridine diphosphate xylose by preparations from rat liver, *Biochem. J.*, 129, 635, 1972.

480. Jansen, P. L. M., Mono- and diglucuronidation of bilirubin, *Folia Med. Neerl.*, 15, 207, 1972.

481. Lathe, G. H., Arias, I. M., and Javitt, N. B., Discussions in *Liver and Bile (Falk Symposium 23)*, Bianchi, L., Gerok, W., and Sickinger, K., Eds., MTP Press, Lancaster, England, 1977, 397.

481a. Jansen, P. L. M., Chowdhury, J. R., and Arias, I. M., Hepatic conversion of bilirubin monoglucuronide to diglucuronide in UDP-glucuronyltransferase-deficient rats, in *Conjugation Reactions in Drug Biotransformation*, Aitio, A., Ed., Elsevier, Amsterdam, 153.

482. Fevery, J., Blanckaert, N., Heirwegh, K. P. M., Preaux, A.-M., and Berthelot, P., Unconjugated bilirubin and an increased portion of bilirubin monoconjugates in the bile of patients with Gilbert's syndrome and Crigler-Najjar disease, *J. Clin. Invest.*, 60, 970, 1977.

483. Chowdhury, J. R., Jansen, P. L. M., Fischberg, E. B., Daniller, A., and Arias, I. M., Hepatic conversion of bilirubin monoglucuronide to diglucuronide in uridine diphosphate-glucuronyltransferase-deficient man and rat by bilirubin glucuronoside glucuronosyltransferase, *J. Clin. Invest.*, 62, 191, 1978.

483a. Wolkoff, A. W., Ketley, J. N., Waggoner, J. G., Berk, P. D., and Jakoby, W. B., Hepatic accumulation and intracellular binding of conjugated bilirubin, *J. Clin. Invest.*, 61, 142, 1978.

483b. Campbell, M. T. and Dutton, G. J., The formation and distribution of bilirubin monoglucuronide and diglucuronide in rat liver slices, *Biochem. J.*, 179, 473, 1979.

484. Ostrow, J. D. and Boonyapisit, S. T., Inaccuracies in measurement of conjugated and unconjugated bilirubin in bile with ethyl anthranilate diazo and solvent-partition methods, *Biochem. J.*, 173, 263, 1978.

485. **Blumenthal, S. G., Taggart, D. B., Ikeda, R. M., and Ruelsner, B. H.**, Conjugated and unconjugated bilirubin in bile of humans and rhesus monkeys. Structure of adult human and rhesus monkey bilirubins compared with dog bilirubins, *Biochem. J.*, 167, 535, 1977.

486. **Heirwegh, K. P. M., Van Hees, G. P., Leroy, P., Van Roy, F. P., and Jansen, F. H.**, Heterogeneity of bile pigment conjugates as revealed by chromatography of their ethyl anthranilate azopigments, *Biochem. J.*, 120, 877, 1970.

487. **Kuenzle, C. C.**, Bilirubin conjugates of human bile. Isolation of phenylazo derivatives of bile bilirubin, *Biochem. J.*, 119, 387, 1970.

488. **Kuenzle, C. C.**, Bilirubin conjugates of human bile. Nuclear-magnetic resonance, infrared and optical spectra of model compounds, *Biochem. J.*, 119, 395, 1970.

489. **Kuenzle, C. C.**, Bilirubin conjugates of human bile. Excretion of bilirubin as the acyl glycosides of aldobiouronic acid, pseudoaldobiouronic acid and hexuronosylhexuronic acid, with a branched chain hexuronic acid as one of the components of hexuronosylhexuronide, *Biochem. J.*, 119, 411, 1970.

490. **Kuenzle, C. C.**, Bilirubin conjugates of human bile, in *Metabolic Conjugation and Metabolic Hydrolysis*, Vol. 3, Fishman, W. H., Ed., Academic Press, New York, 1973, 351.

491. **Noir, B. A.**, Bilirubin conjugates in bile of man, rat and dog. Semi-quantitative analysis of bile composition by thin-layer chromatography, *Biochem. J.*, 155, 365, 1976.

492. **Kuenzle, C. C.**, Microheterogeneity of complex glycosides of bilirubin from human bile, *Experientia*, 31, 626, 1975.

493. **Kuenzle, C. C.**, Bilirubin conjugates in bile, in *The Hepato-Biliary System. Fundamental and Pathological Mechanisms*, Talor, W., Ed., Plenum Press, New York, 1976, 309.

493a. **Blanckaert, N., Compernolle, F., Leroy, P., Van Houtte, R., Fevery, J., and Heirwegh, K. P. M.**, The fate of bilirubin-IXα glucuronide in cholestasis and during storage *in vitro*. Intramolecular rearrangement to positional isomers of glucuronic acid, *Biochem. J.*, 178, 203, 1978.

493b. **Compernolle, F.**, Structure revision of disaccharide conjugates of bilirubin-IXα in human bile and identification of phenylazo derivatives B₄, B₅, and B₆ as 2-, 3- and 4-*O*-acylglucuronides, *Biochem. J.*, 175, 1095, 1978.

494. **Noir, B. A., Garay, E. R., and Royer, M.**, Separation and properties of conjugated biliverdin, *Biochim. Biophys. Acta*, 100, 403, 1965.

494a. **Heirwegh, K. P. M.**, Formation, metabolism and significance of bilirubin-IX glycosides, in *Conjugation Reactions in Drug Biotransformation*, Aitio, A., Ed., Elsevier, Amsterdam, 1978, 67.

495. **Garay, E. R., Noir, B., and Royer, M.**, Biliverdin pigments in green bile, *Biochim. Biophys. Acta*, 100, 411, 1965.

495a. **Chowdhury, J. R., Chowdhury, N. R., Bhargava, M. M., and Arias, I. M.**, Purification and partial characterisation of rat liver bilirubin glucuronoside glucuronosyltransferase, *J. Biol. Chem.*, 254, 8336, 1979.

495b. **Blanckaert, N., Gollan, J., and Schmid, R.**, Bilirubin diglucuronide synthesis by a UDP-glucuronic acid-dependent enzyme system in rat liver microsomes, *Proc. Natl. Acad. Sci., Washington*, 76, 2037, 1979.

496. **Lin, G. L., Himes, J. A., and Cornelius, C. E.**, Bilirubin and biliverdin excretion by the chicken, *Am. J. Physiol.*, 226, 881, 1974.

496a. **Bedford, C. T., Logan, C. J. and Smith, E. H.**, Mechanistic studies of glucuronidation. I. The determination of the glycosylating reactivity of UDPG and UDPGA, in *Conjugation Reactions in Drug Biotransformation*, Aitio, A., Ed., Elsevier, Amsterdam, 1978, 491.

497. **Zakim, D. and Vessey, D. A.**, Techniques for the characterization of UDP-glucuronyltransferase, glucose-6-phosphatase, and other tightly bound microsomal enzymes, *Methods Biochem. Analy.*, 21, 1, 1973.

498. **Isselbacher, K. J. and Axelrod, J.**, Enzymatic formation of corticosteroid glucurinides, *J. Am. Chem. Soc.*, 77, 1070, 1955.

499. **Sanchez, E., del Villar, E., and Tephly, T. R.**, Structural requirements in the reaction of morphine-UDP-glucuronyltransferase with opioid substances, *Biochem. J.*, 169, 173, 1978.

500. **Dutton, G. J. and Ko, V.**, The synthesis of *o*-aminophenyl glucuronide in several tissues of the domestic fowl, *Gallus gallus*, during development, *Biochem. J.*, 99, 550, 1966.

501. **Wong, K. P. and Sourkes, T. L.**, Determination of UDPG and UDPGA in tissues, *Anal. Biochem.*, 21, 444, 1967.

502. **Fyffe, J. and Dutton, G. J.**, Induction of UDPglucose dehydrogenase during development, organ culture and exposure to phenobarbital. Its relation to levels of UDPglucuronic acid and overall glucuronidation in chicken and mouse, *Biochim. Biophys. Acta*, 411, 41, 1975.

503. **Balduini, E., Brovelli, A., de Luca, G., Galligani, L., and Castellani, A. A.**, Uridine diphosphate glucose dehydrogenase from cornea and epiphysial-plate cartilage, *Biochem. J.*, 133, 243, 1973.

504. **Zhivkov, V.**, Measurement of uridine diphosphate glucuronic acid concentrations in animal tissues, *Biochem. J.*, 120, 505, 1970.

505. Zhivkov, V., Measurement of the relative rates of synthesis of uridine diphosphate sugars in some tissues of vertebrates by using [1-^{14}C] glucose, *Int. J. Biochem.*, 3, 258, 1972.

506. Zhivkov, V., Tosheva, R., and Zhivkova, Y., Concentration of uridine diphosphate sugars in various tissues of vertebrates, *Comp. Biochem. Physiol.*, 51B, 421, 1975.

507. Handley, C. J. and Phelps, C. F., The biosynthesis *in vitro* of chondroitin sulphate in neonatal rat epiphysial cartilage, *Biochem. J.*, 126, 417, 1972.

508. Handley, C. J. and Phelps, C. F., The biosynthesis *in vitro* of keratin sulphate in bovine cornea, *Biochem. J.*, 128, 205, 1972.

509. Gainey, P. A. and Phelps, C. F., Uridine diphosphate glucuronic acid production and utilization in various tissues actively synthesizing glycosaminoglycans, *Biochem. J.*, 128, 215, 1972.

510. Losman, M. J. and Harley, E. H., Synthesis of UDP-glucuronic acid in an established hepatoma cell line, *FEBS Lett.*, 85, 57, 1978.

511. Flodgaard, H. J. and Brodersen, R., Bilirubin glucuronide formation in developing guinea pig liver, *Scand. J. Clin. Lab. Invest.*, 19, 149, 1967.

511a. Moldéus, P., Andersson, B., and Norling, A., Interaction of ethanol oxidation with glucuronidation in isolated hepatocytes, *Biochem. Pharmacol.*, 27, 2583, 1978.

512. Müller-Oerlinghausen, B. and Künzel, B., Sex differences in the activity of rat liver UDPG-dehydrogenase, *Life Sci.*, 7, 1129, 1968.

513. Wong, K. P. and Sourkes, T. L., Uridine nucleotides in brain, *J. Neurochem.*, 15, 201, 1968.

514. Winsnes, A. and Rugstad, H. E., Different properties of microsomal UDP-glucuronyl transferase in Buffalo rat liver and a clonal strain of rat hepatoma cells derived from the same rat strain, *Acta Pharmacol. Toxicol.*, 33, 161, 1973.

515. Dalessandro, G. and Northcote, D. H., Changes in enzymic activities of nucleoside diphosphate sugar interconversion during differentiation of cambium to xylem in sycamore and poplar, *Biochem. J.*, 162, 267, 1977.

516. Dalessandro, G. and Northcote, D. H., Changes in enzymic activities of nucleoside diphosphate sugar interconversion during differentiation of cambium to xylem in pine and fir, *Biochem. J.*, 162, 281, 1977.

517. Strominger, J. L., Maxwell, E. S., Axelrod, J., and Kalckar, H. M., Enzymatic formation of uridine diphosphoglucuronic acid, *J. Biol. Chem.*, 224, 79, 1957.

518. Schiller, J. G., Bowser, A. M., and Feingold, D. S., Studies on the mechanism of action of UDP-D-glucose dehydrogenase from beef liver, *Carbohydr. Res.*, 21, 249, 1972.

519. Simonart, P. C., Salo, W. L., and Kirkwood, S., The mechanism of action of UDPG-dehydrogenase, *Biochem. Biophys. Res. Commun.*, 24, 120, 1966.

520. Molz, R. J. and Danishefsky, I., Uridine diphosphate glucose dehydrogenase of rat tissue, *Biochim. Biophys. Acta*, 250, 6, 1971.

521. Nelselstuen, G. L. and Kirkwood, S., The mechanism of action of uridine diphosphoglucose dehydrogenase. Uridine diphosphohexodialdoses as intermediates, *J. Biol. Chem.*, 246, 3828, 1971.

522. Ridley, W. P. and Kirkwood, S., The stereospecificity of hydrogen abstraction by uridine diphosphoglucose dehydrogenase, *Biochem. Biophys. Res. Commun.*, 54, 955, 1973.

523. Ridley, W. P., Houchins, J. P., and Kirkwood, S., Mechanism of action of uridine diphosphoglucose dehydrogenase. Evidence for a second reversible dehydrogenation step involving an essential thiol group, *J. Biol. Chem.*, 250, 8761, 1975.

524. Ordman, A. O. and Kirkwood, S., Mechanism of action of uridine diphosphoglucose dehydrogenase Evidence for an essential lysine residue at the active site, *J. Biol. Chem.*, 252, 1320, 1977.

525. Goldberg, N. D., Dahl, J. L., and Parks, R. E., Jr., Uridine diphosphate glucose dehydrogenase. pH dependence of the reactions with the 5-fluorouracil and 6-azauracil analogues of uridine diphosphate glucose, *J. Biol. Chem.*, 238, 3109, 1963.

526. Roy-Burman, P., Roy-Burman, S., and Visser, D. W., Uridine diphosphate glucose dehydrogenase, studies with 5-hydroxyuridine diphosphate glucose and 5,6-dihydrouridine diphosphate glucose, *J. Biol. Chem.*, 243, 1692, 1968.

527. Druzhinina, T. N., Kusov, Y. Y., Shibaev, V. N., and Kochetkov, N. K., Interaction of uridine diphosphate glucose with calf liver uridine diphosphate glucose dehydrogenase. Significance of hydroxyl groups at C-3, C-4 and C-6 of hexosyl residue, *Biochim. Biophys. Acta*, 403, 1, 1975.

528. Graham, T. L. and Whistler, R. L., Enzymatic synthesis and reactions of uridine 5′-(5-thio-α-D-glucopyranosyl pyrophosphate), *Arch. Biochem. Biophys.*, 171, 721, 1975.

529. Franzen, J. G., Ishman, R., and Feingold, D. S., Half-of-the-sites reactivity of bovine liver uridine diphosphoglucose dehydrogenase toward iodoacetate and iodacetamide, *Biochemistry*, 15, 5665, 1976.

530. Geren, C. R., Oloman, C. M., Primrose, D. C., and Ebner, K. E., Purification of UDP glucose dehydrogenase by hydrophobic and affinity chromatography, *Prep. Biochem.*, 7, 19, 1977.

531. **Wilson, D.,** Purification of uridine diphosphoglucose dehydrogenase from calf liver, *Anal. Biochem.,* 10, 472, 1965.
532. **Zalitis, J. and Feingold, D. S.,** Purification and properties of UDPG dehydrogenase from beef liver, *Arch. Biochem. Biophys.,* 132, 457, 1969.
533. **Gainey, P. A., Pestell, T. C., and Phelps, C. F.,** A study of the subunit structure and the thiol reactivity of bovine liver uridine diphosphate glucose dehydrogenase, *Biochem. J.,* 129, 821, 1972.
534. **Maxwell, E. S., Kalckar, H. M., and Strominger, J. L.,** Some properties of uridine diphosphoglucose dehydrogenase, *Arch. Biochem. Biophys.,* 65, 2, 1956.
535. **Strominger, J. L. and Mapson, R.,** Uridine diphosphoglucose dehydrogenase of pea seedlings, *Biochem. J.,* 66, 567, 1957.
536. **Neufeld, E. F. and Hall, E. W.,** Inhibition of UDP-D-glucose dehydrogenase by UDP-D-xylose: a possible regulatory mechanism, *Biochem. Biophys. Res. Commun.,* 19, 456, 1965.
537. **Gainey, P. A. and Phelps, C. F.,** The binding of oxidised and reduced nicotinamide-adenine dinucleotides to bovine liver uridine diphosphate glucose dehydrogenase, *Biochem. J.,* 141, 667, 1974.
538. **Gainey, P. A. and Phelps, C. F.,** Interactions of uridine diphosphate glucose dehydrogenase with the inhibitor uridine diphosphate xylose, *Biochem. J.,* 145, 129, 1975.
539. **Chen, A., Marchetti, P., Weingarten, M., Franzen, J., and Feingold, D. S.,** Binding studies with bovine liver UDP-D-glucose dehydrogenase, *Carbohydr. Res.,* 37, 155, 1974.
540. **De Luca, G., Speziale, P., Balduini, C., and Castellani, A.,** Biosynthesis of glycosaminoglycans: uridine diphosphate glucose 4'-epimerase from cornea and epiphysial-plate cartilage, *Conn. Tiss. Res.,* 3, 39, 1975.
541. **Grygiel, I. H. and Kratzsch, E.,** Nachweis der Uridindiphosphatglucose-Dehydrogenase, einen spezifischen Enzym der Glucuronsäurebildung, in der menschlichen Placenta, *Arch. Gynaekol.,* 217, 173, 1974.
542. **Fan, D. F. and Troen, P.,** Formation of steroid conjugates in the human. I. Biosynthesis of uridinediphosphate glucuronic acid, *Metabolism,* 23, 125, 1975.
543. **Miettinen, T. A. and Leskinen, E.,** Glucuronic acid pathway, in *Metabolic Conjugation and Metabolic Hydrolysis,* Fishman, W. H., Ed., Vol. 1, Academic Press, New York, 1970, 157.
544. **Solms, J. and Hassid, W. Z.,** Isolation of uridine diphosphate N-acetylglucosamine and uridine diphosphate glucuronic acid from mung bean seedlings, *J. Biol. Chem.,* 228, 357, 1957.
545. **Feingold, D. S., Neufeld, E. F., and Hassid, W. Z.,** Enzymic synthesis of uridine diphosphate glucuronic acid and uridine diphosphate galacturonic acid with extracts from *Phaseolus aureus* seedlings, *Arch. Biochem. Biophys.,* 78, 401, 1958.
546. **Marsh, C. A.,** Reactions of uridine diphosphate sugars catalysed by enzymes of French-bean leaves, *Biochim. Biophys. Acta,* 44, 359, 1960.
547. **Roberts, R. M.,** The formation of uridine diphosphate-glucuronic acid in plants. Uridine diphosphate-glucuronic acid pyrophosphorylase from barley seedlings, *J. Biol. Chem.,* 246, 4995, 1971.
548. **Smith, E. E. B. and Mills, G. T.,** Uridine nucleotide compounds of liver, *Biochim. Biophys. Acta,* 13, 386, 1954.
549. **Touster, O. and Reynolds, V. H.,** The synthesis and properties of β-D-glucuronic acid-1-phosphate, *J. Biol. Chem.,* 197, 863, 1952.
550. **Arias, I. M., Lowy, B. A., and London, I. M.,** Studies of glucuronide synthesis and of glucuronyl transferase activity in liver and serum, *J. Clin. Invest.,* 37, 875, 1958.
551. **Shirai, Y., Wakabayishi, M., Ohishi, K., Hagi, K., and Mayuzumi, Y.,** Studies on the metabolism of glucuronic acid. Comparison of the biological functions of glucuronate, glucuronolactone and of ethyl glucuronate, *J. Biochem. (Tokyo),* 45, 893, 1958.
552. **Jeffrey, P. L. and Rienits, K. G.,** Involvement of UDP derivatives in the transformation of [¹⁴C] glucose into chondroitin sulphate by extracts of embryonic cartilage, *Biochim. Biophys. Acta,* 208, 147, 1970.
553. **Tsuyuki, H. and Idler, D. R.,** The metabolism of inositol in salmon. II. The role of nucleotides in relation to the metabolism of 2-C¹⁴-myoinositol in Coho liver, *Canad. J. Biochem. Physiol.,* 38, 1177, 1960.
554. **Forrest, R. J. and Hansen, R. G.,** A comparative study of nucleotide metabolism in fish liver, *Can. J. Biochem. Physiol.,* 37, 751, 1959.
555. **Burns, J. J. and Conney, A. H.,** Metabolism of glucuronic acid and its lactone, in *Glucuronic Acid,* Dutton, G. J., Ed., Academic Press, New York, 1966, 365.
556. **Loewus, F., Chen, M.-S., and Loewus, H. W.,** The myoinositol pathway to cell wall polysaccharides in *Biogenesis of Plant Cell Wall Polysaccharides,* Loewus, F., Ed., Academic Press, New York, 1973, 1.
557. **Neufeld, E. F., Feingold, D. S., and Hassid, W. Z.,** Enzymic phosphorylation of D-glucuronic acid by extracts from seedlings of *Phaseolus aureua, Arch. Biochem. Biophys.,* 83, 96, 1959.

558. **Pogell, B. M. and Krisman, C. R.**, Enzymic hydrolysis of uridine diphosphate glucuronate in rat skin, *Biochim. Biophys. Acta*, 41, 349, 1960.

559. **Conney, A. H. and Burns, J. J.**, Metabolism of uridine diphosphate glucuronic acid by liver and kidney, *Biochim. Biophys. Acta*, 54, 369, 1961.

560. **Jacobson, B. and Davidson, E. A.**, Biosynthesis of uronic acids by skin enzymes. II. Uridine diphosphate-D-glucuronic acid-5-epimerase, *J. Biol. Chem.*, 237, 638, 1962.

561. **Lindahl, U., Jacobsson, J., Höök, M., Backström, G. and Feingold, D. S.**, Biosynthesis of heparin. Loss of C-5 hydrogen during conversion of D-glucuronic to L-iduronic residues, *Biochem. Biophys. Res. Commun.*, 70, 492, 1976.

562. **Smith, E. E. B., Mills, G. T., Bernheimer, H. P., and Austrian, R.**, The presence of an uronic acid epimerase in a strain of pneumococcus type 1, *Biochim. Biophys. Acta*, 29, 640, 1958.

563. **Leventer, L. L., Buchanan, J. L., Ross, J. E., and Tapley, D. F.**, Solubilisation of N-glucuronyl transferase, *Biochim. Biophys. Acta*, 110, 428, 1965.

564. **Feingold, D. S., Neufeld, E. F., and Hassid, W. Z.**, The 4-epimerization and decarboxylation of uridine diphosphate D-glucuronic acid by extracts from *Phaseolus aureus* seedlings, *J. Biol. Chem.*, 235, 910, 1960.

565. **Gaunt, M. A., Maitra, U. S., and Ankel, H.**, UDP-Galacturonate 4-epimerase from the blue green alga *Anabaena flos-aquae*, *J. Biol. Chem.*, 249, 2366, 1974.

566. **Schliselfeld, H., van Eys, J., and Touster, O.**, The purification and properties of a nucleotide pyrophosphatase of rat liver nuclei, *J. Biol. Chem.*, 240, 811, 1965.

567. **Evans, W. H., Hood, D. O., and Gurd, J. W.**, Purification and properties of a mouse liver plasma-membrane glycoprotein hydrolysing nucleotide pyrophosphate and phosphodiester bonds, *Biochem. J.*, 135, 819, 1973.

568. **Lau, Y. K. and Wong, K. P.**, A comparative study of nuclear and microsomal UDP-glycuronic pyrophosphatase of rat liver, *Biochim. Biophys. Acta*, 334, 431, 1974.

569. **Puhakainen, E., Saarinen, A., and Hänninen, O.**, UDPglucuronic acid pyrophosphatase assay with the aid of alkaline phosphatase, *Acta Chem. Scand.*, B31, 125, 1977.

570. **Hollman, S. and Touster, O.**, Alterations in tissue levels of uridine diphosphate glucose dehydrogenase, uridine diphosphate glucuronic acid pyrophosphatase and glucuronyl transferase induced by substances influencing the production of ascorbic acid, *Biochim. Biophys. Acta*, 62, 338, 1962.

571. **Puhakainen, E.**, Nucleotide pyrophosphatase as a competitor of drug hydroxylation and conjugation in rat liver microsomes, in *Biological Reactive Intermediates*, Jollow, D. J., Kocsis, J. J., Snyder, R., and Vainio, H., Eds., Plenum Press, New York, 1977, 264.

572. **Ginsburg, W., Weissbach, A., and Maxwell, E. S.**, Formation of glucuronic acid from uridinediphosphate glucuronic acid, *Biochim. Biophys. Acta*, 28, 649, 1958.

573. **Stevenson, I. H. and Dutton, G. J.**, Mechanism of glucuronide synthesis in skin, *Biochem. J.*, 77, 19 P, 1960.

574. **Stevenson, I. H. and Dutton, G. J.**, Glucuronide synthesis in kidney and gastrointestinal tract, *Biochem. J.*, 82, 330, 1962.

575. **Jacobson, K. B. and Kaplan, N. O.**, A reduced pyridine nucleotide pyrophosphatase, *J. Biol. Chem.*, 226, 427, 1957.

576. **Touster, O., Aronson, N. N., Jr., Dulaney, J. T., and Hendrickson, H.**, Isolation of rat liver plasma membranes. Use of nucleotide pyrophosphatase and phosphodiesterase I as marker enzymes, *J. Cell Biol.*, 47, 604, 1970.

577. **Bischoff, E., Liersch, M., Keppler, D., and Decker, K.**, Fate of intravenously administered UDP-glucose, *Hoppe Seylers Z. Physiol. Chem.*, 351, 729, 1970.

578. **Bischoff, E., Tran-Thi, T.-A., and Decker, K. F. A.**, Nucleotide pyrophosphatase of rat liver. A comparative study on the enzymes solubilized and purified from plasma membrane and endoplasmic reticulum, *Eur. J. Biochem.*, 51, 353, 1975.

579. **Puhakainen, E. and Hänninen, O.**, Pyrophosphatase and glucuronosyltransferase in microsomal UDPglucuronic-acid metabolism in the rat liver, *Eur. J. Biochem.*, 61, 165, 1976.

580. **Puhakainen, E.**, Formation of free D-Glucuronic Acid from UDPGlucuronic Acid, Ph.D. thesis, University of Kuopio, Finland, 1977.

581. **Adlard, B. P. F.**, The Interaction of Steroids and Bilirubin Glucuronyl Transferase in Relation to Newborn Jaundice, Ph.D. thesis, University of Leeds, England, 1969.

582. **Neufeld, E. F., Feingold, D. S., and Hassid, W. Z.**, Enzymatic conversion of uridine diphosphate-D-glucuronic acid to uridine diphosphate galacturonic acid, uridine diphosphate xylose and uridine diphosphate arabinose, *J. Am. Chem. Soc.*, 80, 4430, 1958.

583. **Castanera, E. G. and Hassid, W. Z.**, Properties of uridine diphosphate D-glucuronic acid decarboxylase from wheat germ, *Arch. Biochem. Biophys.*, 110, 462, 1965.

584. **Grebner, E. E., Hall, C. W., and Neufeld, E. F.**, Glycosylation of serine residues by a uridine diphosphate-xylose: protein xylosyltransferase from mouse mastocytoma, *Arch. Biochem. Biophys.*, 116, 391, 1966.

585. **Robinson, H. C., Telser, A., and Dorfman, A.**, Studies on biosynthesis of the linkage region of chondroitin sulfate-protein complex, *Proc. Natl. Acad. Sci. U.S.A.*, 56, 1859, 1966.

586. **Wellmann, E. and Grisebach, H.**, Purification and properties of an enzyme preparation from *Lemna minor*, L., catalysing the synthesis of UDP-apiose and UDP-D-xylose from UDP-D-glucuronic acid, *Biochim. Biophys. Acta*, 235, 389, 1971.

587. **Sandermann, H., Jr., Tisue, G. T., and Grisebach, H.**, Biosynthesis of D-apiose. IV. Formation of UDP-apiose from UDP-D-glucuronic acid in cell-free extracts of parsley (*Apium petroselinum*, L.,) and *Lemna minor*, L., *Biochim. Biophys. Acta*, 165, 550, 1968.

588. **Sandermann, H., Jr. and Grisebach, H.**, Biosynthesis of D-apiose. V. NAD$^+$-dependent biosynthesis of UDP-apiose and and UDP-xylose from UDP-D-glucuronic acid with an enzyme preparation from *Lemna minor*, L., *Biochim. Biophys. Acta*, 208, 173, 1970.

589. **Mendicino, J. and Hussein, A-Z.**, Conversion of UDP-D-glucuronic acid to UDP-D-apiose and UDP-D-xylose by an enzyme isolated from *Lemna minor*, *Biochim. Biophys. Acta*, 364, 159, 1974.

590. **Kelleher, W. J. and Grisebach, H.**, Hydride transfer in the biosynthesis of uridine diphospho-apiose from uridine diphospho-D-glucuronic acid with an enzyme preparation of *Lemna minor*, *Eur. J. Biochem.*, 23, 136, 1971.

591. **Marselos, M.**, Glucaric acid synthesis in the hepatic glucuronic acid pathway, Ph.D. thesis, University of Kuopio, Finland, 1976.

592. **Rugstad, H. E. and Dybing, E.**, Competition between *p*-aminophenol, *p*-nitrophenol and bilirubin for glucuronidation in cultures of rat hepatoma cells and homogenates of the same cells, *Acta Pharmacol. Toxicol.*, 34, 65, 1974.

593. **Keppler, D. O. R. and Decker, K.**, Studies on the mechanism of galactosamine hepatitis: accumulation of galactosamine-1-phosphate and its inhibition of UDP-glucose pyrophosphorylase, *Eur. J. Biochem.*, 10, 219, 1969.

594. **Levvy, G. A. and Conchie, J.**, β-Glucuronidase and the hydrolysis of glucuronides, in *Glucuronic Acid*, Dutton, G. J., Ed., Academic Press, New York, 1966, 301.

595. **Das, I., Wentworth, M. A., Ide, H., Sie, H-G., and Fishman, W. H.**, Uridine diphosphate β-glucuronic acid, a new substrate for β-glucuronidase, *Biochim. Biophys. Acta*, 201, 375, 1970.

596. **Fishman, W. H.**, Determination of β-glucuronidases, *Methods Biochem. Anal.*, 15, 77, 1967.

597. **Silbert, J. E.**, Metabolism of polysaccharides containing glucuronic acid, in *Glucuronic Acid*, Dutton, G. J., Ed., Academic Press, New York, 1966, 385.

598. **Eisenberg, F., Jr.**, The mechanism of action of β-glucuronidase, *Fed. Proc. Fed. Am. Soc. Exp. Biol.*, 18, 221, 1959.

599. **Stoeber, F.**, Sur la biosynthèse induite de la β-glucuronidase chez *Escherichia coli*, *C. R. Acad. Sci.*, 244, 950, 1957.

600. **Colucci, D. F. and Buyske, D. A.**, The biotransformation of a sulfonamide to a mercaptan and to mercapturic acid and glucuronide conjugates, *Biochem. Pharmacol.*, 14, 457, 1965.

601. **Hètu, C. and Gianetto, R.**, Synthetic 1-thio-β-D-glucosiduronic acids as substrates for rat liver β-glucuronidase, *Can. J. Biochem.*, 48, 799, 1970.

602. **Marsh, C. A. and Levvy, G. A.**, The relationship between glucuronidase and galacturonidase activity in the limpet and in mammalian tissues, *Biochem. J.*, 68, 610, 1958.

603. **Stahl, P. D. and Touster, O.**, β-Glucuronidase of rat liver lysosomes. Purification, properties, subunits, *J. Biol. Chem.*, 246, 5398, 1971.

604. **Owens, J. W. and Stahl, P.**, Purification and characterization of rat liver microsomal β-glucuronidase, *Biochim. Biophys. Acta*, 438, 474, 1976.

605. **Brot, F. E., Glaser, J. H., Roosen, K. J., Sly, W., and Stahl, P.**, *In vitro* correction of deficient human fibroblasts by β-glucuronidase from different human sources, *Biochem. Biophys. Res. Commun.*, 57, 1, 1971.

606. **Paigen, K., Swank, R. T., Tomino, S., and Ganschow, R. E.**, The molecular genetics of mammalian glucuronidase, *J. Cell Physiol.*, 85, 379, 1975.

607. **Chern, C. J.**, Assignment of the structural gene for human β-glucuronidase to chromosome 7 and tetrameric association of subunits in the enzyme molecule, *Am. J. Hum. Genet.*, 28, 350, 1976.

608. **Francke, U.**, The human gene for β-glucuronidase is on chromosome 7, *Am. J. Hum. Genet.*, 28, 357, 1976.

609. **Tomino, S., Paigen, K., Tulsiani, D. R. P., and Touster, O.**, Purification and chemical properties of mouse liver β-glucuronidase L-form, *J. Biol. Chem.*, 250, 8503, 1975.

610. **Swank, R. T. and Paigen, K.**, Biochemical and genetic evidence for a macromolecular β-glucuronidase complex in microsomal membranes, *J. Mol. Biol.*, 77, 371, 1973.

611. **Walker, P. G.**, The preparation and properties of β-glucuronidase. III. Fractionation and activity of homogenates in isotonic media, *Biochem. J.*, 51, 223, 1952.

612. **De Duve, C. B. C., Preesman, R., Gianetto, R., Wattiaux, R., and Appelmans, F.**, Tissue fractionation studies. 6. Intracellular distribution studies of enzymes in rat liver tissue, *Biochem. J.*, 60, 604, 1955.

613. **Owens, J. W., Gammon, K. L., and Stahl, P. D.**, Multiple forms of β-glucuronidase in rat liver lysosomes and microsomes, *Arch. Biochem. Biophys.*, 166, 258, 1975.

614. **Mandell, B. and Stahl, P.**, Effects of di-isopropyl phosphofluoridate on rat liver microsomal and lysosomal β-glucuronidase, *Biochem. J.*, 164, 549, 1977.

615. **Marsh, C. A., Lin, C.-W., and Fishman, W. H.**, Golgi β-glucuronidase of androgen-stimulated mouse kidney, *Biochem. J.*, 142, 491, 1974.

616. **Owerbach, D. and Lusis, A. T.**, Phenobarbital induction of egasyn: availability of egasyn *in vivo* determines β-glucuronidase binding to membrane, *Biochem. Biophys. Res. Commun.*, 69, 628, 1976.

617. **Kiyomoto, A., Harigaya, S., Oshima, S., and Morita, T.**, Studies on glucosaccharolactones - I. Glucosaccharo - 1:4-lactone, *Biochem. Pharmacol.*, 12, 105, 1963.

618. **Mulder, G. J.**, Effect of phenobarbital treatment on lysosomal enzyme activity in rat liver, *Biochem. Pharmacol.*, 20, 1328, 1971.

619. **Lucier, G. W. and McDaniel, O. S.**, Alterations in rat liver microsomal and lysosomal β-glucuronidase by compounds which induce hepatic drug metabolizing enzymes, *Biochim. Biophys. Acta*, 261, 168, 1972.

620. **Platt, D. and Katzemeier, U.**, Untersuchungen zur Frage der Induzierbarkeit lysosomaler Enzyme der Rattenleber durch Phenobarbital, *Arzneim. Forsch.*, 20, 258, 1970.

621. **Marsh, C. A., Alexander, F., and Levvy, G. A.**, Glucuronide decomposition in the digestive tract, *Nature, (London)*, 170, 163, 1953.

622. **Boyland, E. and Williams, R. T.**, The biochemistry of aromatic amines. 7. The enzymic hydrolysis of aminonaphthyl glucosiduronic acids, *Biochem. J.*, 76, 388, 1960.

623. **Hartiala, K. and Hakkinen, I.**, Inhibition of cinchophen ulcer in chicks with boiled potassium hydrogen saccharate, *Acta Physiol. Scand.*, 49, 92, 1960.

624. **Matsuchiro, T.**, Identification of glucaro-1,4-lactone in bile as a factor responsible for inhibitory effect of bile on bacterial β-glucuronidase, *Tohuku J. Exp. Med.*, 85, 330, 1965.

625. **Woods, L. A.**, Distribution and fate of morphine in non-tolerant and tolerant dogs and rats, *J. Pharm. Exp. Therap.*, 112, 158, 1954.

626. **Clark, A. G., Fischer, F. J., Millburn, P., Smith, R. L., and Williams, R. T.**, The role of gut flora in the enterohepatic circulation of stilboestrol in the rat, *Biochem. J.*, 112, 17 P, 1969.

627. **Marselos, M., Dutton, G., and Hänninen, O.**, Evidence that glucaro-1→4-lactone shortens the pharmacological action of drugs being disposed *via* the bile as glucuronides, *Biochem. Pharmacol.*, 24, 1855, 1975.

628. **Williams, C. H.**, β-Glucuronidase activity in the serum and liver of rats administered pesticides and hepatotoxic agents, *Toxicol. Appl. Pharmacol.*, 14, 283, 1969.

629. **Iida, R., Nagata, S., Kakimoto, M., Akaike, H., Watanabe, H., and Shioya, A.**, 2,5-Di-*O*-acetyl-D-glucosaccharo-(1→4) (6→3)-dilactone, a new potent β-glucuronidase inhibitor, *Jpn. J. Pharmacol.*, 15, 88, 1965.

630. **Levvy, G. A.**, The preparation and properties of β-glucuronidase. 4. Inhibition by sugar acids and their lactones, *Biochem. J.*, 52, 464, 1952.

631. **Tomašić, J. and Keglević, D.** The kinetics of hydrolysis of synthetic glucuronic esters and glucuronic ethers by bovine liver and *Escherichia coli* β-glucuronidase, *Biochem. J.*, 133, 789, 1973.

632. **El-Zoghby, S. M., El-Kholy, Z. A., Abdel-Tawab, G. A., Hamoud, F., and Kandil, G.**, Studies on the effectiveness of liver and bacterial β-glucuronidase enzymes in the hydrolysis of urinary β-glucosiduronides of some tryptophan metabolites, *Invest. Urol.*, 13, 154, 1975.

633. **Levvy, G. A., McAllan, A., and Marsh, C. A.**, Purification of β-glucuronidase from the preputial gland of female rat, *Biochem. J.*, 69, 22, 1958.

634. **Hudson, B., Sheath, J., and Duimanis, A.**, Hydrolysis of steroid glucuronides with β-glucuronidase from preputial glands of the female rat, *Endocrinology*, 70, 189, 1962.

634a. **Faed, E. M.**, Origin of an unusual glucuronide conjugate of clofibric acid — a possible intramolecular rearrangement, in *Conjugation Reactions in Drug Biotransformation*, Aitio, A., Ed., Elsevier, Amsterdam, 1978, 496.

635. **Dutton, G. J.**, Commentary. Control of UDP-glucuronyltransferase activity, *Biochem. Pharmacol.*, 24, 1835, 1975.

636. **Dutton, G. J.**, Developmental aspects of drug conjugation, *Ann. Rev. Pharmacol. Toxicol.*, 18, 17, 1978.

636a. **Dutton, G. J.**, Factors controlling onset of perinatal drug-metabolizing enzymes, *Progr. Drug. Res.*, 25, 1980, in press.

636b. **Onishi, S., Kawade, N., Itoh, S., Isobe, K., and Sugiyama, S.**, Postnatal development of UDP-glucuronyltransferase activity towards bilirubin and 2-aminophenol in human liver, *Biochem. J.*, 184, 705, 1979.

636c. **Campbell, M. T. and Wishart, G. J.**, The effect of premature and delayed birth on the development of UDP-glucuronosyltransferase activities towards bilirubin, morphine and testosterone in the rat, *Biochem. J.*, 186, 617, 1980.

636d. **Leakey, J. E. A., Dutton, G. J., and Wishart, G. J.**, Differential stimulation of hepatic mono-oxygenase and glucuronidating systems in chick embryo and neonatal rat by glucocorticoids, *Med. Biol.*, 57, 256, 1979.

636e. **Kapitulnik, J., Tshershedsky, M., and Barenholz, Y.**, Fluidity of the rat liver microsomal membrane: increase at birth, *Science*, 206, 843, 1979.

637. **Neims, A. H., Warner, M., Loughnan, P. M., and Aranda, J. V.**, Developmental aspects of the hepatic cytochrome P-450 monooxygenase system, *Ann. Rev. Pharmacol. Toxicol.*, 16, 427, 1976.

637a. **Short, C. R., Kinden, D. A., and Stith, R.**, Fetal and neonatal development of the microsomal monooxygenase system, *Drug Metab. Rev.*, 5, 1, 1976.

638. **Pelkonen, O.**, Transplacental transfer of foreign compounds and their metabolism by the foetus, *Progr. Drug. Metab.*, 2, 119, 1977.

639. **Wilson, J. T.**, Developmental pharmacology: a review of its application to clinical and basic science, *Ann. Rev. Pharmacol.*, 12, 423, 1972.

640. **Yaffe, S. J. and Juchau, M. R.**, Perinatal pharmacology, *Ann. Rev. Pharmacol.*, 14, 219, 1974.

641. **Done, A. K.**, Developmental pharmacology, *Clin. Pharmacol. Ther.*, 5, 432, 1964.

642. **Vest, M. F. and Rossier, R.**, Detoxification in the newborn: the ability of the newborn infant to form conjugates with glucuronic acid, glycine, acetate and glutathione, *Ann. N.Y. Acad. Sci.*, 111, 183, 1963.

643. **Vest, M. F. and Salzberg, R.**, Conjugation reactions in the newborn infant: the metabolism of *p*-aminobenzoic acid, *Arch. Dis. Child.*, 40, 97, 1965.

644. **Goldenthal, E. I.**, A compilation of LD_{50} values in newborn and adult animals, *Toxicol. Appl. Pharmacol.*, 18, 185, 1971.

645. **Horning, M. G., Stratton, C., Nowlin, J., Wilson, A., Horning, E. C., and Hill, R. M.**, Placental transfer of drugs, in *Fetal Pharmacology*, Boréus, L., Ed., Raven Press, New York, 1973, 355.

646. **Oh, Y. and Mirkin, B. L.**, Transfer of drugs into the central nervous system and across the placenta: a comparative study utilizing aminopyrine (A), diphenylhydantoin (D), sodium salicylate (S), and mecamylamine (M), *Fed. Proc. Fed. Am. Soc. Exp. Biol.*, 30, Abstr. 2034, 1971.

647. **Speirs, J. and Sim, A. W.**, The placental transfer of pancuronium bromide, *Br. J. Anaesth.*, 44, 370, 1972.

648. **Juchau, M. R. and Dyer, D. C.**, Pharmacology of the placenta, *Pediatr. Clin. North Am.*, 19, 65, 1972.

649. **Gartner, L. M., Lee, K., Vaisman, S., Lane, D., and Zarafu, I.**, Development of bilirubin transport and metabolism in the newborn rhesus monkey, *J. Pediatr.*, 90, 513, 1977.

650. **Greengard, O.**, Enzymic differentiation of human liver: comparison with the rat model, *Pediatr. Res.*, 11, 669, 1977.

651. **Hietanen, E.**, Developmental pattern of hepatic uridine diphosphoglucuronosyl transferase activity in growing male rats, *Biol. Neonate*, 31, 135, 1977.

652. **O'Toole, K. and Pollak, J. K.**, Changes in free and membrane bound ribosomes during the development of chick liver. A new cell fractionation approach, *Biochem. J.*, 138, 359, 1974.

653. **Leskes, A., Siekevitz, P., and Palade, G. E.**, Differentiation of endoplasmic reticulum in hepatocytes. II. Glucose-6-phosphatase in rough microsomes, *J. Cell Biol.*, 49, 288, 1971.

654. **Koga, A.**, Morphogenesis of intrahepatic bile duct of the human foetus, *Z. Anat. Entwickluugsgesch.*, 135, 156, 1971.

655. **Mikhail, G., Wiqvist, N., and Diczfalusy, E.**, Oestriol metabolism in the pre-viable human foetus, *Acta Endocrinol.*, 42, 519, 1963.

656. **Diczfalusy, E., Franksson, G., Lisboa, B. P., and Martinsen, B.**, Formation of oestrone glucosiduronate by the human intestinal tract, *Acta Endocrinol.*, 40, 537, 1962.

657. **Solomon, S., Bird, C. E., Ling, W., Iwamiya, M., and Young, P. C. M.**, Formation and metabolism of steroids in the fetus and placenta, *Recent Progr. Horm. Res.*, 23, 297, 1967.

658. **Diczfalusy, E. and Levitz, M.**, Formation, metabolism and transport of estrogen conjugates, in *Chemical and Biological Aspects of Steroid Conjugation*, Springer, Berlin, 1970, 291.

659. **Bird, C. E., Solomon, S., Wiqvist, N., and Diczfalusy, E.**, Formation of C-21 steroid sulphates and glucuronides by pre-viable human foetuses perfused with [4-^{14}C] progesterone, *Biochim. Biophys. Acta*, 104, 623, 1965.

660. **Ling, W., Coutts, J. R. T., MacNaughton, M. C., and Solomon, S.**, Formation of conjugated metabolites after injection of [4-^{14}C]17α-hydroxyprogesterone into the umbilical vein of the human foetus at midpregnancy, *J. Endocrinol.*, 58, 477, 1973.

661. **Huhtaniemi, I.**, Endogenous steroid sulphates and glucuronides in the gallbladder bile from early and midterm human foetuses, *J. Endocrinol.*, 59, 503, 1973.

662. **Tikkanen, M. J. and Adlercreutz, H.**, Oestriol conjugates in amniotic fluid. Qualitative and quantitative aspects, including preliminary studies in Rh isoimmunization, *Acta Endocrinol.*, 73, 555, 1973.

663. Schweitzer, M., Klein, G. P. and Giroud, C. J. P., Characterization of 17-deoxy and 17-hydroxy corticosteroids in human liquor amnii, *J. Clin. Endocrinol. Metab.*, 33, 605, 1971.

664. Schweitzer, M. and Giroud, C. J. P., A comparison in the pattern of steroid glucuronides and sulfates in maternal plasma, umbilical cord plasma and amniotic fluid, *J. Clin. Endocrinol. Metab.*, 33, 793, 1971.

665. Young, B. K., Jirku, H. and Levitz, M., Estriol conjugates in amniotic fluid at midpregnancy and term, *J. Clin. Endocrinol. Metab.*, 35, 208, 1972.

666. Kinsella, R. A. and Francis, F. E., Steroids and sterols in meconium, *J. Clin. Endocrinol. Metab.*, 32, 801, 1971.

667. Gustafsson, J.-Å. and Stenberg, Å., Steroids in meconium from male and female newborn infants, *Eur. J. Biochem.*, 22, 246, 1971.

668. Goebelsmann, U., Wiqvist, N., and Diczfalusy, E., Placental transfer of oestriol glucosiduronates, *Acta Endocrinol.*, 59, 426, 1968.

669. Goebelsmann, U., Roberts, J. M. and Jaffe, R. B., Placental transfer of ³H-oestriol-3-sulphate-16-glucosiduronate and ³H-oestriol-16-glucosiduronate-¹⁴C, *Acta Endocrinol.*, 70, 132, 1972.

670. O'Donoghue, S. E. F., Distribution of pethidine and chlorpromazine in maternal, foetal and neonatal biological fluids, *Nature (London)*, 229, 124, 1971.

671. Rane, A., Urinary excretion of diphenylhydantoin metabolites in newborn infants, *J. Pediatr.*, 85, 543, 1974.

672. Done, A. K., Cohen, S. N., and Strebel, L., Pediatric clinical pharmacology and the "therapeutic orphan", *Ann. Rev. Pharmacol. Toxicol.*, 17, 561, 1977.

673. Eriksson, M. and Yaffe, S. J., Drug metabolism in the newborn, *Ann. Rev. Med.*, 24, 29, 1973.

674. Gladtke, E. and Rind, H., Bilirubinstoffwechsel beim Neugeborenen, *Monatsschr. Kinderheilkd.*, 115, 231, 1967.

675. Di Toro, R., Lupi, L., and Ansanelli, V., Glucuronidation of the liver in premature babies, *Nature (London)*, 219, 265, 1968.

676. Krawczynska, H., Zachmann, M., and Pradev, A., Urinary testosterone glucuronide and sulphate in newborns and young infants, *Acta Endocrinol.*, 82, 842, 1976.

677. Stern, L., Khanna, N. N., Levy, G., and Yaffe, S. J., Effect of phenobarbital on hyperbilirubinemia and glucuronide formation in newborns, *Am. J. Dis. Child.*, 120, 26, 1970.

678. Heirwegh, K. P. M., Blanckaert, N., and Fevery, J., Methods of determination and nature of diazo-positive bile pigments present in body fluids in neonatal jaundice and related conditions, *Birth Defects Orig. Artic. Ser.*, 12, 293, 1976.

679. Wilson, J. T., Hojer, B., and Rane, A., Loading and conventional dose therapy with phenytoin in children: kinetic profile of parent drug and main metabolite in plasma, *Clin. Pharm. Ther.*, 20, 48, 1976.

680. Karunairatnam, M. C., Kerr, L. M. H., and Levvy, G. A., The glucuronide synthesizing system in the mouse and its relation to β-glucuronidase, *Biochem. J.*, 45, 496, 1949.

681. Brown, A. K. and Zuelzer, W. W., Studies on the neonatal development of the glucuronide conjugating system, *J. Clin. Invest.*, 37, 332, 1958.

682. Dutton, G. J., Glucuronide synthesis in foetal liver and other tissues, *Biochem. J.*, 71, 141, 1959.

683. Lathe, G. H. and Walker, M., The synthesis of bilirubin glucuronide in animal and human liver, *Biochem. J.*, 70, 705, 1958.

684. Schenker, S. and Schmid, R., Excretion of C¹⁴-bilirubin in newborn guinea pigs, *Proc. Soc. Exp. Biol. Med.*, 115, 446, 1964.

685. Grodsky, G. M., Kolb, H. J., Fanska, R. E., and Nemechek, E., Effect of age on rate of development of hepatic carriers for bilirubin: a possible explanation for physiologic jaundice and hyperbilirubinemia of the newborn, *Metabolism*, 19, 246, 1970.

686. Levi, A. J., Gatmaitan, Z., and Arias, I. M., Deficiency of organic anion-binding protein, impaired organic anion-binding protein, impaired organic anion uptake by liver and 'physiologic' jaundice in newborn monkeys, *N. Engl. J. Med.*, 283, 115, 1970.

687. De Wolf-Peeters, C., Moens Bullens, A.-M., Van Assche, A., and Desmet, V. J., Conjugated bilirubin in fetal liver in eryhtroblastosis, *Lancet*, i, 471, 1969.

688. De Wolf-Peeters, C., De Vos, R., and Desmet, V., Histochemical evidence of a cholestatic period in neonatal rats, *Pediatr. Res.*, 5, 704, 1971.

689. Feuer, G. and Liscio, A., Origin of delayed development of drug metabolism in the newborn rat, *Nature (London)*, 223, 68, 1969.

690. Marsh, C. A. and Carr, A. J., Changes in enzyme activity, related to D-glucaric acid synthesis, with age, pregnancy and malignancy, *Clin. Sci.*, 28, 209, 1965.

691. Lucier, G. W., Sonawane, B. R., and McDaniel, O. S., Glucuronidation and deglucuronidation reactions in hepatic and extrahepatic tissues during perinatal development, *Drug Metab. Dispos.*, 5, 279, 1977.

692. **Leakey, J. E. A. and Donald, A. M.,** Activation *versus* induction of foetal enzymes during culture: a problem illustrated by increased uridine diphosphate glucuronic acid-5-hydorxytryptamine glucuronosyltransferase activity through organ culture on rafts and paper strips, *Biochem. Soc. Trans.,* 4, 1071, 1976.

693. **G. J. Wishart and G. J. Dutton,** Regulation of onset of development of UDP-glucuronyltransferase activity towards *o*-aminophenol by glucocorticoids in late foetal rat liver *in utero,Biochem. J.,* 168, 507, 1977.

694. **Krasner, J., Juchau, M. R., and Jaffe, S. J.,** Postnatal developmental changes in hepatic bilirubin UDP-glucuronyl transferase. Studies on the solubilized enzyme, *Biol. Neonate,* 23, 381, 1973.

695. **Felsher, B. F., Maidman, J. E., Carpio, N. M., Van Couvering, K., and Woolley, M. M.,** Reduced hepatic bilirubin uridine diphosphate glucuronyl transferase and uridine diphosphate glucose dehydrogenase activity in the human fetus, *Pediatr. Res.,* 12, 838, 1978.

696. **Brodersen, R., Jacobsen, J., Hertz, H., Rebbe, H., and Sørensen, B.,** Bilirubin conjugation in the human fetus, *Scand. J. Clin. Lab. Invest.,* 20, 41, 1967.

697. **Short, C. R. and Davis, L. E.,** Perinatal development of drug-metabolizing enzyme activity in swine, *J. Pharmacol. Exp. Ther.,* 174, 185, 1970.

698. **Schröter, W.,** Intracellularer Transport und Membran des endoplasmatischen Reticulums der Leberzelle: neue Aspekte in der Genese der Transitorischen Neugeborenenbilirubinämie, *Monatsschr. Kinderheilk.,* 120, 119, 1972.

699. **Kandall, S. R., Thaler, M. M., and Erickson, R. P.,** Intestinal development of lysosomal and microsomal beta glucuronidase and bilirubin uridine diphosphoglucuronyl transferase in normal and jaundiced rats, *J. Pediatr.,* 82, 1013, 1973.

700. **Wu, C.-H., Archer, D. F., Flickinger, G. L., and Touchstone, J. C.,** *In vitro* estrogen biosynthesis in foetal organs of mid-pregnancy, *Acta Endocrinol.,* 65, 675, 1970.

701. **Rane, A., Sjöqvist, F., and Orrenius, S.,**Drugs and fetal metabolism, *Clin. Pharm. Ther.,* 14, 666, 1973.

702. **Flint, M., Lathe, G. H., Ricketts, T. R., and Silman, G.,** Development of glucuronyl transferase and other enzyme systems in the newborn rabbit, *Q. J. Exp. Physiol.,* 49, 66, 1964.

703. **Stevens, L.,** A comparative study of enzymes in foetal, young and adult rat, *Comp. Biochem. Physiol.,* 6, 129, 1962.

704. **Dutton, G. J.,** Neonatal drug toxicity caused by defective glucuronide synthesis: unsuitability of the rat as a test animal?, in *Proc. European Society for the Study of Drug Toxicity,* Vol. 4, Davey, D. G., Ed., Excerpta Medica, Amsterdam, 121, 1964.

705. **Greengard, O.,** Enzymic differentiation in mammalian tissues, *Essays Biochem.,* 7, 159, 1971.

706. **Thaler, M. M.,** Substrate-induced conjugation of bilirubin in genetically-deficient newborn rats, *Science,* 170, 550, 1970.

707. **Campbell, M. T. and Wishart, G. J.,** The effect of birth and related hormonal events on the neonatal development of UDP-glucuronosyltransferase activity towards bilirubin and other substrates in rat liver, *Biochem. Soc. Trans.,* 6, 175, 1978.

708. **Vaisman, S. L., Lee, K.-S., and Gartner, L. M.,** Xylose, glucose and glucuronic acid conjugation in the newborn rat, *Pediatr. Res.,* 10, 967, 1976.

709. **Ducharme, J. R., Limal, J.-M., Autic, B., and Sandor, T.,** C_{21} steroid metabolism and conjugation in the human premature neonate. II. Tissue metabolism and conjugation of exogenous cortisol, 11-deoxycortisol, corticosterone and pregnenolone in the presence and absence of ACTH *in vitro, Acta Endocrinol.,* 72, 63, 1973.

710. **Short, C. R., Maines, M. D., and Westfall, B. A.,** Postnatal development of drug-metabolizing enzyme activity in liver and extrahepatic tissues of swine, *Biol. Neonate,* 21, 54, 1972.

711. **Leakey, J. E. A., Wishart, G. J., and Dutton, G. J.,** Precocious development of UDP-glucuronyltransferase activity in chick embryo liver following administration of adrenocorticotrophic hormone and of certain 11β-hydroxy-corticosteroids, *Biochem. J.,* 158, 419, 1976.

712. **Jakowicki, J., Evvast, H.-S., and Adlercreutz, H.,** Gas chromatographic determination and mass spectrometric identification of oestrogens in normal human placental tissue at term, *J. Steroid Biochem.,* 4, 181, 1973.

713. **Smith, S. W. and Axelrod, L. R.,** Studies on the metabolism of steroid hormones and their precursors by the human placenta at various stages of gestation. I. *In vitro* metabolism of 1,3,5 (10)-estratriene-3,17β-diol, *J. Clin. Endocrinol. Metab.,* 29, 85, 1969.

714. **Chin, B. H., Eldridge, J. M., and Sullivan, L. J.,** Metabolism of carbaryl by selected human tissues using an organ-maintenance technique, *Clin. Toxicol.,* 7, 37, 1974.

715. **Chakraborty, J., Hopkins, R., and Parke, D. V.,** Biological oxygenation of drugs and steroids in the placenta, *Biochem. J.,* 125, 15P, 1971.

716. **Kyegombe, D., Franklin, C., and Turner, P.,** Drug-metabolizing enzymes in the human placenta, their induction and repression, *Lancet,* i, 405, 1973.

717. **Aitio, A.**, UDP glucuronyltransferase activity in various rat tissues, *Int. J. Biochem.*, 5, 325, 1974.
718. **Berté, F., Manzo, L., de Bernardi, M., and Benzi, G.**, Ability of the placenta to metabolize oxazepam and aminopyrine before and after drug stimulation, *Arch. Int. Pharmacodyn.*, 182, 182, 1969.
719. **Lester, R. and Schmid, R.**, Bile pigment excretion in amphibia, *Nature (London)*, 190, 452, 1961.
720. **Arias, I. M.**, Studies of bile pigment metabolism in the dogfish (Squalus acanthia) and the goosefish (Lophius piscatorius), *Bull. Mt. Desert Isl. Biol. Lab.*, 10, 614, 1962.
721. **Yaffe, S. J., Levy, G., Matsuzawe, T., and Baliah, T.**, Enhancement of glucuronide conjugating capacity in a hyperbilirubinemic infant due to apparent enzyme induction by phenobarbital, *N. Engl. J. Med.*, 275, 461, 1966.
722. **Ramboer, C., Thompson, R. P. H., and Williams, R.**, Controlled trials of phenobarbitone therapy in neonatal jaundice, *Lancet*, i, 966, 1969.
723. **Boréus, L. O., Jalling, B., and Køllberg, N.**, Clinical pharmacology of phenobarbital in the neonatal period, in *Basic and Therapeutic Aspects of Perinatal Pharmacology*, Morselli, P. L., Garattini, S., and Sereni, F., Raven Press, New York, 1975, 331.
724. **Nathenson, G., Cohen, M. I., Flitt, I., and McNamara, H.**, The effect of maternal heroin addiction on neonatal jaundice, *J. Pediatr.*, 81, 899, 1972.
725. **Idéo, G., de Franchis, R., del Minno, E., Cocucci, C., and Dioguardi, N.**, Increase of some rat liver microsomal enzymes as a consequence of prolonged alcohol intake: comparison with the effect of phenobarbitone, *Enzyme*, 12, 473, 1971.
726. **Waltman, R., Bonura, F., Nigrin, G., and Pipat, C.**, Ethanol in prevention of hyperbilirubinaemia in the newborn. A controlled trial, *Lancet*, ii, 1265, 1969.
727. **Remmer, H.**, Vermehrte Glucuronidierung von Sulfamethoxin während und nach Phenobarbitalbehandlung bei Ratten, *Naunyn Schmiedebergs Arch. Exp. Pathol. Pharmakol.*, 247, 461, 1964.
728. **Catz, C. and Yaffe, S. J.**, Pharmacological modification of bilirubin conjugation in the newborn, *Am. J. Dis. Child.*, 104, 516, 1962.
729. **Arias, I. M., Fleischner, G., Listowsky, I., Kamisaka, K., Mishkin, S., and Gatmaitan, Z.**, On the structure and function of ligandin and Z protein, in *The Hepatobiliary System*, Taylor, W., Ed., Plenum Press, New York, 1976, 81.
730. **Hales, B. F. and Neims, A. H.**, Developmental aspects of glutathione *S*-transferase B (ligandin) in rat liver, *Biochem. J.*, 160, 231, 1976.
731. **Kuenzig, W., Kamm, J. J., Boublik, M., Jenkins, F., and Burns, J. J.**, Perinatal drug metabolism and morphological changes in the hepatocytes of normal and phenobarbital treated guinea pigs, *J. Pharmacol. Exp. Ther.*, 191, 32, 1974.
732. **Arias, I. M., Gartner, L., Furman, M., and Wolfson, S.**, Effect of several drugs and chemicals on hepatic glucuronide formation in newborn rats, *Proc. Soc. Exp. Biol. Med.*, 112, 1037, 1963.
733. **Inscoe, J. K. and Axelrod, J.**, Some factors affecting glucuronide formation *in vitro*, *J. Pharmacol. Exp. Ther.*, 129, 128, 1960.
734. **Parke, D. V., Rahman, K. M. Q., and Walker, R.**, Effect of linalool on hepatic drug-metabolizing enzymes in the rat, *Biochem. Soc. Trans.*, 2, 615, 1974.
734a. **Lucier, G.**, Unique structure-activity relationships for the perinatal pharmacology of the polychlorinated and polybrominated biphenyls (PCBS and PBBBS), *Abstr. 7th Internat. Congr. Pharmacol.*, Abstr. 2555, Pergamon Press, Oxford, 1978, 826.
734b. **Lucier, G.**, Perinatal development of UDPglucuronyltransferase: pharmacological and toxicological considerations, in *Conjugation Reactions in Drug Biotransformation*, Aitio, A., Ed., Elsevier, Amsterdam, 1978, 167.
735. **Burchell, B., Dutton, G. J., and Nemeth, A. M.**, Development of phenobarbital-sensitive control mechanisms for uridine diphosphate glucuronyltransferase activity in chick embryo liver, *J. Cell Biol.*, 55, 448, 1972.
736. **Aarts, E. M.**, Differentiation of the barbiturate stimulation of the glucuronic acid pathway from *de novo* enzyme synthesis, *Biochem. Pharmacol.*, 15, 1469, 1966.
737. **Zeidenberg, P., Orrenius, S., and Ernster, L.**, Increase in levels of glucuronylating enzymes and associated rise of mitochondrial oxidative enzymes upon phenobarbital administration in the rat, *J. Cell Biol.*, 32, 528, 1967.
738. **Ko, V., Dutton, G. J., and Nemeth, A. M.**, Development of uridine diphosphate glucuronyltransferase activity in cultures of chick-embryo liver, *Biochem. J.*, 104, 991, 1967.
739. **Sherer, G. K.**, Tissue interaction in chick liver development: a reevaluation. II. Parenchymal differentiation: mesenchymal influence and morphogenetic independence, *Dev. Biol.*, 46, 296, 1975.
740. **Skea, B. R. and Nemeth, A. M.**, Factors influencing premature induction of UDP-glucuronyltransferase activity in cultured chick embryo liver cells, *Proc. Natl. Acad. Sci. U.S.A.*, 64, 795, 1969.
741. **Benzo, C. A. and Nemeth, A. M.**, The ultrastructural development of early chick embryo liver during organ culture in a simple medium, *J. Cell Biol.*, 48, 235, 1971.

742. **Coutalik, A. and Monder, C.**, Metabolism of cortisol by fetal rat liver explants, *Endocrinology*, 95, 466, 1974.

743. **Grieninger, G. and Granick, S.**, Synthesis and differentiation of plasma proteins in cultured embryonic chicken liver cells: a system for study of regulation of protein synthesis, *Proc. Natl. Acad. Sci. U.S.A.*, 72, 5007, 1975.

744. **Goodridge, A. G.**, Hormonal regulation of the activity of the fatty acid synthesizing system and of the malic enzyme concentration in liver cells, *Fed. Proc. Fed. Am. Soc. Exp. Biol.*, 34, 117, 1975.

745. **Benzo, C. A.**, The annulate lamellae of chick embryo liver cells in organ culture, *Anat. Rec.*, 174, 399, 1972.

745a. **Bakken, A. F.**, Bilirubin excretion in newborn human infants. I. Unconjugated bilirubin as a possible trigger for bilirubin excretion, *Acta Pediatr. Scand.*, 59, 148, 1970.

746. **Bakken, A. F.**, Effects of unconjugated bilirubin on bilirubin-UDP-glucuronyl transferase activity in liver of newborn rats, *Pediatr. Res.*, 3, 205, 1969.

747. **Bratlid, D. and Winsnes, A.**, Determination of conjugated and unconjugated bilirubin by methods based on direct spectrophotometry and chloroform extraction. A reappraisal, *Scand. J. Clin. Lab. Invest.*, 28, 41, 1971.

748. **Winsnes, A. and Bratlid, D.**, Effects of bilirubin loading of pregnant rats on hepatic UDP-glucuronyltransferase in the offspring, *Biol. Neonate*, 22, 367, 1973.

749. **Cusworth, R. B. and Simons, J. A.**, Is the ontogeny of glucuronyltransferase adaptive? *Nature (London), New Biol.*, 229, 216, 1971.

750. **Dutton, G. J., Hanson, K. B., and Burchell, B.**, Effect of phenobarbital and other compounds on induction or maintenance of liver uridine diphosphate glucuronyltransferase during organ culture, *Biochem. J.*, 120, 15P, 1970.

751. **Klinger, W., Zwacka, G., and Ankermann, H.**, Untersuchungen zum Mechanismus der Enzyminduktion. VIII. Die Übertragung eines zytoplasmatischen Hemmfaktors der Entwicklung mikrosomaler Leberenzyme aus der Leber von Rattenfoeten auf infantile Ratten, *Acta Biol. Med., Ger.*, 20, 137, 1968.

752. **Soyka, L. F. and Long, R. J.**, In vitro inhibition of drug metabolism by metabolites of progesterone, *J. Pharmacol. Exp. Ther.*, 182, 320, 1972.

753. **Wilson, J. T.**, The effect of a pituitary mammotropic tumor on hepatic microsomal drug metabolism in the rat, *Biochem. Pharmacol.*, 17, 1449, 1968.

754. **Wishart, G. J. and Dutton, G. J.**, Precocious development of detoxicating enzymes following pituitary graft, *Nature (London)*, 252, 408, 1974.

755. **Wishart, G. J. and Dutton, G. J.**, Precocious development of glucuronidating and hydroxylating enzymes in chick embryos treated with pituitary grafts, *Biochem. J.*, 152, 325, 1975.

756. **Brasch, M. and Betz, T. W.**, The hormonal activities associated with the cephalic and caudal regions of the cockerel *pars distalis*, *Gen. Comp. Endocrinol.*, 16, 241, 1971.

757. **Leakey, J. E. A. and Dutton, G. J.** Precocious development *in vivo* of UDP-glucuronyltransferase and aniline hydroxylase by corticosteroids and ACTH, using a simple new continuous flow technique, *Biochem. Biophys. Res. Commun.*, 66, 250, 1975.

758. **Wishart, G. J., Leakey, J. E. A., and Dutton, G. J.**, Differential effects of hormones on precocious yolk sac retraction in chick embryos following administration by a new technique, *Gen. Comp. Endocrinol.*, 31, 373, 1977.

759. **Leakey, J. E. A. and Wishart, G. J.**, Differential stimulation of monooxygenase and uridine diphosphate glucuronosyltransferase activities in chick liver during natural development and following treatment *in ovo* with corticosterone, *Biochem. Soc. Trans.*, 4, 1072, 1976.

760. **Wishart, G. J. and Dutton, G. J.**, Precocious development of UDP-glucuronyltransferase activity in cultured fetal rat liver brought about by glucocorticoids and requiring amino acid incorporation into protein, *Biochem. Biophys. Res. Commun.*, 73, 960, 1976.

761. **Dupouy, J. P., Coffigny, H., and Magre, S.**, Maternal and foetal corticosterone lvels during pregnancy in rats, *Endocrinology*, 65, 347, 1975.

762. **Zarrow, M. X., Philpott, J. E., and Denenberg, V. H.**, Passage of ^{14}C-4-corticosterone from the rat mother to the foetus and neonate, *Nature (London)*, 226, 1058, 1970.

763. **Wishart, G. J. and Dutton, G. J.**, Precocious development of foetal glucuronidating enzymes induced by glucocorticoids in culture and *in utero*, *Nature (London)*, 266, 183, 1977.

764. **Wishart, G. J. and Dutton, G. J.**, Precocious development *in utero* of certain UDP-glucuronyltransferase activities in rat fetuses exposed to glucocorticoids, *Biochem. Biophys. Res. Commun.*, 75, 125, 1977.

765. **Greengard, O.**, Effects of hormones on development of fetal enzymes, *Clin. Pharmacol. Ther.*, 14, 721, 1973.

766. **Wong, M. D., Thompson, M. J., and Burton, A. F.**, Metabolism of natural and synthetic corticosteroids in relation to their effects on mouse fetuses, *Biol. Neonate*, 28, 12, 1976.

767. **Holt, P. G. and Oliver, I. T.**, Plasma corticosterone concentrations in the perinatal rat, *Biochem. J.*, 108, 339, 1968.

768. **Cohen, A. and Brault, S.**, Modifications de la concentration en corticosterone des surrénales et du plasma chez le foetus de rat après surénalectomie maternelle, *C. R. Acad. Sci.*, 228, 1755, 1974.

769. **Milkovic, S., Milkovic, K., and Paunovic, J.**, The initiation of fetal adrenocorticotrophic activity in the rat, *Endocrinology*, 92, 380, 1973.

770. **Kamoun, A.**, Activité cortico-surrénale au cours de la lactation et du developpement pré et postnatal chez le rat. I. Concentration et cinétique de disparition de la corticosterone, *J. Physiol. (Paris)*, 62, 5, 1970.

771. **Jacquot, R. L., Plas, C., and Nagel, J.**, Two examples of physiological maturations in rat fetal liver, *Enzyme*, 15, 296, 1973.

772. **Parvez, S., Parvez, H., and Youdim, M. B. H.**, Adrenal-pituitary implications for maintenance of tissue glycogen stores in cyclic and pregnant rats, *J. Steroid Biochem.*, 7, 653, 1976.

773. **Murphy, B. E. P. and Diez d'Aux, R. C.**, Steroid levels in the human fetus: cortisol and cortisone, *J. Clin. Endocrinol. Metab.*, 35, 678, 1972.

774. **Giannopoulos, G.**, Levels and subcellular distribution of endogenous corticosterone in rat liver during development, *Steroids*, 29, 539, 1977.

775. **Cake, M. H., Ghisalberti, A. V., and Oliver, I. T.**, Cytoplasmic binding of dexamethasone and induction of tyrosine aminotransferase in neonatal rat liver, *Biochem. J.*, 54, 983, 1973.

776. **Quattropani, S. L., Stenger, V. G., and Dvorchik, B. H.**, Ultrastructure of the developing fetal hepatocyte of *Macaca arctoides*: a proposed model for studies of fetal drug metabolism, *Anat. Rec.*, 182, 103, 1975.

777. **Schmucker, D. L., Mooney, J. S., and Jones, A. L.**, Age-related changes in the hepatic endoplasmic reticulum: a quantitative analysis, *Science*, 197, 1005, 1977.

778. **Lagoguey, M., Dray, F., Chauffournier, T., and Reinberg, A.**, Circadian and circannual rhythms of urine testosterone and epitestosterone glucuronides in healthy adult men, *Int. J. Chronobiol.*, 1, 91, 1973.

779. **Crooks, J., O'Malley, K., and Stevenson, I. H.**, Pharmacokinetics in the elderly, *Clin. Pharmacokinet.*, 1, 280, 1976.

780. **Graham, A. B., Woodcock, B. G., and Wood, G. C.**, The phospholipid dependence of uridine diphosphate glucuronyltransferase. Effect of protein deficiency on the phospholipid composition and enzyme activity of rat liver microsomal fraction, *Biochem. J.*, 137, 567, 1974.

781. **Bock, K. W., Fröhling, W., and Remmer, H.**, Influence of fasting and hemin on microsomal cytochromes and enzymes, *Biochem. Pharmacol.*, 22, 1557, 1973.

782. **Wood, G. C. and Woodcock, B. G.**, Effects of dietary protein deficiency on conjugation of foreign compounds in rat liver, *J. Pharm. Pharmacol.*, (Suppl.), 22, 60S, 1970.

783. **Woodcock, B. G. and Wood, G. C.**, Effect of protein-free diet on UDP-glucuronyltransferase and sulphotransferase in rat liver, *Biochem. Pharmacol.*, 20, 2703, 1971.

784. **Eakins, M. N. and Slater, T. F.** Enhanced bilirubin clearance in protein-deprived rats, *Biochem. Soc. Trans.*, 1, 931, 1973.

785. **Smith, J. A., Butler, T. C., and Poole, D. T.**, Effect of protein depletion in guinea pigs on glucuronate conjugation of chloramphenicol by liver microsomes, *Biochem. Pharmacol.*, 22, 981, 1973.

786. **Sachon, D. S.**, Effects of low and high protein diets on the induction of microsomal drug-metabolizing enzymes in rat liver, *J. Nutr.*, 105, 1631, 1975.

787. **Dickerson, J. W. T. and Walker, R.**, Nutrition, age and drug metabolism, *Proc. Nutr. Soc.*, 33, 191, 1974.

788. **Okolicsanyi, L. and Scremin, S.**, Effect of uridine-diphosphate-glucose on liver glucuronyl transferase activity in normal and fasting rats, *Enzyme*, 14, 366, 1972—1973.

789. **Marselos, M. and Laitinen, M.**, Starvation and phenobarbital treatment effects on drug hydroxylation and glucuronidation in the rat liver and small intestinal mucosa, *Biochem. Pharmacol.*, 24, 1529, 1975.

790. **Hietanen, E. and Hänninen, O.**, Variable activities of drug-metabolizing enzymes in the gut of the rat after feeding with different pelleted diets, *Comp. Gen. Pharmacol.*, 5, 255, 1974.

791. **Chadwick, R., Peoples, A., and Cramner, M.**, The effect of protein quality and ascorbic acid deficiency on stimulation of hepatic microsomal enzymes in guinea pigs, *Toxicol. Appl. Pharmacol.*, 24, 603, 1973.

792. **Laitinen, M.**, Dietary Lipids on the Control of Drug Metabolizing Activity in the Rat, Ph.D. thesis, University of Kuopio, Kuopio, 1977.

793. **Laitinen, M., Hietanen, E., Vainio, H., and Hänninen, O.**, Dietary fats and properties of endoplasmic reticulum. I. Dietary lipid induced changes in composition of microsomal membranes in liver and gastroduodenal mucosa of rats, *Lipids*, 10, 461, 1975.

794. **Hietanen, E., Laitinen, M., Vainio, H., and Hänninen, O.** Dietary fats and properties of endoplasmic reticulum. II. Dietary lipid induced changes in activities of drug metabolizing enzymes in liver and duodenum of rats, *Lipids,* 10, 467, 1975.

795. **Laitinen, M.,** Hepatic and duodenal drug metabolism in the rat during fat deficiency, *Gen. Pharmacol.,* 7, 263, 1976.

796. **Laitinen, M.,** Enhancement of hepatic drug metabolism with dietary cholesterol in the rat, *Acta Pharmacol. Toxicol.,* 39, 241, 1976.

797. **Hietanen, E., Laitinen, M., Lang, M., and Vainio, H.,** Inducibility of mucosal drug-metabolizing enzymes of rats fed on a cholesterol-rich diet by polychlorinated biphenyl, 3-methylcholanthrene and phenobarbital, *Pharmacology,* 13, 287, 1975.

798. **Laitinen, M., Lang, M., Hietanen, E., and Vainio, H.,** Enhancement of hepatic drug biotransformation rate by polychlorinated biphenyls in rats fed cholesterol-rich diet, *Toxicology,* 5, 79, 1975.

799. **Lang, M.,** Dietary cholesterol caused modification in the structure and function of rat hepatic microsomes, studied by fluorescent probes, *Biochim. Biophys. Acta,* 455, 947, 1976.

800. **Gollan, J. L., Bateman, C., and Billing, B. H.,** Effect of dietary composition on the conjugated hyperbilirubinaemia of Gilbert's syndrome, *Gut,* 17, 335, 1976.

801. **Gronwall, R. and Mia, A. S.,** Fasting hyperbilirubinemia in horses, *Am. J. Dig. Dis.,* 17, 473, 1972.

802. **Felsher, B. F., Rickard, D., and Redecker, A. G.** The reciprocal relation between caloric intake and the degree of hyperbilirubinemia in Gilbert's syndrome, *N. Engl. J. Med.,* 283, 170, 1970.

803. **Barrett, P. V. D.,** The effect of diet and fasting on the serum bilirubin concentration in the rat, *Gastroenterology,* 60, 572, 1971.

804. **Owens, D. and Sherlock, S.,** Diagnosis of Gilbert's syndrome: role of reduced caloric intake test, *Br. Med. J.,* 3, 559, 1973.

805. **Felsher, B. F. and Carpio, N. M.,** Caloric intake and unconjugated hyperbilirubinemia, *Gastroenterology,* 69, 42, 1975.

806. **Bakken, A. F., Thaler, M. M., and Schmid, R.,** Metabolic regulation of heme catabolism and bilirubin production, *J. Clin. Invest.,* 51, 530, 1972.

807. **Bloomer, J. R., Barrett, P. V., Rodkey, F. L., and Berlin, N. I.,** Studies on the mechanism of fasting hyperbilirubinemia, *Gastroenterology,* 61, 479, 1971.

808. **Felsher, B. F., Craig, J. R., and Carpio, N.,** Hepatic glucuronidation in Gilbert's syndrome, *J. Lab. Clin. Med.,* 81, 829, 1973.

809. **Gollan, J. L., Hatt, K. J., and Billing, B. H.,** The influence of diet on unconjugated hyperbilirubinaemia in the Gunn rat, *Clin. Sci. Mol. Med.,* 49, 229, 1975.

810. **Oyama, J. H.,** Hyperbilirubinemia in healthy males on acutely restricted dietary intakes receiving parenteral nutrition, *Am. J. Clin. Nutr.,* 25, 459, 1972.

811. **Gollan, J. L., Huang, S. N., Billing, B. H., and Sherlock, S.,** Prolonged survival in three brothers with severe type 2 Crigler-Najjar syndrome: ultrastructural and metabolic studies, *Gastroenterology,* 68, 1543, 1975.

812. **Duvaldestin, P., Mahu, J.-L., and Berthelot, P.,** Effects of fasting on substrate specificity of UDP-glucuronosyltransferase, *Biochim. Biophys. Acta,* 384, 811, 1975.

813. **Freedland, A.,** Effects of thyroid hormones on metabolism, *Endocrinology,* 77, 19, 1965.

814. **Phelps, C., Stevens, R., Young, A., and Luscombe, M.,** The control of the early steps in polysaccharide synthesis, in *Protides of the Biological Fluids,* 22nd Colloquium, Peeters, H., Ed., Pergamon Press, Oxford, 1975, 205.

815. **Flint, M., Lathe, G. H., and Ricketts, T. R.,** The effect of undernutrition and other factors on the development of glucuronyl transferase activity in the newborn rabbit, *Ann. N.Y. Acad. Sci.,* 111, 295, 1963.

816. **Basu, T. K., Dickerson, J. W. T., and Parke, D. V.,** Effect of underfeeding suckling rat on the activity of hepatic drug metabolizing enzymes, *Biol. Neonate,* 23, 109, 1973.

817. **Catz, C. S., Winick, M., and Yaffe, S. J.,** The effect of early malnutrition on drug metabolism, Proc. 12th Int. Congr. Pediatr. Wien, Österreich, 24, 17, 1971.

818. **Felsher, B. F., Carpio, N. M., Woolley, M. M., and Asch, M. J.,** Hepatic bilirubin glucuronidation in neonates with unconjugated hyperbilirubinemia and congenital gastrointestinal obstruction, *J. Lab. Clin. Med.,* 83, 90, 1974.

819. **Adlard, B. P. and Lathe, G. H.,** Breast milk jaundice: effect of 3α,20β-pregnanediol on bilirubin conjugation by human liver, *Arch. Dis. Child.,* 45, 186, 1970.

820. **Arias, I. M. and Gartner, L. M.,** Production of unconjugated hyperbilirubinaemia in full term newborn human infants following administration of pregnane-3α,20β-diol, *Nature (London),* 203, 1291, 1964.

821. **Arias, I. M., Gartner, L. M., Seifter, S., and Furman, M.,** Prolonged neonatal unconjugated hyperbilirubinemia associated with breast feeding and a steroid, pregnane-3α,20β-diol, in maternal milk that inhibits glucuronide formation *in vitro, J. Clin. Invest.,* 43, 2037, 1964.

822. **Rosenfeld, K. S., Arias, I. M., Gartner, L. M., Hellman, L., and Gallagher, T. F.,** Studies of urinary pregnane-3α,20β-diol during pregnancy, post-partum, lactation and progesterone ingestion, *J. Clin. Endocrinol.,* 27, 1705, 1967.

823. **Bevan, B. R. and Holton, J. B.,** Inhibition of bilirubin conjugation in rat liver slices by free fatty acids, with relevance to the problem of breast milk jaundice, *Clin. Chim. Acta,* 41, 101, 1972.

824. **Brewington, C. R., Park, O. W., and Schwartz, D. P.,** Conjugated compounds in cow's milk, *J. Agric. Food Chem.,* 21, 38, 1973.

825. **Brewington, C. R., Park, O. W., and Schwartz, D. P.,** Conjugated compounds in cow's milk II, *J. Agric. Food Chem.,* 21, 293, 1974.

826. **Rao, L. G. S. and Taylor, W.,** The metabolism of progesterone by animal tissues *in vitro.* Sex and species differences in conjugate formation during the metabolism of [4-¹⁴C]progesterone by liver homogenates, *Biochem. J.,* 96, 172, 1965.

827. **Sever, P. S., Dring, L. G., and Williams, R. T.,** The metabolism of hydroxyamphetamines in man and animals, *Biochem. Soc. Trans.* 1, 1158, 1973.

828. **Schriefers, H., Ghraf, R., and Lax, E. R.,** Sex-specific aglucone patterns of testosterone metabolism in rat liver and their alteration following interference with the sexual differentiation, *Hoppe Seylers Z. Physiol. Chem.,* 353, 371, 1972.

829. **Chhabra, R. S. and Fouts, J. R.,** Sex differences in the metabolism of xenobiotics by extrahepatic tissue in rats, *Drug Metab. Dispos.,* 2, 375, 1974.

830. **Hietanen, E.,** Effect of sex and castration on hepatic and intestinal activity of drug metabolizing enzymes, *Pharmacology,* 12, 84, 1974.

831. **Künzel, B. and Müller-Oerlinghausedn, B.,** Wirkung von Testosteron und einem Anti-Androgen (Cyproteronacetat) auf die Glucuronbildung in der Rattenleber, *Naunyn Schmidiedbergs Arch. Exp. Pathol. Pharmakol.,* 262, 112, 1969.

832. **Vilnikka, L. and Janne, O.,** Urinary excretion of neutral steroid glucuronides with reference to the menstrual cycle, *Clin. Chim. Acta,* 49, 277, 1973.

833. **Kato, R.,** Sex-related differences in drug metabolism, *Drug Metab. Rev.,* 3, 1, 1974.

834. **Skett, P. and Gustafsson, J.-Å.,** Imprinting of enzyme systems of xenobiotic and steroid metabolism, *Rev. Biochem. Toxicol.,* 1, 27, 1979.

835. **Chung, L. W. K.,** Characteristics of neonatal androgen-induced imprinting of rat hepatic microsomal monooxygenases, *Biochem. Pharmacol.,* 26, 1979, 1977.

836. **Tabone, D. and Tabone, J.,** L'uridine diphosphate facteur de la synthèse de l'ester glucosidique de l'acide authranilique, *C. R. Acad. Sci.,* 242, 302, 1956.

837. **Renwick, A. G.,** Microbial metabolism of drugs, in *Drug Metabolism from Microbe to Man,* Parke, D. V. and Smith, R. L., Eds., Taylor & Francis, London, 1977, 169.

838. **Baldwin, B. C.,** Xenobiotic metabolism in plants, in *Drug Metabolism from Microbe to Man,* Parke, D. V. and Smith, R. L., Eds., Taylor & Francis, London, 1977, 191.

839. **Marsh, C. A.,** Glucuronide metabolism in plants. 2. The isolation of flavone glucosiduronic acids from plants, *Biochem. J.,* 59, 58, 1955.

840. **Dutton, G. J. and Duncan, A. M.,** Mechanism of conjugation of *o*-aminophenol in plants and an insect, *Biochem. J.,* 77, 18 P, 1960.

841. **Yamaha, T. and Cardini, C. E.,** The biosynthesis of plant glycosides. I. Monoglucosides, *Arch. Biochem. Biophys.,* 86, 127, 1960.

842. **Yamaha, T. and Cardini, C. E.,** The biosynthesis of plant glycosides. II. Gentobiosides, *Arch. Biochem. Biophys.,* 86, 133, 1960.

843. **Douch, P. G. C. and Blair, S. S. B.,** The metabolism of foreign compounds in the cestode *Moniezia expansa* and the nematode *Ascaris lumbricoides, Xenobiotica,* 5, 279, 1975.

844. **Allsop, T. F.,** Comparative Detoxification of Phenols by some Marine Invertebrates, M.Sc. thesis, University of Wellington, N.Z., 1965.

845. **Dutton, G. J.,** Uridine diphosphate glucose and the synthesis of phenolic glucosides by mollusks, *Arch. Biophys. Biochem.,* 116, 399, 1966.

846. **Leakey, J. E. A. and Dutton, G. J.,** Effect of phenobarbital on UDP-glucosyltransferase activity and phenolic glucosidation in the mollusc *Arion ater, Comp. Biochem. Physiol.,* 51 C, 215, 1975.

847. **Puyear, R. L.,** The glucuronic acid pathway in the blue crab *Callinectes sapidus* Rathbun: the enzymes of the UDPglucose to glucuronic acid portion of the pathway, *Comp. Biochem. Physiol.,* 20, 499, 1967.

848. **Dutton, G. J.,** The mechanism of *o*-aminophenol glucoside formation in *Periplaneta americana, Comp. Biochem. Physiol.,* 7, 39, 1962.

849. **Smith, J. N. and Turbert, H. B.,** Enzymic glucoside synthesis in locusts, *Nature (London),* 189, 600, 1961.

850. **Gessner, T. and Acara, M.,** Metabolism of thiols: *S*-glucosylation, *J. Biol. Chem.,* 243, 3142, 1968.

851. **Dutton, G. J. and Ko, V.,** The apparent absence of uridine diphosphate glucuronyltransferase for detoxication in *Musca domestica, Comp. Biochem. Physiol.,* 11, 269, 1964.

852. **Heenan, M. and Smith, J. N.**, Water-soluble metabolites of *p*-nitrophenol and 1-naphthyl-*N*-methylcarbamate in flies and grass grubs: formation of glucose-phosphate and phosphate conjugates, *Biochem. J.*, 144, 303, 1974.

853. **Koolman, J., Hoffman, J. A., and Carlson, P.**, Sulphate esters as inactivation products of ecdysone in *Locusta migratoria*, *Z. Physiol. Chem.*, 354, 1043, 1973.

854. **Sieber, S. M. and Adamson, R. H.**, The metabolism of xenobiotics by fish, in *Drug Metabolism from Microbe to Man*, Parke, D. V. and Smith, R. L., Eds., Taylor & Francis, London, 1977, 233.

855. **Adamson, R. H. and Guarino, A. M.**, The effect of foreign compounds on elasmobranchs and the effect of elasmobranchs on foreign compounds, *Comp. Biochem. Physiol.*, 42 A, 171, 1972.

856. **Dutton, G. J. and Montgomery, J. P.**, Glucuronide synthesis in fish and the influence of temperature, *Biochem. J.*, 70, 17 P, 1958.

857. **Nagayama, F., Yamada, T., and Tauti, M.**, Activities of UDPglucose dehydrogenase, UDPglucuronyltransferase, and some glycosidases of freshwater fish during the adaptation to salt water, *Bull. Jpn. Soc. Sci. Fish.*, 34, 950, 1968.

858. **Adamson, R. H.**, Drug metabolism in marine vertebrates, *Fed. Proc. Fed. Am. Soc. Exp. Biol.*, 26, 1047, 1967.

859. **Burger, J. W.**, Some aspects of liver function in the spiny dogfish *Squalus acanthias*, in *Sharks, Skate and Rays*, Gilbert, P. W., Mathewson, R. F., and Rall, D. P., Eds., Johns Hopkins Press, Baltimore, 1967, 293.

860. **Guarino, A. M. and Anderson, J. B.**, Excretion of phenol red and its glucuronide in the dogfish shark, *Xenobiotica*, 6, 1, 1976.

861. **Statham, C. N., Pepple, S., and Lech, J. J.**, Biliary excretion products of 1-[1-^{14}C]-naphthyl-*N*-methylcarbamate (Carbaryl) in rainbow trout (*Salmo gairdneri*), *Drug Metab. Dispos.*, 3, 400, 1975.

862. **Huang, K. C. and Collins, S. F.**, Conjugation and excretion of aminobenzoic acid isomers in marine fishes, *J. Cell Comp. Physiol.*, 60, 49, 1962.

863. **Akitake, H. and Kobayashi, K.**, Studies on the metabolism of chlorophenols in fish. III. Isolation and identification of a conjugated PCP excreted by goldfish, *Bull. Jpn. Soc. Sci. Fish.*, 41, 321, 1975.

864. **Dewaide, J. H.**, Metabolism of Xenobiotics — Comparative and Kinetic Studies as a Basis for Environmental Pharmacology, Ph.D. thesis, University of Nijmegen, The Netherlands, 1971.

865. **Grajcer, D. and Idler, D. R.**, Conjugated testosterone in the blood and testes of spawned Fraser-river Sockeye salmon (*Oncorhyncus nerka*), *Can. J. Biochem.*, 41, 23, 1963.

866. **Sinclair, D. A.**, Iodothyronine-glucuronide conjugates in the bile of brook trout, *Salvelinus fontinalis* (Mitchellii) and other freshwater teleosts, *Gen. Comp. Endocrinol.*, 19, 552, 1972.

867. **Lech, J. J.**, Isolation and identification of 3-trifluoromethyl-4-nitrophenyl glucuronide from bile of rainbow trout exposed to 3-trifluoroethyl-4-nitrophenol, *Toxicol. Appl. Pharmacol.*, 24, 114, 1973.

868. **Statham, C. N. and Lech, J. J.**, Metabolism of 2',5-dichloro-4'-nitrosalicylanilide (Bayer 73) in rainbow trout (Salmo gairdneri), *J. Fish. Res. Board Can.*, 32, 515, 1975.

869. **Lech, J. J., Pepple, S., and Anderson, M.**, Role of glucuronide formation in the selective toxicity of 3-trifluoromethyl-4-nitrophenol (TFM) for the sea lamprey: comparative aspects of TFM uptake and conjugation in sea lamprey and rainbow trout, *Toxicol. Appl. Pharmacol.*, 31, 150, 1975.

870. **Lech, J. J., Pepple, S., and Anderson, M.**, Effects of novobiocin on the acute toxicity, metabolism and biliary excretion of 3-trifluoromethyl-4-nitrophenol in rainbow trout, *Toxicol. Appl. Pharmacol.*, 25, 542, 1973.

871. **Lech, J. J.**, Glucuronide formation in rainbow trout — effect of salicylamide on the acute toxicity, conjugation and excretion of 3-trifluoromethyl-4-nitrophenol, *Biochem. Pharmacol.*, 23, 2403, 1974.

872. **Ozon, R. and Breuer, H.**, Studies on the metabolism of steroid hormones in vertebrates. VII. Enzymatic glucuronidation of estrogens by *Pleurodeles waltlii* Micah (Amphibia: Urodele) *in vivo* and *in vitro*, *Gen. Comp. Endocrinol.*, 6, 295, 1966.

873. **Paulson, G. D., Doktor, M. M., Jakobson, A. M., and Zaylskie, R. G.**, Isopropyl carbanilate (Propham) metabolism in the chicken: balance studies and isolation and identification of excreted metabolites, *J. Agric. Food Chem.*, 20, 867, 1972.

874. **Gillett, S. W. and Arscott, G. H.**, Microsomal epoxidation in Japanese quail: induction by dietary dieldrin, *Comp. Biochem. Physiol.*, 30, 589, 1969.

875. **Cornelius, C. E., Kelly, K. C., and Himes, J. A.**, Heterogeneity of bilirubin conjugates in several animal species, *Cornell Vet.*, 65, 90, 1974.

876. **Sova, Z. and Nemeč, Z.**, Values of free bilirubin, bilirubin monoglucuronides and bilirubin diglucuronides in the serum of four goose breeds kept on Czechoslovak farms, *Vet. Med. (Prague)*, 17, 469, 1972.

877. **Tenhunen, R.**, The green colour of avian bile: biochemical explanation, *Scand. J. Clin. Lab. Invest.*, *Suppl.*, 116, 9, 1971.

878. **Robinson, A. A., Henneberry, G. O., and Common, R. H.,** Steroid estrogen conjugates of hen's urine: identification of radioactive estrone β-glucuronide, estradiol-17α-3β-glucuronide, estrone sulphate, estradiol-17β-3-sulphate and estradiol-17α-3-sulphate as metabolites of injected [¹⁴C]estrone, *Biochim. Biophys. Acta,* 326, 93, 1973.

879. **Newcomer, W. S. and Heninger, R. W.,** Glucuronic conjugation of steroids in the avian adrenal gland, *Proc. Soc. Exp. Biol. Med.,* 105, 32, 1960.

880. **Gorrod, J. W.,** Species differences in the formation of 4-amino-3-biphenyl-β-D-glucosiduronate *in vitro, Biochem. J.,* 121, 29 P, 1971.

881. **Wong, K. P.,** Bilirubin glucosyl and glucuronyltransferases. A comparative study and the effect of drugs — on rat, mouse, cat, guinea pig, hamster and man, *Biochem. Pharmacol.,* 21, 1485, 1972.

882. **Andoh, B. Y. A., Renwick, A. G., and Williams, R. T.,** The excretion of [³⁵S] Dapsone and its metabolites in the urine, faeces and bile of the rat, *Xenobiotica,* 4, 571, 1974.

883. **Hirom, P. C., Idle, J. R., and Millburn, P.,** Comparative aspects of the biosynthesis and excretion of xenobiotic conjugates by nonprimate mammals, in *Drug Metabolism from Microbe to Man,* Parke, D. V. and Smith, R. L., Eds., Taylor & Francis, London, 1977, 299.

884. **Williams, R. T.,** Inter-species variations in the metabolism of xenobiotics, *Biochem. Soc. Trans.,* 2, 13, 1974.

885. **Hartiala, K. J. V.,** Studies on detoxication mechanisms; glucuronide synthesis of various organs with special reference to detoxifying capacity of mucous membrane of alimentary canal, *Ann. Med. Exp. Biol. Fenn.,* 33, 239, 1955.

886. **Dutton, G. J. and Greig, C. G.,** Observations on the distribution of glucuronide synthesis in tissues, *Biochem. J.,* 66 52 P, 1958.

887. **Robinson, D. and Williams, R. T.,** Do cats form glucuronides?, *Biochem. J.,* 68, 23 P, 1958.

888. **Capel, I. D., French, M. R., Millburn, P., Smith, R. L., and Williams, R. T.,** The fate of [¹⁴C]phenol in various species, *Xenobiotica,* 2, 25, 1972.

889. **Capel, I. D., Millburn, P., and Williams, R. T.,** Monophenyl phosphate, a new conjugate of phenol in the cat, *Biochem. Soc. Trans.,* 2, 305, 1974.

890. **Yeh, S. Y., Chernov, H. I., and Woods, L. A.,** Metabolism of morphine by cats, *J. Pharm. Sci.,* 60, 469, 1971.

891. **Miller, J. J., Powell, G. M., Olavesen, A. H., and Curtis, C. G.** The metabolism and toxicity of phenols in cats, *Biochem. Soc. Trans.,* 1, 1163, 1973.

892. **Weisburger, J. H., Grantham, P. H., and Weisburger, E. K.,** The metabolism of *N*-2-fluorenylacetamide in the cat: evidence for glucuronic acid conjugates, *Biochem. Pharmacol.,* 13, 469, 1964.

893. **Bridges, J. W., French, M. R., Smith, R. L., and Williams, R. T.,** The fate of benzoic acid in various species, *Biochem. J.,* 18, 47, 1970.

894. **Wong, K. P.,** Species differences in the conjugation of 4-hydroxy-3-methoxy-phenylethanol with glucuronic acid and sulphuric acid, *Biochem. J.,* 158, 33, 1976.

895. **Caldwell, J., French, M. R., Idle, J. R., Renwick, A. G., Bassir, O., and Williams, R. T.,** Conjugation of foreign compounds in the elephant and hyaena, *FEBS Lett.,* 60, 391, 1975.

896. **McChesney, E. W.,** On glucuronide formation in the cat, *Biochem. Pharmacol.,* 13, 1366, 1964.

897. **Schillings, R. T., Sisenwine, S. F., Schwartz, M. H., and Ruelius, H. W.,** Lorazepam: glucuronide formation in the cat, *Drug Metab. Dispos.,* 3, 85, 1975.

898. **Myant, N. B.,** Excretion of the glucuronide of thyroxine in cat bile, *Biochem. J.,* 99, 341, 1966.

899. **Taylor, W. and Scratcherd, I.,** The metabolism of [¹⁴C]progesterone in the cat: Biliary and urinary excretion of conjugated metabolites, *Biochem. J.,* 81, 398, 1961.

900. **Archer, S. E. H., Scratcherd, T., and Taylor, W.,** Steroid metabolism in the cat: biliary and urinary excretion of metabolites of [4-¹⁴C]testosterone, *Biochem. J.,* 94, 778, 1965.

901. **Karim, M. F. and Taylor, W.,** Steroid metabolism in the cat. Biliary and urinary excretion of metabolites of [4-¹⁴C]oestradiol, *Biochem. J.,* 117, 267, 1970.

902. **Potrepka, R. F. and Spratt, J. L.,** Effect of phenobarbital and 3-methylcholanthrene pretreatment on guinea pig hepatic microsomal bilirubin glucuronyltransferase activity, *Biochem. Pharmacol.,* 20, 861, 1971.

903. **Spector, W. S.,** *Handbook of Toxicology,* Vol. 1, W. B. Saunders, Philadelphia, 1956.

904. **Hietanen, E. and Vainio, H.,** Interspecies variation in small intestinal and hepatic drug hydroxylation and glucuronidation, *Acta Pharmacol. Toxicol.,* 33, 57, 1973.

905. **Marsh, C. A.,** The mode of glucuronic acid excretion in marsupials, *Proc. Aust. Biochem. Soc.,* 2, 8, 1969.

906. **Davies, D. S.,** Drug metabolism in man, in *Drug Metabolism from Microbe to Man,* Parke, D. V. and Smith, R. L., Eds., Taylor & Francis, London, 1977, 357.

907. **Williams, R. T.,** Future developments, in *Drug Metabolism from Microbe to Man,* Parke, D. V. and Smith, R. L., Eds., Taylor & Francis, London, 1977, 433.

908. **Smith, R. L. and Williams, R. T.**, Comparative metabolism of drugs in man and monkeys, *J. Med. Primatol.*, 3, 138, 1974.

909. **Owens, I. S. and Nebert, D. W.**, Genetically-mediated UDPglucuronyltransferase induction by polycyclic aromatic compounds associated with the "*Ah* locus" in the mouse, *Pharmacologist*, 17, 217, 1975.

910. **Owens, I. S.**, Genetic regulation of UDP-glucuronosyltransferase: induction by polycyclic aromatic compounds in mice, *J. Biol. Chem.*, 252, 2827, 1977.

910a. **Owens, I. S., Koteen, G. M., Pelkonen, O. and Legraverend, C.**, Activation of certain benzo(a)pyrene phenols and the effect of some conjugating enzyme activities, in *Conjugation Reactions in Drug Biotransformation*, Aitio, A., Ed., Elsevier, Amsterdam, 1978, 39.

911. **Gunn, C. K.**, Hereditary acholuric jaundice, *J. Hered.*, 29, 137, 1938.

912. **Schmid, R., Axelrod, J., Hammaker, L., and Swarm, K. L.**, Congenital jaundice of rats, due to a defect in glucuronide formation, *J. Clin. Invest.*, 37, 1123, 1958.

913. **Arias, I. M. and Johnson, L.**, Studies of bilirubin excretion in normal and Gunn rats, *Clin. Res.*, 7, 291, 1959.

914. **Arias, I. M., Johnson, L., and Wolfson, S.**, Biliary excretion of injected conjugated and unconjugated bilirubin by normal and Gunn rats, *Am. J. Physiol.*, 200, 1091, 1961.

915. **Puukka, R., Tanner, P., and Hänninen, O.**, Enzymes of the glucuronic acid pathway in Wistar and Gunn rats and their heterozygotes, *Biochem. Genet.*, 9, 343, 1973.

916. **Schmid, R. and McDonagh, A. F.**, Hyperbilirubinemia, in *The Metabolic Basis of Inherited Disease*, 4th ed., Stanbury, J. B., Wyngaarden, J. B., and Fredrickson, D. S., Eds., McGraw-Hill, New York, 1978, 1221.

917. **Black, M. and Billing, B. H.**, Hepatic bilirubin UDP-glucuronyltransferase activity in liver disease and Gilbert's syndrome, *N. Engl. J. Med.*, 280, 1266, 1969.

918. **Arias, I. M., Furman, M., Tapley, D. F., and Ross, J. E.**, Glucuronide formation and transport of various compounds by Gunn rat intestine *in vitro, Nature (London)*, 197, 1109, 1963.

919. **Arias, I. M., Wolfson, S., and Johnson, L.**, Studies of extrahepatic glucuronide formation in normal and Gunn rats, *Clin. Res.*, 8, 28, 1960.

920. **Yeary, R. A. and Grothaus, R. H.**, The Gunn rat as an animal model in comparative medicine, *Lab. Anim. Sci.*, 21, 362, 1971.

921. **Cornelius, E. E. and Arias, I. M.**, Animal model of human disease. Crigler-Najjar syndrome animal model: hereditary nonhemolytic unconjugated hyperbilirubinemia in Gunn rats, *Am. J. Pathol.*, 69, 369, 1972.

922. **Guarino, A. M., Reagan, R. L., and Gram, T. E.**, Submicrosomal distribution of hepatic drug-metabolizing components in the Gunn rat, *Pharmacology*, 10, 257, 1973.

923. **Bastomsky, C. H.**, The biliary excretion of thyroxine and its glucuronic acid conjugate in normal and Gunn rats, *Endocrinology*, 92, 35, 1973.

924. **Bock, K. W., Clausbruch, U. C. V., and Ottenwalder, H.**, UDP-Glucuronyltransferase in perfused rat liver and in microsomes. V. Studies with Gunn rats, *Biochem. Pharmacol.*, 27, 369, 1978.

925. **Vainio, H. and Hietanen, E.**, Drug metabolism in Gunn rat — inability to increase bilirubin glucuronidation by phenobarbital treatment, *Biochem. Pharmacol.*, 23, 3405, 1974.

926. **Yeary, R. A. and Fleming, M.**, Developmental changes in *p*-nitrophenol glucuronidation in the Gunn rat, *Biochem. Pharmacol.*, 20, 375, 1971.

927. **Yeary, R. A., Gerken, D., and Davis, D. R.**, Postnatal development of hepatic drug-metabolizing enzymes in the Gunn rat, *Biol. Neonate*, 23, 371, 1973.

928. **Auclair, C., Hakim, J., Boivin, P., Troube, H., and Boucherot, J.**, Bilirubin and paranitrophenyl glucuronyl transferase activities of the liver in patients with Gilbert's syndrome, *Enzyme*, 21, 97, 1976.

929. **Marniemi, J., Vainio, H., and Parkki, M.**, Drug conjugation in Gunn rats: reduced UDP-glucuronyltransferase and UDP-glucosyltransferase activities with increased glycine-N-acyltransferase activity, *Pharmacology*, 13, 492, 1975.

930. **Blanckaert, N., Fevery, J., Heirwegh, K. P. M., and Compernolle, F.**, Characterization of the major diazo-positive pigments in bile of homozygous Gunn rats, *Biochem. J.*, 164, 237, 1977.

931. **Blanckaert, N., Compernolle, F., Heirwegh, K. P. M., De Groote, J., and Fevery, J.**, Bile pigments in serum and urine of normals and in patients with unconjugated hyperbilirubinemia, *Digestion*, 10, 315, 1974.

932. **De Leon, A., Gartner, L. M., and Arias, I. M.**, Effect of phenobarbital on hyperbilirubinemia in UDP-glucuronyltransferase-deficient rats, *J. Lab. Clin. Med.*, 70, 273, 1967.

933. **Robinson, S.**, Increased bilirubin conjugation in heterozygous Gunn rats treated with phenobarbital, *Nature (London)*, 222, 990, 1969.

934. **Robinson, S. H.**, Production and excretion of bilirubin in Gunn rats treated with phenobarbital, *Proc. Soc. Exp. Biol. Med.*, 138, 281, 1971.

935. **Robinson, S. H., Yannoni, C., and Nagasawa, S.,** Bilirubin excretion in rats with normal and impaired bilirubin conjugation: effect of phenobarbital, *J. Clin. Invest.,* 50, 2606, 1971.
936. **Garay, E. A. R., Morisoli, L. S., and Spetale, M. del R.,** Increased bilirubin conjugation in the liver and intestinal mucosa of phenobarbital-treated rats, *Experientia,* 28, 944, 1972.
937. **Blaschke, T. F. and Berk, P. D.,** Augmentation of bilirubin UDP glucuronyltransferase activity in rat liver homogenates by glutethimide, *Proc. Soc. Exp. Biol. Med.,* 140, 1315, 1972.
938. **Rugstad, H. E., Robinson, S. H., Yannoni, C., and Tashjian, A. H.,** Transfer of bilirubin uridine diphosphate-glucuronyltransferase to enzyme-deficient rats, *Science,* 170, 553, 1970.
939. **Foliot, A., Christoforov, B., Petite, J. P., Etienne, J. P., Housset, E., and Dubois, M.,** Bilirubin UDP glucuronyltransferase activity of Wistar rat kidney, *Am. J. Physiol.,* 229, 340, 1975.
940. **Mukherjee, A. B. and Krasner, J.,** Induction of an enzyme in genetically-deficient rats after grafting of normal liver, *Science,* 182, 68, 1973.
941. **van Houwelingen, C. A. J. and Arias, I. M.,** Attempts to induce hepatic uridine diphosphate glucuronyl transferase in genetically deficient Gunn rats by grafting of normal liver tissue, *Pediatr. Res.,* 10, 830, 1976.
942. **Crigler, S. F. and Najjar, V. A.,** Congenital familial nonhemolytic jaundice with kernicterus, *Pediatrics,* 10, 169, 1952.
943. **Arias, I. M.,** Chronic unconjugated hyperbilirubinemia without overt signs of hemolysis in adolescents and adults, *J. Clin. Invest.,* 41, 2233, 1962.
944. **Arias, I. M., Gartner, L. M., Cohen, M., Ben Ezzer, I., and Levi, A. J.,** Uniformly decreased 4-methylumbelliferone glucuronyltransferase and o-aminophenol glucuronyltransferase in six patients with chronic non-hemolytic unconjugated hyperbilirubinemia, *Am. J. Med.,* 47, 395, 1969.
945. **Hunter, J., Stewart, D. A., Maxwell, J. D., and Williams, R.,** Urinary D-glucaric acid excretion and total liver content of cytochrome P-450 in guinea pigs: relationship during enzyme induction and following inhibition of protein synthesis, *Biochem. Pharmacol.,* 22, 743, 1973.
946. **Crigler, J. F. and Gold, N. I.,** Effect of sodium phenobarbital on bilirubin metabolism in an infant with congenital, nonhemolytic, unconjugated hyperbilirubinemia, and kernicterus, *J. Clin. Invest.,* 48, 42, 1969.
947. **Szabó, L., Kovács, Z., and Ebrey, P. B.,** Crigler-Najjar syndrome, *Acta Paediatr. Hung.,* 3, 49, 1962.
948. **Billing, B. H., Williams, R., and Richards, T. G.,** Defects in hepatic transport of bilirubin in congenital hyperbilirubinemia: an analysis of plasma bilirubin disappearance curves, *Clin. Sci.,* 27, 245, 1964.
949. **Jansen, F. H., Malvaux, P., Heirwegh, K. P. M., and Devriendt, A.,** Congenital nonhemolytic jaundice: Crigler-Najjar syndrome, *Biol. Neonate,* 14, 53, 1969.
950. **Gilbert, A., Lerebouillet, P., and Herscher, M.,** Les trois cholémies congénitales, *Bull. Soc. Med. Hop. Paris,* 24, 1203, 1907.
951. **Powell, L. W., Hemingway, E., Billing, B. H., and Sherlock, S.,** Idiopathic unconjugated hyperbilirubinemia (Gilbert's Syndrome). A study of 42 families, *N. Engl. J. Med.,* 277, 1108, 1967.
952. **Owens, D. and Evans, J.,** Population studies on Gilbert's syndrome, *J. Med. Gen.,* 12, 152, 1975.
953. **Berk, P. D., Bloomer, J. R., Howe, R. B., and Berlin, N. I.,** Constitutional hepatic dysfunction (Gilbert's syndrome): a new definition based on kinetic studies with unconjugated radiobilirubin, *Am. J. Med.,* 49, 296, 1970.
954. **Arias, I. M.,** Transfer of bilirubin from blood to bile, *Semin. Hematol.,* 55, 1972.
955. **Arias, I. M. and London, I. M.,** Bilirubin glucuronide formation *in vitro*; demonstration of a defect in Gilbert's disease, *Science,* 126, 563, 1957.
956. **Metge, W. R., Owen, C. A., Jr., Foulk, W. T., and Hoffman, H. N.,** Bilirubin glucuronyl transferase activity in liver disease, *J. Lab. Clin. Med.,* 64, 89, 1964.
957. **Carulli, N., Ponz de León, M., Mauro, D., Manenti, F., and Ferrari, A.,** Alteration in drug metabolism in Gilbert's syndrome, *Gut,* 17, 581, 1976.
958. **Berk, P. D., Blaschke, T. F., and Waggoner, J. G.,** Defective bromosulfophthalein clearance in patients with constitutional hepatic dysfunction (Gilbert's syndrome), *Gastroenterology,* 63, 472, 1972.
959. **Adlercreutz, H. and Tikkanen, M. J.,** Defects in hepatic uptake and transport and biliary excretion of estrogens, *Med. Chir. Dig.,* 2, 59, 1973.
960. **Fevery, J., Heirwegh, K. P. M., and de Groote, J.,** Unconjugated hyperbilirubinemia in achalasia, *Gut,* 15, 121, 1974.
961. **Black, M. and Sherlock, S.,** Treatment of Gilbert's syndrome with phenobarbitone, *Lancet,* i, 1359, 1970.
962. **Berthelot, P., Erlinger, S., Dhumeaux, D., and Preaux, A.-M.,** Mechanism of phenobarbital-induced hypercholeresis in the rat, *Am. J. Physiol.,* 219, 809, 1970.

963. **Blanckaert, N., Heirwegh, K. P. M., and Compernolle, F.,** Synthesis and separation by thin layer chromatography of bilirubin-IX isomers. Their identification as tetrapyrroles and dipyrrolic ethyl anthranilate azo derivatives, *Biochem. J.,* 155, 405, 1976.

964. **Cornelius, C. E.,** Organic anion transport in mutant sheep with congenital hyperbilirubinemia, *Arch. Environ. Health,* 19, 852, 1969.

965. **Mia, A. S., Gronwall, R. R., and Cornelius, C. E.,** Unconjugated and conjugated bilirubin transport in normal and mutant Corriedale sheep with Dubin-Johnson syndrome, *Proc. Soc. Exp. Biol. Med.,* 135, 33, 1970.

966. **Arias, I. M., Bernstein, L., Toffler, R., and Ezzer, J. B.,** Biliary and urinary excretion of 7-H³-epinephrine in mutant Corriedale sheep with hepatic pigmentation, *Gastroenterology,* 48, 495, 1965.

967. **Cornelius, C. E. and Gronwall, R. R.,** Congenital photosensitivity and hyperbilirubinemia in Southdown sheep in the United States, *Am. J. Vet. Res.,* 29, 291, 1968.

968. **Mia, A. S., Gronwall, R. R., and Cornelius, C. E.,** Bilirubin-¹⁴C turnover studies in normal and mutant Southdown sheep with congenital hyperbilirubinemia, *Proc. Soc. Exp. Biol. Med.,* 133, 955, 1970.

969. **Anwer, M. S., Mia, A. S., and Gronwall, R.,** Bilirubin uridyldiphospho-glucuronyl transferase and β-glucuronidase activity in tissues of horse (*Equus caballus*) and sheep (*Ovis aries*), *Comp. Biochem. Physiol.,* 43B, 929, 1972.

969a. **Dodge, R. H., Cerniglia, C. E., and Gibson, D. T.,** Fungal metabolism of biphenyl, *Biochem. J.,* 178, 223, 1979.

970. **Pulkkinen, M. O. and Hartiala, K.,** Studies of the possible role of sex hormones on the change in conjugation capacity at various ages in the rat, *Ann. Acad. Sci. Fenn. Ser. A.,* 106, 13, 1964.

971. **Wong, K. P. and Sourkes, T. L.,** Measurement of endogenous UDPG in tissues, and some hormonal effects on its concentration in liver, *Fed. Proc. Fed. Am. Soc. Exp. Biol.,* 26, 835, 1967.

972. **Müller-Oerlinghausen, B., Jahns, R., Künzel, B., and Hasselblatt, A.,** Die Wirkung von Tolbutamid auf Blutglucose und GlucuronsaureKonjugation in Lebergewebe normaler und adrenalektomierter Mause, *Naunyn Schmiedebergs Arch. Exp. Pathol. Pharmakol.,* 262, 17, 1969.

973. **Simons, J. A.,** The effect of growth hormone, thyroxine, and testosterone on the ontogeny of bilirubin UDPglucuronyltransferase in mouse liver, *Biochem. Med.,* 6, 53, 1972.

974. **Szepesi, B. and Freedland, A.** Effect of thyroid hormones on metabolism. IV. Comparative aspects of enzyme responses, *Am. J. Physiol.,* 216, 1054, 1969.

975. **Marselos, M., Rantanen, T., and Puhakainen, E.,** Comparison of the effect of phenobarbital on the D-glucuronic acid pathway in euthryoid and hypothyroid rats, *Acta Pharmacol. Toxicol.,* 37, 415, 1975.

976. **Mougdal, N. R., Raghupathy, E., and Sarma, P. S.,** Effect of thyroid imbalance on the detoxication of benzoic acid in the rat, *Endocrinology,* 64, 326, 1959.

976a. **Constantopoulos, A. and Matsantiotis, N.,** Augmentation of uridine diphosphate glucuronyltransferase activity in rat liver by adenosine 3′,5′-monophosphate, *Gastroenterology,* 75, 486, 1978.

977. **Tikkanen, M. J. and Adlercreutz, H.,** Recurrent jaundice in pregnancy. III. Quantitative determination of urinary estriol conjugates, including studies in pruritus gravidarum, *Am. J. Med.,* 54, 600, 1973.

978. **Adlercreutz, H., Tikkanen, M. J., Wichmann, K., Svanborg, Å., and Anberg, Å.,** Recurrent jaundice in pregnancy. IV. Quantitative determination of urinary and biliary estrogens, including studies in pruritus gravidarum, *J. Clin. Endocrinol. Med.,* 38, 51, 1974.

979. **Lueders, K. K., Dyer, H. M., Thompson, E. B., and Kuff, E. L.,** Glucuronyltransferase activity in transplantable rat hepatomas, *Cancer Res.,* 30, 274, 1970.

980. **Dybing, E.,** Inhibition of acetaminophen glucuronidation by oxazepam, *Biochem. Pharmacol.,* 25, 1421, 1976.

981. **Dybing, E.,** Chlorpromazine inhibition of *p*-aminophenol glucuronidation by rat hepatoma cells in culture, *Acta Pharmacol. Toxicol.,* 31, 287, 1972.

982. **Dybing, E.,** Effects of membrane stabilizers on glucuronidation and amino acid transport in cultures of rat hepatoma cells, *Acta Pharmacol. Toxicol.,* 32, 481, 1973.

983. **Yeh, S. Y. and and Mitchell,** Potentiation and reduction of the analgesia of morphine in the rat by Pargyline, *J. Pharmacol. Exp. Ther.,* 179, 642, 1971.

984. **Yeh, S. Y. and Mitchell, C. L.,** Effect of monoamine oxidase inhibitors on formation of morphine glucuronide, *Biochem. Pharmacol.,* 21, 571, 1972.

985. **Hargreaves, T.,** The effect of monoamine oxidase inhibitors on conjugation, *Experientia,* 24, 157, 1968.

986. **Hüthwohl, B. and Trüber, E.,** The inhibitory effect of triiodinated radiographic contrast media on *p*-nitrophenol glucuronide formation by Ca-sedimented microsomes, *Naunyn Schmiedebergs Arch. Pharmacol.,* Suppl. R37, 282, 1974.

987. **Cooke, W. J., and Cooke, L.,** Effects of anesthetic agents on the biliary excretion of iodopanoate in the rat, *Drug Metab. Dispos.,* 5, 377, 1977.
988. **Brown, B. R., Jr.,** Effect of inhalation anaesthetics on hepatic glucuronide conjugation: a study of the rat *in vitro, Anesthesiology,* 37, 483, 1972.
989. **Rugstad, H. E. and Bratlid, D.,** The effect of sulfisoxazole (Gantrisin) and albumin on bilirubin conjugation in cultures of a clonal cell line with liver-like function, *Biochem. Pharmacol.,* 23, 1432, 1974.
990. **Hargreaves, T., Piper, R. F., and Cam, J.,** Effect of oral contraceptives on glucuronyl transferase, *Nature (London) New Biol.,* 234, 110, 1971.
991. **Malaka-Zafiriu, K., Tsiouris, I., and Cassimos, C.,** The effect of gentamycin on liver glucuronyl transferase, *J. Pediatr.,* 82, 118, 1973.
992. **Smith, D. S. and Fujimoto, J. U.,** Alterations produced by novobiocin during biliary excretion of morphine, morphine-3-glucuronide and other compounds, *J. Pharmacol. Exp. Ther.,* 188, 504, 1974.
993. **Duvaldestin, P., Mahu, J.-L., Preaux, A.-M., and Berthelot, P.,** Novobiocin-inhibition and magnesium-interaction of rat liver microsomal bilirubin UDP-glucuronyltransferase, *Biochem. Pharmacol.,* 25, 2587, 1976.
994. **Mehendale, H. M. and Dorough, H. W.,** Glucuronidation mechanisms in the rat and their significance in the metabolism of insecticides, *Pest. Biochem. Physiol.,* 1, 307, 1972.
995. **Jollow, D. J., Thorgeirsson, S. S., Potter, W. Z., Hashimoto, M., and Mitchell, J. R.,** Acetaminophen-induced hepatic necrosis. VI. Metabolic disposition of toxic and nontoxic doses of acetaminophen, *Pharmacology,* 12, 251, 1974.
996. **Yuasa, A.,** Experimental studies on glucuronidation. I. The influence of 4-allyl-2-methoxy phenol (eugenol), *Jpn. J. Vet. Sci.,* 34, 49, 1972.
997. **Dingell, J. V., Miller, K. W., Heath, E. C., and Klausner, H. A.,** The intracellular localization of Δ^9-tetrahydrocannabinol in liver and its effect on drug metabolism *in vitro, Biochem. Pharmacol.,* 22, 949, 1973.
998. **Okolicsanyi, L., Frei, J., and Magnenat, P.,** Multiplicity and specificity of UDP-glucuronyltransferase. II. Influence of phenobarbital and cholestasis on the activity of glucuronyl transferase and β-glucuronidase in rat liver, *Enzymol. Biol. Clin.,* 11, 402, 1970.
999. **Stevenson, I. H. and Turnbull, M. J.,** Hepatic drug-metabolising enzyme activity and duration of hexobarbitone anaesthesia in barbitone-dependent and withdrawn rats, *Biochem. Pharmacol.,* 17, 2297, 1968.
1000. **Hänninen, O. and Aitio, A.,** Enhanced glucuronide formation in different tissues following drug administration, *Biochem. Pharmacol.,* 17, 2307, 1968.
1001. **Hänninen, O. and Puukka, R.,** Effect of phenobarbital administration on the hepatic UDP glucuronyltransferase activity in the rat, *Proc. 1st Congr. Hungarian Pharmacol. Society,* Budapest, 1973, 125.
1002. **Black, M., Perrett, R. D., and Carter, A. E.,** Hepatic bilirubin UDP-glucuronyltransferase activity and cytochrome P-450 content in a surgical population and the effects of preoperative drug therapy, *J. Lab. Clin. Med.,* 81, 704, 1973.
1003. **Orme, M. L., Davies, L., and Breckenridge, A.,** Increased glucuronidation of bilirubin in man and rat by administration of antipyrine (Phenazone), *Clin. Sci. Mol. Med.,* 46, 511, 1974.
1004. **Pilcher, C. W. T., Thompson, R. P. H., and Williams, R.,** Effect of phenobarbitone on hepatic microsomal enzymes of the male rat, *Biochem. Pharmacol.,* 21, 129, 1972.
1005. **Roerig, D. l., Hasegawa, A. T., Peterson, R. E., and Wang, R. I. H.,** Effect of chloroquine and phenobarbital on morphine glucuronidation and biliary excretion in the rat, *Biochem. Pharmacol.,* 23, 1331, 1974.
1006. **Henderson, P. Th.,** Phenobarbital induction of UDP-glucuronyl transferase in isolated rat hepatocytes, *Life Sci.,* 10, 655, 1971.
1007. **Adachi, Y. and Yamamoto, T.,** Influence of drugs and chemicals upon hepatic enzymes and proteins. 1. Structure-activity relationship between various barbiturates and microsomal enzyme induction in rat liver, *Biochem. Pharmacol.,* 25, 663, 1976.
1008. **Glazer, R. I. and Sartorelli, A. C.,** The effect of phenobarbital on the synthesis of nascent protein on free and membrane-bound polyribosomes of normal regenerating liver, *Mol. Pharmacol.,* 8, 701, 1972.
1009. **Vainio, H. and Parkki, M. G.,** Enhancement of microsomal monooxygenase, epoxide hydrase and UDP-glucuronyltransferase by aldrin, dieldrin and isosafrole administration in rat liver, *Toxicology,* 5, 279, 1976.
1010. **Notten, W. R. F. and Henderson, P. Th.,** Alteration in urinary D-glucaric acid excretion as an indication of exposition to xenobiotics, Proc. W. H. O. Int. Symp. Rec. Adv. Assessment Health Effects Environ. Pollution, 1975, 2047.

1011. **Wishart, G. J. and Dutton, G. J.**, Effect of phenobarbital and other xenobiotics on the repression *in ovo* of chick embryo uridine diphosphate glucuronyltransferase activity, *Biochem. Soc. Trans.*, 1, 1214, 1973.

1012. **Goldstein, J. A. and Taurog, A.**, Enhanced biliary excretion of thyroxine-glucuronide in rats pretreated with benzpyrene, *Biochem. Pharmacol.*, 17, 1049, 1968.

1013. **Vainio, H. and Aitio, A.**, Influence of hematin, carbon tetrachloride and SKF 525-A on the enhancement of microsomal monooxygenase and UDPglucuronosyltransferase by 3,4-benzpyrene in rat liver, *Acta Pharmacol. Toxicol.*, 37, 23, 1975.

1014. **Aitio, A., Vainio, H., and Hanninen, O.**, Enhancement of drug oxidation and conjugation by carcinogens in different rat tissues, *FEBS Lett.*, 24, 237, 1974.

1015. **Dutton, G. J. and Stevenson, I. H.**, The stimulation by 3,4-benzpyrene of glucuronide synthesis in skin, *Biochim. Biophys. Acta*, 58, 633, 1962.

1016. **Aitio, A.**, Induction of UDPglucuronyltransferase in the liver and extrahepatic organs of the rat, *Life Sci.*, 13, 1705, 1973.

1017. **Laitinen, M., Lang, M., and Hänninen, O.** Changes in the protein lipid interaction in rat liver microsomes after pretreatment of the rat with barbiturates and polycyclic hydrocarbons, *Int. J. Biochem.*, 5, 747, 1974.

1018. **Lake, B. G., Longland, R. C., Hodgson, R. A., Severn, B. J., Gangolli, S. D., and Lloyd, A. G.**, The effect of some inducers of hepatic xenobiotic-metabolizing enzymes on the urinary excretion of D-glucuronic acid metabolites in the rat, *Biochem. Soc. Trans.*, 3, 188, 1975.

1019. **Notten, W. R. F. and Henderson, P. Th.**, Alterations in the glucuronic acid pathway caused by various drugs, *Int. J. Biochem.*, 6, 111, 1975.

1020. **Lucier, G. W., McDaniel, O. S., Bend, J. R., and Faeder, E.**, Effects of hycanthone and two of its chlorinated analogs on hepatic microsomes, *J. Pharmacol. Exp. Ther.*, 186, 416, 1973.

1021. **Hohenwaller, W. and Klima, J.**, *In vivo* activation of glucuronyltransferase in rat liver by eucalyptole, *Biochem. Pharmacol.*, 20, 3463, 1971.

1022. **Potrepka, R. F. and Spratt, J. L.**, Bilirubin glucuronidation by hepatic microsomal subfractions and the effect of 3-methylcholanthrene, *Biochem. Pharmacol.*, 20, 2247, 1971.

1023. **Yuasa, A.**, Experimental studies on glucuronidation. II. The influence of some glucosiduronic drugs, *Jpn. J. Vet. Sci.*, 34, 49, 1972.

1024. **Gessner, T.**, Occurrence of steroid glucuronyltransferases in a hepatoma, *Cancer Res.*, 37, 2275, 1977.

1025. **Bastomsky, C. H. and Murthy, P. V. N.**, Enhanced *in vitro* hepatic glucuronidation of thyroxine in rats following cutaneous application or ingestion of polychlorinated biphenyls, *Can. J. Physiol. Pharmacol.*, 54, 23, 1976.

1026. **Vainio, H.**, Enhancement of microsomal drug oxidation and glucuronidation by an environmental chemical, polychlorinated biphenyl, *Chem. Biol. Interact.*, 9, 379, 1974.

1027. **Ecobichon, D. J. and Corneau, A. M.**, Competitive effects of commercial Arachlors on liver enzyme activities, *Chem. Biol. Interact.*, 9, 341, 1974.

1028. **Grote, W., Schmoldt, A., and Dammann, H. G.**, The metabolism of foreign compounds in rats after treatment with polychlorinated biphenyls (PCBs), *Biochem. Pharmacol.*, 24, 1121, 1975.

1029. **Lake, B. G., Longland, R. C., Gangolli, S. D., and Lloyd, A. G.**, The influence of some foreign compounds on hepatic xenobiotic metabolism and the urinary excretion of D-glucuronic acid metabolites in the rat, *Toxicol. Appl. Pharmacol.*, 35, 113, 1976.

1030. **Marselos, M., Törrönen, R., Alakuijala, P., and MacDonald, E.**, Hepatic hydroxylation and glucuronidation in the rat after subacute pyrazole treatment, *Toxicology*, 8, 251, 1977.

1031. **Bitto, T., Mirabile, V., Tornaghi, G. C., del Ninno, E., Chiesa, G., Portaleone, D., and Ideo, G.**, Effeto della somministrazione cronica di spironolattone etanolo e fenobarbital sulla bilirubin-UDPG-transferasi epatica nell' ittero di Gilbert, *Quad. Sclavo. Diagn.*, 9, 357, 1972.

1032. **Radzialowski, F.**, Effect of spironolactone and pregnenolone 16α-carbonitrile on bilirubin metabolism and plasma levels in male and female rats, *Biochem. Pharmacol.*, 22, 1607, 1973.

1033. **Solymoss, B. and Zsigmond, G.**, Effect of various steroids on the hepatic glucuronidation and biliary excretion of bilirubin, *Can. J. Physiol. Pharmacol.*, 51, 319, 1973.

1034. **Parkki, M., Vainio, H., and Marniemi, J.**, Effects of styrene administration on the detoxifying enzymes in rat liver, Proc. 6th Int. Congr. Pharmacol., Helsinki, 1975, Abstr. 1099, 1975, 459.

1035. **Lucier, G. W., McDaniel, O. S., Hook, G. E. R., Fowler, B. A., Sonawane, B. R., and Faeder, E.**, TCDD-induced changes in rat liver microsomal enzymes, *Environ. Health Perspect.*, 5, 199, 209, 1973.

1036. **Chadwick, R. W., Cranmer, M. F., and Peoples, A. J.**, Metabolic alterations in the squirrel monkey induced by DDT administration and ascorbic acid deficiency, *Toxicol. Appl. Pharmacol.*, 20, 308, 1971.

1037. **Davis, M., Simmons, C. J., Dordoni, B., and Williams, R.,** Urinary D-glucaric acid excretion and plasma antipyrine kinetics during enzyme induction, *Br. J. Clin. Pharmacol.*, 1, 253, 1974.

1038. **Miettinen, T. A. and Leskinen, E.,** The glucuronic acid pathway in thyroid dysfunction, *Ann. Med. Exp. Fenn.*, 45, 80, 1967.

1039. **Gillette, J. R.,** Effect of various inducers on electron transport system associated with drug metabolism by liver microsomes, *Metabolism*, 20, 228, 1971.

1040. **Kamil, I. A., Smith, J. N., and Williams, R. T.,** Studies in detoxication. 50. The isolation of methyl and ethyl glucuronides from the urine of rabbits receiving methanol and ethanol, *Biochem. J.*, 54, 390, 1953.

1041. **Cooke, B. A. and Taylor, W.,** The metabolism of progesterone by animal tissues *in vitro*. 5. Inhibition of conjugate formation by ethanol and propylene glycol during the metabolism of [4-^{14}C] progesterone by rat liver *in vitro*, *Biochem. J.*, 87, 214, 1963.

1041a. **Gessner, T. and Bolanowska, W.,** Drug interactions with acetaminophen metabolism, *Abstr. 7th Int. Congr. Pharmacol.*, *Abstr. 2598*, Pergamon Press, Oxford, 1978, 840.

1042. **Vainio, H.,** On the Topology and Synthesis of Drug Metabolizing Enzymes in Hepatic Endoplasmic Reticulum, thesis, University of Turku, Turku, Finland, 1973.

1043. **Laitinen, M., Lang, M., and Hänninen, O.,** Changes in the protein-lipid interaction in rat liver microsomes after pretreatment of the rat with barbiturates and polycyclic hydrocarbons, *Int. J. Biochem.*, 5, 747, 1974.

1044. **Wishart, G. J.,** Demonstration of functional heterogeneity of hepatic uridine diphosphate glucuronosyltransferase activities after administration of 3-methylcholanthrene and phenobarbital to rats, *Biochem. J.*, 174, 671, 1978.

1045. **Uotila, P. and Marniemi, J.,** Variable effects of cigarette smoking on aryl hydrocarbon hydroxylase, epoxide hydratase and UDP-glucuronosyltransferase activites in rat lung, kidney and intestinal mucosa, *Biochem. Pharmacol.*, 25, 2323, 1976.

1046. **Uotila, P.,** Effects of single and repeated cigarette smoke exposures on the activities of aryl hydrocarbon hydroxylase, epoxide hydratase and UDP glucuronosyltransferase in rat lung, kidney and small intestinal mucosa, *Res. Commun. Chem. Pathol. Pharmacol.*, 17, 101, 1977.

1047. **Hänninen, O.,** Age and exposure factors in drug metabolism, *Acta Pharmacol. Toxicol.*, (Suppl. 11), 36, 3, 1975.

1048. **Vainio, H.,** Enhancement of hepatic microsomal drug hydroxylation and glucuronidation by 1,1,1-trichloro-2,2-bis(p-chlorophenyl)ethane (DDT), *Chem. Biol. Interact.*, 9, 7, 1974.

1049. **Vainio, H., Parkki, M. G., and Marniemi, J.,** Effects of aliphatic chlorohydrocarbons on drug-metabolizing enzymes in rat liver *in vitro*, *Xenobiotica*, 6, 599, 1976.

1050. **Hargreaves, T.,** Inhibition of conjugation by anti-inflammatory drugs, *Nature (London)*, 208, 1101, 1965.

1051. **Windorfer, A.,** Steigerung der Glucuronidierungsrate in der Neugeborenenperiode durch Therapeutische Phenobarbitaldosen, *Z. Kinderheilkd.*, 113, 33, 1972.

1052. **Takemori, A. E.,** Enzymic studies on morphine glucuronide synthesis in acutely and chronically morphinized rats, *J. Pharmacol. Exp. Ther.*, 130, 370, 1960.

1053. **Hänninen, O.,** Effect of salicylamide administration on D-glucuronic acid metabolism in the rat, *Ann. Acad. Sci. Fenn. Ser. A.*, 123, 1966.

1054. **Yuasa, A.,** Enhancement of liver UDP-glucose dehydrogenase caused by an effective glucosidurogenic compound, eugenol, *Jpn. J. Vet. Sci.*, 36, 273, 1974

1055. **Prasannan, K. and Kurup, P. A.,** Oral hypoglycaemic agents and cardiovascular complications of diabetes, *Atherosclerosis*, 18, 459, 1973.

1056. **Pulkkinen, M. and Hartiala, K.,** Effect of 3,4-benzpyrene on the formation and hydrolysis of conjugates, *Nature (London)*, 207, 646, 1965.

1057. **Rantanen, T., Marselos, M., and Puhakainen, E.,** Effects of 1,1,1-trichloro-2,2-bis(p-chlorophenyl)ethane (DDT) administration on the glucuronic acid pathway in the rat liver, in *Environmental Quality and Safety*, Vol. 3, Coulston, F. and Korte, F., Eds., Georg Thieme Verlag, Stuttgart, 1975, 494.

1058. **Marselos, M., Lang, M., and Törrönen, R.,** Modification of drug metabolism by disulfiram and diethylthiocarbamate. II. D-Glucuronic acid pathway, *Chem. Biol. Interact.*, 15, 277, 1976.

1059. **Törrönen, R., Alakuijala, P., and Marselos, M.,** Chelating agents and hepatic drug metabolism in the rat, *Arch. Int. Pharmacodyn.*, 222, 167, 1977.

1060. **Aarts, E. M.,** Evidence for the function of D-glucaric acid as an indicator for drug induced enhanced metabolism through the glucuronic acid pathway in man, *Biochem. Pharmacol.*, 14, 359, 1965.

1061. **Hunter, J. and Chasseaud, L.,** Clinical aspects of microsomal enzyme induction, *Prog. Drug Metab.*, 1, 129, 1976.

1062. **Mowat, A. P.**, Developmental effects on liver D-glucuronolactone dehydrogenase levels and on D-glucaric acid excretion in urine; hormonal effects on D-glucaric acid excretion in urine, *J. Endocrinol.*, 42, 585, 1968.

1063. **Talafant, E., Hoškova, A., and Pojerova, A.**, Glucaric acid excretion as index of hepatic glucuronidation in neonates after phenobarbital treatment, *Pediatr. Res.*, 9, 480, 1975.

1064. **Levy, G., Khanna, N. N., Soda, D. M., Tsuzuki, O., and Stern, L.**, Pharmacokinetics of acetaminophen in the human neonate: formation of acetaminophen glucuronide and sulfate in relation to plasma bilirubin concentration and D-glucaric acid excretion, *Pediatrics*, 55, 818, 1975.

1065. **Notten, W. R. F.**, Alterations in the D-Glucuronic Acid Pathway and Drug Metabolism by Exogenous Compounds, thesis, University of Nijmegen, Netherlands, 1975.

1066. **Jones, A. L. and Fawcett, D. W.**, Hypertrophy of the agranular endoplasmic reticulum in hamster liver induced by phenobarbital, *J. Histochem. Cytochem.*, 14, 215, 1966.

1067. **Notten, W. R. F., Henderson, P. Th., and Kuyper, Ch. A. V.**, Stimulation of the glucuronic acid pathway in isolated rat liver cells by phenobarbital, *Int. J. Biochem.*, 6, 713, 1975.

1068. **Notten, W. R. F. and Henderson, P. Th.**, The interaction of chemical compounds with the functional state of the liver. II. Estimation of changes in D-glucaric acid synthesis as a method for diagnosing exposure to xenobiotics, *Int. Arch. Occup. Environ. Health*, 38, 209, 1977.

1068a. **Baker, T. S., Jennison, K. M., and Kellie, A. E.**, The direct radioimmunoassay of oestrogen glucuronides in human female urine, *Biochem. J.*, 177, 729, 1979.

1068b. **Ricci, G. L. and Fevery, J.**, Stimulation by secretion of bilirubin UDP-glycosyltransferase activities and of cytochrome *P*-450 concentration in rat liver, *Biochem. J.*, 182, 881, 1979.

1068c. **Wishart, G. J. and Campbell, M. T.**, Demonstration of two functionally heterogeneous groups within the activities of UDP-glucuronosyltransferase towards a series of 4-alkyl-substituted phenols, *Biochem. J.*, 178, 443, 1979.

1069. **Powell, G. M., Miller, I. J., Olavesen, A. H., and Curtis, C. G.** Liver as major organ of phenol detoxication?, *Nature (London)*, 252, 234, 1974.

1070. **Aitio, A.**, Extrahepatic Microsomal Drug Metabolism, thesis, University of Turku, Finland, 1973.

1071. **Støa, K. F. and Levitz, M.**, Comparison of the conjugated metabolites of intravenously and intraduodenally administered oestriol, *Acta Endocrinol.*, 57, 657, 1968.

1072. **Aitio, A.**, Glucuronide synthesis in the rat and guinea pig lung, *Xenobiotica*, 3, 13, 1973.

1073. **Hobkirk, R., Green, R. N., Nilsen, M., and Jennings, B. A.**, Formation of estrogen glucosiduronates by human kidney homogenates, *Can. J. Biochem.*, 52, 9, 1974.

1074. **De Schepper, J. and Van der Stock, J.**, Increase of the total amount of conjugated and unconjugated bilirubin in the normothermic perfused isolated dog kidney system, *Pflugers Arch. Gesamte Physiol.*, 333, 62, 1972.

1075. **Mellor, J. D. and Hobkirk, R.**, *In vitro* synthesis of estrogen glucuronides and sulfates by human renal tissue, *Can. J. Biochem.*, 53, 779, 1975.

1076. **Quebbemann, A. J. and Anders, M. W.** Renal tubular conjugation and excretion of phenol and *p*-nitrophenol in the chicken: differing mechanisms of renal transfer, *J. Pharmacol. Exp. Ther.*, 184, 695, 1975.

1077. **Weiner, I. M. and Mudge, G. H.**, Renal tubular mechanisms for excretion of organic acids and bases, *Am. J. Med.*, 36, 743, 1964.

1078. **Kellie, A. E. and Smith, E. R.**, Renal clearance of 17-oxo steroid conjugates found in human peripheral plasma, *Biochem. J.*, 66, 490, 1957.

1079. **Sperber, I.**, Secretion of organic anions in the formation of urine and bile, *Pharmacol. Rev.*, 11, 109, 1959.

1080. **Hartiala, K.**, Metabolism of hormones, drugs and other substances by the gut, *Physiol. Rev.*, 53, 496, 1973.

1081. **Zini, F.**, Sulla glicurono-coniugazione dell'o aminofenolo da fettine d'organi di ratto, *Sperimentale*, 102, 40, 1952.

1082. **Shirai, Y. and Ohkubo, T.**, Synthesis of glucuronides by tissue slices. I., *J. Biochem. (Tokyo)*, 41, 341, 1954.

1083. **Hänninen, O., Aitio, A., and Hartiala, K.**, Gastrointestinal distribution of glucuronide synthesis and the relevant enzymes in the rat, *Scand. J. Gastroenterol.*, 3, 461, 1968.

1084. **Bock, K. W. and Winne, D.**, Glucuronidation of 1-naphthol in the rat intestinal loop, *Biochem. Pharmacol.*, 24, 859, 1975.

1085. **Josting, D., Winne, D., and Bock, K. W.**, Glucuronidation of paracetamol, morphine and 1-naphthol in the rat intestinal loop, *Biochem. Pharmacol.*, 25, 613, 1976.

1086. **Aitio, A., Hietanen, E., and Hänninen, O.**, Mucosal drug metabolism and drug-induced ulcer, in *Experimental Ulcer, Models, Methods, Clinical Validity*, Gheorghiu, T., Ed., Witzstrock, Baden Baden, 1975, 17.

1087. **Lehtinen, A., Nurmikko, V., and Hartiala, K.,** Duodenal glucuronide synthesis. II. Identification of estradiol glucuronide as a conjugation product of estradiol by the rat duodenal mucosa. Quantitative studies, *Acta Chem. Scand.,* 12, 1585, 1958.

1088. **Franco, D., Preaux, A.-M., Bismuth, H., and Berthelot, P.,** Extra hepatic formation of bilirubin glucuronides in the rat, *Biochim. Biophys. Acta,* 286, 55, 1972.

1089. **Royer, M., Noir, B. A., Sfarcich, D., and Nanet, H.,** Extrahepatic bilirubin formation and conjugation in the dog, *Digestion,* 10, 423, 1974.

1090. **Poland, R. L. and Odell, G. B.,** Physiologic jaundice: the enterohepatic circulation of bilirubin, *N. Engl. J. Med.,* 284, 1, 1971.

1091. **Windorfer, A., Jr., Künzer, W., Bolze, H., Ascher, K., Wilcken, F., and Hochne, K.,** Studies on the effect of orally administered agar on the serum bilirubin level of premature infants and mature newborns, *Acta Paediatr. Scand.,* 62, 699, 1975.

1092. **Koldowsky, O. and Palmieri, M.,** Cortisone-evoked decrease of acid β-galactosidase, β-glucuronidase, N-acetyl-β-glucosaminidase and arylsulphatase in the ileum of suckling rats, *Biochem. J.,* 125, 697, 1971.

1093. **Schachter, D., Kass, D. J., and Lannon, T. J.,** The biosynthesis of salicyl glucuronides by tissue slices of various organs, *J. Biol. Chem.,* 234, 201, 1959.

1094. **Gingell, R., Bridges, J. W., and Williams, R. T.,** Intestinal azo-reduction and glucuronide conjugation of prontosil, *Xenobiotica,* 3, 599, 1973.

1095. **Taylor, W.,** The hepatobiliray system: retrospect and prospect. A personal view, in *The Hepato-Biliary System,* Taylor, W., Ed., Plenum Press, New York, 1976, 1.

1095a. **Levine, W. G.,** Biliary excretion of drugs and other xenobiotics, *Ann. Rev. Pharmacol. Toxicol.,* 18, 81, 1978.

1096. **Williams, R. T., Millburn, P., and Smith, R. L.,** The influence of enterohepatic circulation on toxicity of drugs, *Ann. N.Y. Acad. Sci.,* 123, 110, 1965.

1097. **Mulder, G. J.,** The rate-limiting step in the biliary elimination of some substrates of uridine diphosphate glucuronyltransferase in the rat, *Biochem. Pharmacol.,* 22, 1751, 1973.

1098. **Millburn, P.,** Excretion of xenobiotic compounds in bile, in *The Hepato-Biliary System,* Taylor, W., Ed., Plenum Press, New York, 1976, 109.

1099. **Uesugi, T., Ikeda, M., Kanei, Y., Hori, R., and Arita, T.,** Studies on the biliary excretion mechanisms of drugs. II. Biliary excretion of thiamphenicol, chloramphenicol and their glucuronides in the rat, *Biochem. Pharmacol.,* 23, 2315, 1974.

1100. **Gessner, T. and Hamada, N.,** High biliary and systemic excretion of a glucuronide during its hepatic synthesis, *Life Sci.,* 15, 83, 1974.

1101. **Levine, W. G. and Singer, R. W.,** Hepatic intracellular distribution of foreign compounds in relation to their biliary excretion, *J. Pharm. Exp. Ther.,* 183, 411, 1972.

1102. **Hillier, A. P.,** Transport of thyroxine glucuronide into bile, *J. Physiol.,* 227, 195, 1972.

1103. **Fleischner, G. and Arias, I. M.,** Recent advances in bilirubin formation, transport, metabolism and excretion, *Am. J. Med.,* 49, 576, 1970.

1104. **Cooke, W. J. and Munck, L.,** Effect of phenobarbital on biliary excretion of p-nitrophenol in the rat, *Abstr. 6th Int. Congr. Pharmacol.,* Helsinki, 1975, 644.

1105. **Cooke, W. S. and Cooke, L.,** Biliary excretion of iopanoate glucuronide by the rat, *Drug Metab. Dispos.,* 5, 368, 1977.

1106. **Okolicsanyi, L., Vassanelli, P., Scremin, S., and Stella, G. D.,** Modifications of the biliary excretion of bilirubin glucuronide in rats with portacaval shunt: the role of UDP-glucuronyltransferase, *Biomedicine,* 19, 253, 1973.

1107. **Inoue, N., Sandberg, A. A., Graham, J. B., and Slaunwhite, W. R., Jr.,** Studies on phenolic steroids in human subjects. VIII. Metabolism of estriol-16α-glucosiduronate, *J. Clin. Invest.,* 48, 380, 1969.

1108. **Inoue, N., Sandberg, A. A., Graham, J. B., and Slaunwhite, W. R., Jr.,** Studies on phenolic steroids in human subjects. IX. Role of the intestine in the conjugation of estriol, *J. Clin. Invest.,* 48, 390, 1969.

1109. **Støa, K. F. and Skulstad, P. A.,** Biliary and urinary metabolites of intravenously and intraduodenally administered 17β-oestradiol and oestriol, *Steroids Lipid Res.,* 3, 299, 1972.

1110. **Tikkanen, M. J., Pulkkinen, M. O., and Adlercreutz, H.,** Effect of Ampicillin treatment on the urinary excretion of estriol conjugated in pregnancy, *J. Steroid Biochem.,* 4, 439, 1973.

1111. **Sandberg, A. A., Slaunwhite, W. R., Jr., and Kirdani, R. Y.,** Metabolic conjugation and hydrolysis of estrogens and progesterone in the enterohepatic circulation, in *Metabolic Conjugation and Metabolic Hydrolysis,* Vol. 2, Fishman, W. H., Ed., Academic Press, New York, 1970, 123.

1112. **Musey, P. J., Green, R. N., and Hobkirk, R.,** The role of an enterohepatic system in the metabolism of 17β-estradiol-17-glucosiduronate in the human female, *J. Clin. Endocrinol. Metab.,* 35, 448, 1972.

1113. **Hearse, D. J., Powell, G. M., Olavesen, A. H., and Dodgson, K.,** The influence of some physico-chemical factors on the biliary excretion of a series of structurally related aryl sulphate esters, *Biochem. Pharmacol.,* 18, 181, 1969.

1113a. **Back, D. J. and Breckenridge, A. M.,** The effect of antibiotics on the enterohepatic circulation of ethinlestradiol and norethistrone in the rat, *J. Steroid Biochem.,* 9, 527, 1978.

1114. **Smith, R. L.,** The role of the gut flora in the conversion of inactive compounds to active metabolites, in *Mechanisms of Toxicity,* Aldridge, W. N., Ed., Macmillan, London, 1971, 229.

1115. **Strand, L. P. and Scheline, R. R.,** The metabolism of vanillin and isovanillin in the rat, *Xenobiotica,* 5, 49, 1975.

1116. **Harper, K. H.,** The intermediary metabolism of 3:4-benzpyrene: the biosynthesis and identification of the X_1 and X_2 metabolites, *Br. J. Cancer,* 12, 645, 1958.

1117. **Harper, K. H. and Calcutt, G.,** Conjugation of 3:4 benzpyrenols in mouse skin, *Nature (London),* 186, 80, 1960.

1118. **De Bernardi, M., Ferrara, A., and Manzo, L.,** Aspetto del farmacometabolismo polmonare nel coniglio adulto, *Boll. Soc. Ital. Biol. Sper.,* 48, 102, 1972.

1119. **El-Shourbagy, N. A. and Dorough, H. W.,** Glycoside conjugative activity in different insect and vertebrate species, *J. Econ. Entomol.,* 67, 344, 1974.

1120. **Litterst, C. L., Minnaugh, E. G., and Gram, T. E.,** Comparative alterations in extrahepatic drug metabolism by factors known to affect hepatic activity, *Biochem. Pharmacol.,* 26, 749, 1977.

1121. **Aitio, A., Hartiala, J., and Uotila, P.,** Glucuronide synthesis in the isolated perfused rat lung, *Biochem. Pharmacol.,* 25, 1919, 1976.

1121a. **Gessner, T., Dresner, J. H., Freedman, H. J., Gurtoo, H. L., and Paigen, B.,** Conjugation pathways in humans: glucuronyltransferase activity in human lymphocytes, in *Conjugation Reactions in Drug Biotransformation,* Aitio, A., Ed., Elsevier, Amsterdam, 1978, 77.

1121b. **Grafström, R., Moldéus, P., Andersson, B., and Orrenius, S.,** Xenobiotic metabolism by isolated rat small intestinal cells, *Med. Biol.,* 57, 287, 1979.

1121c. **Ball, L. M., Plummer, J. L., Smith, B. R., and Bend, J. R.,** Benzo(a)pyrene oxidation, conjugation and disposition in the isolated perfused rabbit lung: role of the glutathione S-transferases, *Med. Biol.,* 57, 298, 1979.

1122. **Fry, J. R. and Bridges, J. W.,** The metabolism of xenobiotics in cell suspensions and cell cultures, *Progr. Drug Metab.,* 2, 71, 1977.

1123. **Sandström, B.,** Maintenance in vitro of functionally active adult human liver, *Acta Hepato Gastroenterol.,* 20, 19, 1973.

1124. **Chessebeuf, M., Olsson, A., Bournot, P., Deggres, J., Guiget, M., Maume, G., Maume, B. F., Perissel, B., and Padieu, P.,** Long term cell culture of rat liver epithelial cells retaining some hepatic functions, *Biochimie,* 56, 1365, 1974.

1125. **Burchell, B., Wishart, G. J., and Dutton, G. J.,** Relation between the induction of glucuronidation and of hydroxylation in chick liver, *FEBS Lett.,* 43, 323, 1974.

1126. **Dybing, E. and Rugstad, H. E.,** Para-aminophenol metabolism in an established cell line with liver-like functions, *Acta Pharmacol. Toxicol.,* 31, 153, 1972.

1127. **Novikoff, A. B.,** Enzyme localization in tumor cells, in *Cell Physiology of Neoplasia,* Hsu, T., Ed., University of Texas Press, Austin, 1960, 219.

1128. **Novikoff, A. B. and Biempica, L.,** Cytochemical and electron microscopical examination of Morris 5123 and Reuber H 35 hepatomas after several years of transplantation, in *Biological and Biochemical Evaluation of Malignancy in Experimental Hepatomas,* Gann Monograph 1, Yoshida, T., Ed., Jpn. Fdn. Canc, Res., Jpn. Canc. Assoc., Tokyo, 1965, 67.

1129. **Dybing, E. and Rugstad, H. E.,** The inhibitory effect of diethylaminoethyl diphenylvalerate (SKF 525-A) on glucuronidation by cultures of rat hepatoma cells, *Acta Pharmacol. Toxicol.,* 32, 113, 1973.

1130. **Rugstad, H. E., Robinson, J. H., Yannoni, C., and Tashjian, A. H.,** Metabolism of bilirubin by a clonal strain of rat hepatoma cells, *J. Cell Biol.,* 47, 703, 1970.

1131. **Rugstad, H. E. and Bratlid, D.,** The effect of sulfisoxazole (Gantrisin) and albumin on bilirubin conjugation in cultures of a clonal cell line with liver-like functions, *Biochem. Pharmacol.,* 23, 1432, 1974.

1132. **Zimmerman, J. J., Gorski, J. P., and Kasper, C. B.,** Quantitative relationship of UDP-glucuronosyltransferase to the NADPH and NADH electron-transport systems in Morris hepatomas with varying growth rates, *Drug Metab. Dispos.,* 5, 572, 1977.

1133. **Gessner, T.,** Studies of glucuronidation and sulphation in tumor-bearing rats, *Biochem. Pharmacol.,* 23, 1809, 1974.

1133a. **Bock, K. W., Lorch, F., and van Ackeren, G.,** Activation and induction of microsomal UDP-glucuronyltransferases in rat liver and Morris hepatomas, *Hoppe Seylers Zeit. Physiol. Chem.,* 335, 1177, 1974.

1134. **Falck, B., Hansson, C., Kennedy, B.-M., and Rosengren, E.,** Conjugated catechol derivatives in a transplantable islet cell tumor of the golden hamster, *Acta Physiol. Scand.*, 99, 217, 1977.

1135. **Vadi, H., Jernström, B., and Orrenius, S.,** Recent studies on benzo(a)pyrene metabolism in rat liver and lung, in *Carcinogenesis*, Vol. 1, Freudenthal, R. I. and Jones, P. W., Eds., Raven Press, New York, 321, 1976.

1135a. **Hesse, S., Mezger, M., and Wolff, T.,** Activation of [¹⁴C] chlorobiphenyls to protein-binding metabolites by rat liver microsomes, *Chem. Biol. Interact.*, 20, 355, 1978.

1135b. **Nemoto, N.,** Glucuronidation in the metabolism of benzo(a)pyrene, in *Conjugation Reactions in Drug Biotransformation*, Aitio, A., Ed., Elsevier, Amsterdam, 1978, 17.

1135c. **Kinoshita, N. and Gelboin, H.,** β-Glucuronidase catalysed hydrolysis of benzo(a)pyrene-3-glucuronide and binding to DNA, *Science*, 199, 307, 1978.

1135d. **Goldman, P.,** Biochemical pharmacology of the intestinal flora, *Ann. Rev. Pharmacol. Toxicol.*, 18, 523, 1978.

1136. **Bauer, C. and Reuter, W.,** Inhibition of UDPglucose dehydrogenase with galactosamine-1-phosphate and UDPgalactosamine, *Biochim. Biophys. Acta*, 293, 11, 1973.

1137. **Decker, K., Keppler, D., Rudigier, J., and Domschke, J.,** Cell damage by trapping of biosynthetic intermediates. The role of uracil nucleotides in experimental hepatitis *Hoppe Seylers Z. Physiol. Chem.*, 352, 412, 1971.

1138. **Quick, A. J.,** The production of conjugated glycuronic acids in depancreatized dogs, *J. Biol. Chem.*, 70, 59, 1926.

1139. **Müller-Oerlinghausen, B., Hasselblatt, A., and Jahns, R.,** Vermehrte Bildung von Bilirubinglucuronid in der Leber der Insulin- und Solfonylharnstoff-Hypoglykamie, *Naunyn-Schmiedebergs Arch. Exp. Pathol. Pharmakol.*, 160, 254, 1968.

1140. **Benzo, C. A. and de la Haba, G.,** Development of chick embryo liver during organ culture: requirement for zinc insulin, *J. Cell. Physiol.*, 79, 53, 1972.

1141. **Adlercreutz, H. and Tenhunen, R.,** Some aspects of the interaction between natural and synthetic female sex hormones and the liver, *Am. J. Med.*, 49, 630, 1970.

1142. **Schriefers, H., Keck, B., and Otto, M.,** Biosynthese von Steroidglucuroniden bei verschiedenen Stoffwechselzustanden, *Acta Endocrinol.*, 50, 25, 1965.

1143. **Woolley, M. M., Felsher, D. F., Asch, J., Carpio, N., and Isaacs, H.,** Jaundice, hypertrophic pyloric stenosis and hepatic glucuronyltransferase, *J. Pediatr. Surg.*, 9, 359, 1974.

1144. **Blanckaert, N., Compernolle, F., Leroy, P., Van Houtte, R., Fevery, J., and Heirwegh, K. P. M.,** The fate of bilirubin-IXα glucuronide in cholestasis and during storage *in vitro*. Intramolecular rearrangement to positional isomers of glucuronic acid, *Biochem. J.*, 171, 203, 1978.

1145. **Desmet, V. J., Bullens, A.-M., and de Groote, J.,** A clinical and histochemical study of cholestasis, *Gut*, 11, 516, 1970.

1146. **Thompson, R. P. H. and Hofmann, A. F.,** Separation of bilirubin and its conjugates by thin layer chromatography, *Clin. Chim. Acta*, 35, 517, 1971.

1147. **Kanie, F.,** Histochemical and enzymatic studies on bilirubin metabolism in experimental and clinical liver disorders, *Nagoya Med. J.*, 17, 123, 1972.

1148. **Reid, E.,** Significant biochemical effects of carcinogens in the rat. (A review), *Cancer Res.*, 22, 398, 430, 1962.

1149. **Dohnálek, J., Eysselt, M., Martinek, K., and Kadlecova, D.,** Unsere Erfahrungen mit dem OIB-Test (ein Leberfunktionstest durch Verfolgen des o-Jodbenzoesäure-¹³¹J-Stoffwechsels), *Nucl. Med.*, 6, 76, 1967.

1150. **Muting, D.,** Detoxication capacity of the diseased liver, *Germ. Med. Mth.*, 8, 198, 1963.

1151. **Hammar, C.-H., Prellwitz, W., and Gempp-Friedrich, W.,** Glucuronyl Transferase bei chronischen Leberkrankheiten, *Verh. Dtsch. Ges. Inn. Med.*, 75, 350, 1969.

1152. **Adachi, Y., Wakisaka, G., and Yamamoto, T.,** Salicylamide glucuronide formation in liver disease and its change by drugs, *Gastroenterol. Jpn.*, 10, 120, 1975.

1153. **Nilius, R. and Rath, F. W.,** Glucuronidase bei akuter Tetrachlorkohlenstoff-schadigung der Rattenleber, *Z. Gesamte Exp. Med.*, 144, 157, 1967.

1154. **Lentz, P. E. and van Lancker, J. L.,** The transfer of β-glucuronidase from the microsomal to the small mitochondrial fraction in hypoxic livers, *Am. J. Pathol.*, 52, 4a, 1968.

1155. **Bauer, C. H., Hassels, B. F., and Reutter, W. G.,** Galactose metabolism in regenerating liver, *Biochem. J.*, 154, 141, 1976.

1156. **Kozaryn, I., Chodera, A., Szczawinska, K., Cenijek, D., and Radola, P.,** Activity of enzymes participating in the synthesis and degradation of glucuronides in postirradiative diseases, *Acta Physiol. Pol.*, 26, 189, 1975.

1157. **Pridham, J. B.,** Biosynthesis of phenolic glucosides in the absence of UDPG, *Chem. Ind.*, 1172, 1961.

1158. **Hopkinson, S. and Pridham, J. B.,** Enzymic glycosylation of phenols, *Biochem. J.*, 105, 655, 1967.

1159. **Pridham, J. B.,** The phenol glucosylation reaction in the plant kingdom, *Phytochemistry,* 3, 493, 1964.
1160. **Cardini, C. E. and Yamaha, T.,** Biosynthesis of plant glycosides from uridine diphosphate glucose, *Nature (London),* 182, 1946, 1958.
1161. **Jacobelli, G., Tabone, M. J., and Tabone, J.,** La synthèse enzymatique de l'ester β glucosidique de l'acide anthranilique: ses relations avec les processus de synthèse et de dégradation de l'uridine diphosphate glucose, *Bull. Soc. Chim. Biol.,* 40, 955, 1958.
1162. **Barber, G. A.,** Enzymic glycosylation of quercetin to rutin, *Biochemistry,* 1, 463, 1962.
1163. **Sutter, A. and Grisebach, H.,** Free reversibility of the UDP-glucose: flavonol 3-O-glucosyltransferase reaction, *Arch. Biochem. Biophys.,* 167, 444, 1975.
1164. **Staver, M. J., Glick, K., and Baisted, D. J.,** Uridine diphosphate glucose-sterol glucosyltransferase and nucleoside diphosphatase activities in etiolated pea seedlings, *Biochem. J.,* 169, 297, 1978.
1165. **Trivelloni, J. C.,** Biosynthesis of glucosides and glycogen in the locust, *Arch. Biochem. Biophys.,* 89, 149, 1960.
1166. **Gessner, T. and Vollmer, C. A.,** Glucosylation by mouse liver microsomes, *Fed. Proc. Fed. Am. Soc. Exp. Biol.,* 28, 545, 1969.
1167. **Gessner, T., Jacknowitz, A., and Vollmer, C. A.,** Studies of mammalian glucoside conjugation, *Biochem. J.,* 132, 249, 1973.
1168. **Gessner, T. and Hamada, N.,** Identification of *p*-nitrophenyl glucoside as a urinary metabolite, *J. Pharmaceut. Sci.,* 59, 1528, 1970.
1169. **Duggan, D. E., Baldwin, J. J., Anson, B. H., and Rhodes, R. E.,** *N*-Glucoside formation as a detoxification mechanism in mammals, *J. Pharmacol. Exp. Ther.,* 190, 563, 1974.
1170. **Williamson, D. G., Collins, D. C., Layne, D. S., Conrow, R. B., and Bernstein, S.,** Isolation of 17α-estradiol 17-β-D-glucopyranoside from rabbit urine, and its synthesis and characterization, *Biochemistry,* 8, 4299, 1969.
1171. **Collins, D. C., Williamson, D. G., and Layne, D. S.,** Enzymatic synthesis by a partially purified transferase from rabbit liver microsomes, *J. Biol. Chem.,* 245, 873, 1970.
1172. **Williamson, D. G., Collins, D. C., and Layne, D. S.,** Steroid estrogen glycosides. Formation of glucosides and galactosides by human liver and kidney, *J. Biol. Chem.,* 247, 3286, 1972.
1173. **Wong, K. P.,** Formation of bilirubin glucoside, *Biochem. J.,* 125, 929, 1971.
1173a. **Layne, D. S., Labow, R. S., and Williamson, D. G.,** The formation and metabolism of glucosides of phenolic steroids: an effect on intracellular transport, in *Conjugation Reactions in Drug Biotransformation,* Aitio, A., Ed., Elsevier, Amsterdam, 1978, 59.
1174. **Fevery, J., Van Hees, G. P., Leroy, P., Compernolle, F., and Heirwegh, K. P. M.,** Excretion in dog bile of glucose and xylose conjugates of bilirubin, *Biochem. J.,* 125, 803, 1971.
1175. **Gordon, E. R., Dadoun, M., Goresky, C. A., Chan, T.-H., and Perlin, A. S.,** The isolation of an azobilirubin β-D-monoglucoside from dog gall-bladder bile, *Biochem. J.,* 143, 97, 1974.
1176. **Basu, S., Kaufman, B., and Roseman, S.,** Enzymatic synthesis of ceramide-glucose and ceramide-lactose by glycosyltransferases from embryonic chicken brain, *J. Biol. Chem.,* 243, 5802, 1968.
1177. **Jirku, H. and Layne, D. S.,** The formation of estradiol-3-glucuronoside-17α-*N*-acetylglucosaminide by rabbit liver homogenate, *Biochemistry,* 4, 2126, 1965.
1178. **Zatta, P., Zakim, D., and Vessey, D. A.,** Incorporation of *N*-acetyl glucosamine into lipid linked oligosaccharides, *Biochem. Biophys. Res. Commun.,* 70, 1015, 1976.
1179. **Vessey, D. A., Lysenko, N., and Zakim, D.,** Evidence for multiple enzymes in the dolichol utilizing pathway of glycoprotein biosynthesis, *Biochim. Biophys. Acta,* 428, 138, 1976.
1180. **Silbert, J. E. and Reppucci, A. C.,** The biosynthesis of chondroitin sulfate: independent addition of glucuronic acid and *N*-acetylgalactosamine to oligosaccharides, *J. Biol. Chem.,* 251, 3942, 1976.
1181. **Vessey, D. A. and Zakim, D.,** The identification of a unique *p*-nitrophenol conjugating enzyme in guinea pig liver microsomes, *Biochim. Biophys. Acta,* 315, 43, 1973.
1182. **Berthillier, G., Azzar, G. J.-C., and Got, R.,** Étude de l'activité de transfert de glucose, a partir d'UDP-glucose dans les membranes microsomiques des hépatocytes de rat, *Eur. J. Biochem.,* 51, 275, 1975.
1182a. **Layne, D. S., Labow, R. S., Paquet, A., and Williamson, D. G.,** Transfer of xylose to steroids by rabbit liver microsomes, *Biochemistry,* 15, 1268, 1976.
1183. **Dodgson, K. S.,** Conjugation with sulphate, in *Drug Metabolism from Microbe to Man,* Parke, D. V. and Smith, R. L., Eds., Taylor & Francis, London, 1977, 91.
1184. **Roy, A. B.,** Sulphate conjugation enzymes, *Handb. Exp. Pharmacol.,* 28(2), 536, 1971.
1185. **Dodgson, K. S. and Rose, F. A.,** Sulfoconjugation and sulfohydrolysis, in *Metabolic Conjugation and Metabolic Hydrolysis,* Vol. 1, Fishman, W. H., Ed., Academic Press, New York, 1970, 239.
1186. **Powell, G. M. and Roy, A. B.,** Sites of sulfation and the fates of sulfate esters, *Extrahepatic Metabolism of Drugs and other Foreign Compounds,* Gram, T. E., Ed., Spectrum Publications, Jamaica, New York, 1978.

1187. **Bock, K. W.**, Dual role of glucuronyl- and sulfotransferases converting xenobiotics into reactive or biologically inactive and easily excretable compounds, *Arch. Toxicol.*, 39, 77, 1977.

1187a. **Mulder, G. J. and Meerman, J. H. N.**, Glucuronidation and sulphation *in vivo* and *in vitro*: selective inhibition of sulphation by drugs and deficiency of inorganic sulphate, in *Conjugation Reactions in Drug Biotransformation*, Aitio, A., Ed., Elsevier, Amsterdam, 1978, 389.

1188. **Flynn, T. G., Dodgson, K. S., Powell, G. M., and Rose, F. A.**, The metabolism of dipotassium 2-hydroxy-5-nitrophenyl [^{35}S] sulphate, a substrate for lysosomal arylsulphatases A and B, *Biochem. J.*, 105, 1003, 1967.

1189. **Gatehouse, P. W., Roy, A. B., Dodgson, K. S., Powell, G. M., Lloyd, A. G., and Olavesen, A. H.**, The metabolism of sodium cortisone 21-[^{35}S]-sulphate in the rat, *Biochem. J.*, 127, 661, 1972.

1190. **Cresswell, D.**, Catalytic Properties of Sulfotransferases, Ph.D. thesis, University of Wales, 1975.

1191. **Pasqualini, J.**, Metabolic conjugation and hydrolysis of steroid hormones in the fetoplacental unit, in *Metabolic Conjugation and Metabolic Hydrolysis*, Vol. 2, Fishman, W. H., Ed., Academic Press, New York, 1970, 153.

1192. **Yaffe, S. J., Krasner, J., and Catz, C. S.**, Variations in detoxicating enzymes during mammalian development, *Ann. N.Y. Acad. Sci.*, 151, 887, 1968.

1193. **Carroll, J. and Armstrong, L. M.**, Regulation of hepatic sulphotransferases, *Biochem. Soc. Trans.*, 4, 871, 1976.

1193a. **Powell, G. M. and Curtis, C. G.**, Sites of sulphation and the fate of sulphate esters, in *Conjugation Reactions in Drug Biotransformation*, Aitio, A., Ed., Elsevier, Amsterdam, 1978, 409.

1194. **Bray, H. G., Humphris, B. G., Thorpe, W. V., White, K., and Wood, P. B.**, Kinetic studies of the metabolism of foreign organic compounds. 4. The conjugation of phenols with sulphuric acid, *Biochem. J.*, 52, 419, 1952.

1195. **Levy, G. and Matsuzawa, T.**, Pharmacokinetics of salicylamide elimination in man, *J. Pharmacol. Exp. Therap.*, 156, 285, 1967.

1196. **Slotkin, T. A., Distefano, V., and Au, W. Y. W.**, Blood levels and urinary excretion of harmine and its metabolites in man and rats, *J. Pharmacol. Exp. Therap.*, 173, 26, 1970.

1197. **Ninck, K., Schupp, R. R., Illing, H. P. A., Kahl, G. F. and Netter, K. J.**, Interrelation between demethylation of *p*-nitroanisole and conjugation of *p*-nitrophenol in rat liver, *Naunyn-Schmiedebergs Arch. Pharmacol.*, 279, 347, 1973.

1198. **Mulder, G. J., Hagen-Keulemans, K., and Sluiter, N. E.**, UDP glucuronyltransferase and phenolsulfotransferase from rat liver *in vivo* and in vitro, *Biochem. Pharmacol.*, 4, 103, 1975.

1199. **Mulder, G. J. and Hagedoorn, A. H.**, UDP glucuronyltransferase and phenolsulfotransferase *in vivo* and *in vitro*. Conjugation of harmol and harmalol, *Biochem. Pharmacol.*, 23, 2101, 1974.

1200. **Mulder, G. J. and Scholtens, E.**, Phenol sulphotransferase and uridine diphosphate glucuronyltransferase from rat liver *in vivo* and *in vitro*. 2,6-Dichloro-4-nitrophenol as selective inhibitor of sulphation. *Biochem. J.*, 165, 553, 1977.

1201. **Mulder, G. J.**, A method for comparison of the properties of UDP glucuronyltransferase and phenolsulfotransferase from rat liver, with a joint substrate, *Anal. Biochem.*, 64, 350, 1975.

1201a. **Levy, G.**, Pharmacokinetic and toxicologic implications of glucuronide and sulfate conjugation of certain non-narcotic analgesics in man, in *Conjugation Reactions and Drug Biotransformation*, Aitio, A., Ed., Elsevier, Amsterdam, 1978, 469.

1201b. **Grafström, R., Moldéus, P., Andersson, B., and Orrenius, S.**, Oxidative and conjugative metabolism of drugs and carcinogens studied in isolated intestinal epithelial cells, in *Conjugation Reactions in Drug Biotransformation*, Aitio, A., Ed., Elsevier, Amsterdam, 1978, 497.

1202. **Powell, G. M., Gregory, P. A., Olavesen, A. H., and Jones, J. G.**, Studies on the mechanism of biliary excretion of aryl sulfates in the rat, *Biochem. Soc. Trans.*, 1, 1165, 1973.

1203. **Holler, M., Grochtmann, W., Napp, M., and Breuer, H.**, Studies on the metabolism of oestrone sulphate, *Biochem. J.*, 166, 363, 1977.

1204. **Barford, P. A., Olavesen, A. H., Curtis, C. G., and Powell, G. M.**, Biliary excretion of some anionic derivatives of diethylstilbestrol and phenolphthalein in the guinea pig, *Biochem. J.*, 168, 373, 1977.

1205. **Capel, I. D., Millburn, P., and Williams, R. T.**, Route of administration and the conjugation of phenol in hens, *Biochem. Soc. Trans.*, 2, 875, 1974.

1206. **Remmer, H.**, Induction of drug-metabolizing enzymes in the liver, *Eur. J. Biochem.*, 5, 116, 1972.

1207. **Pelkonen, O. and Karki, N. T.**, Drug metabolism in human fetal tissues, *Life Sci.*, 13, 1163, 1973.

1208. **von Bahr, C., Rane, A., Orrenius, S., and Sjoqvist, F.**, Metabolism of desmethylimipramine in human foetal and adult liver microsomes, *Acta Pharmacol. Toxicol.*, 34, 58, 1974.

1209. **Knowles, R. G. and Burchell, B.**, A simple method for purification of epoxide hydratase from rat liver, *Biochem. J.*, 163, 381, 1977.

1210. **Albert, K. S., Hallmark, M. R., Sakmar, E., Weidler, D. I., and Wagner, J. G.**, Plasma concentrations of diphenylhydantoin, its para-hydroxylated metabolite, and corresponding glucuronide in man, *Res. Commun. Chem. Path. Pharmacol.*, 9, 463, 1974.

1211. **Gram, L. F., Christiansen, J., and Fredricson Øvero, K.**, Interaction between neuroleptics and tricyclic antidepressants, in *Drug Interactions*, Morselli, P. L., Garattini, S., and Cohen, S. N., Eds., Raven Press, New York, 1974, 271.

1212. **Vainio, H., Aitio, A., and Hanninen, O.**, Action of transcription and translation inhibitors on the enhancement of drug hydroxylation and glucuronidation by 3-methylcholanthrene and phenobarbital, *Int. J. Biochem.*, 5, 193, 1974.

1213. **Poland, A., Glover, E., and Keude, A. S.**, Stereospecific, high affinity binding of 2,3,7,8-tetrachlorodibenzo-*p*-dioxin by hepatic cytosol. Evidence that the binding species is receptor for induction of aryl hydrocarbon hydroxylase, *J. Biol. Chem.*, 251, 4936, 1976.

1214. **Franklin, M. R.**, Simultaneous hepatic microsomal oxidation of *p*-nitroanisole and glucuronide conjugation of *p*-nitrophenol, *Biochem. Soc. Trans.*, 2, 891, 1974.

1215. **Wiebkin, P., Fry, J. R., Jones, C. A., Lowing, R., and Bridges, J. W.**, The metabolism of biphenyl by isolated viable rat hepatocytes, *Xenobiotica*, 6, 725, 1976.

1216. **Oesch, F.**, Epoxide hydratase, *Progr. Drug. Metab.*, 3, 253, 1979.

1217. **Oesch, F.**, Transplacental control of epoxide-forming and inactivating enzymes of rat fetal liver by clinically-used drugs and by environmental chemicals, in *Basic and Therapeutic Aspects of Perinatal Pharmacology*, Morselli, P. L., Garattini, S., and Sereni, F., Eds., Raven Press, New York, 1975, 53.

1218. **Oesch, F., Commentary.** Metabolic transformation of clinically-used drugs to epoxides — new perspectives in drug-drug interactions, *Biochem. Pharmacol.*, 25, 1935, 1976.

1219. **Temple, A. R., George, D. J., and Done, A. K.**, Reduced nicotinamide adenine dinucleotide phosphate (NADPH) enhancement of *p*-nitrophenol glucuronidation, *Biochem. Pharmacol.*, 20, 1718, 1971.

1219a. **Lind, C., Vadi, H., and Ernster, L.**, A possible role of diaphorase in the formation of glucuronyl conjugates of benzo(a)pyrene metabolites, in *Conjugation Reactions in Drug Biotransformation*, Aitio, A., Ed., Elsevier, Amsterdam, 1978, 505.

1219b. **Tomasič, J.**, Analysis of glucuronic acid conjugates, in *Drug Fate and Metabolism*, Vol. 2, Garrett, E. R. and Hirtz, J. L., Eds., Marcel Dekker, New York, 1978, 281.

1220. **Kellie, A. E., Samuel, V. K., Riley, W. J., and Robertson, D. M.**, Steroid glucuronoside-BSA complexes as antigens: the radioimmunoassay of steroid conjugates, *J. Steroid Biochem.*, 3, 275, 1972.

1221. **Kellie, A. E.**, The radioimmunoassay of steroid conjugates, *J. Steroid Biochem.*, 6, 277, 1975.

1222. **Samarajeeva, P. and Kellie, A. E.**, The radioimmunoassay of steroid glucuronides. The oestrogen C-3 glucuronides as haptens, *Biochem. J.*, 151, 369, 1975.

1223. **Soares, J. R., Zimmerman, E., and Gross, S. J.**, Direct radioimmunoassay of 16-glucosiduronate metabolites of estriol in human plasma and urine, *FEBS Lett.*, 61, 263, 1976.

1224. **Lehtinen, T. and Adlercreutz, H.**, Solid-phase radioimmunoassay of estriol-16α-glucuronide in urine and pregnancy plasma, *J. Steroid Biochem.*, 8, 99, 1977.

1225. **Davis, S. E. and Loriaux, D. L.**, A simple specific assay for estriol in maternal urine, *J. Clin. Endocrinol. Metab.*, 40, 895, 1975.

1226. **Soares, J. R., Gross, S. J., and Bashore, R.**, Evaluation of azoestriol antisera for estriol measurements in pregnancy plasma directly and after extraction, *J. Clin. Endocrinol. Metab.*, 40, 970, 1975.

1227. **Koida, M., Takahashi, M., and Kaneto, H.**, The morphine 3-glucuronide directed antibody: its immunological specificity and possible use for radioimmunoassay of morphine in urine, *Jpn. J. Pharmacol.*, 24, 707, 1974.

1228. **Mattox, V. R., Litwiller, R. D., and Goodrich, J. E.**, Extraction of steroidal glucosiduronic acids from aqueous solutions by anionic liquid ion exchangers, *Biochem. J.*, 126, 533, 1972.

1229. **Mattox, V. R., Litwiller, R. D., and Goodrich, J. E.**, Recovery of steroidal glucosiduronic acids from organic solvents containing anionic liquid ion exchangers, *Biochem. J.*, 126, 545, 1972.

1230. **Mattox, V. R., Goodrich, J. E., and Litwiller, R. D.**, Liquid ion exchangers in paper chromatography of steroidal glucosiduronic acids, glucosiduronic esters and free steroids. Influence of concentration of exchanger and counterion, *J. Chromatogr.*, 108, 23, 1975.

1231. **Fransson, B. and Schill, G.**, Isolation of acidic conjugates by ion pair extraction. I. Extraction of glucuronic, sulphuric and glycine conjugates, *Acta Pharm. Suec.*, 12, 107, 1975.

1231a. **Schill, G., Borg, K. O., Modin, R., and Persson, B. A.**, Ion-pair extraction methods, *Progr. Drug. Metab.*, 2, 219, 1977.

1232. **Assandri, A. and Perazzi, A.**, Separation of phenolic O-glucuronides and phenolic sulphate esters by multiple liquid-liquid partition, *J. Chromatogr.*, 95, 213, 1974.

1233. **Van der Wal, Sj. and Huber, J. F. K.**, High pressure liquid chromatography with ion-exchange celluloses and its application to the separation of estrogen glucuronides, *J. Chromatogr.*, 102, 353, 1974.

1234. **Bakke, J. E.**, Recent advances in the isolation and identification of glucuronide conjugates, in *Bound and Conjugated Pesticide Residues*, Ser. 29, Kaufman, D. D., Still, G. G., Paulson, G. D., and Bandal, S. K., Eds., American Chemical Soc., N.Y., 1976, 55.

1235. **Tamura, Z. and Imanari, T.**, Gas chromatography of glucuronides, *Chem. Pharm. Bull.*, 12, 1386, 1964.

1236. **Imanari, T. and Tamura, T.**, Gas chromatography of glucuronides, *Chem. Pharm. Bull.*, 15, 1677, 1967.

1237. **Horning, E. C., Horning, M. C., Ikekawa, M., Chombaz, E. M., Jaakonmaki, P. I., and Brooks, C. J. W.**, Studies of analytical separations of human steroids and steroid glucuronides, *J. Gas Chromatogr.*, 5, 283, 1967.

1238. **Billets, S., Lietman, P. S., and Fenselau, C.**, Mass spectral analysis of glucuronides, *J. Med. Chem.*, 16, 30, 1973.

1239. **Knaack, J. B., Eldridge, J. M., and Sullivan, L. J.**, Systematic approach to preparation and identification of glucuronic acid conjugates, *J. Agric. Food Chem.*, 15, 605, 1967.

1240. **Paulson, G. D., Zaylskie, R. G., and Doktor, M. M.**, Characterization of aryl glucuronic acid conjugates by derivatization and mass spectral analysis, *Anal. Chem.*, 45, 21, 1973.

1241. **Mrochek, J. R. and Rainey, W. T., Jr.**, Gas chromatography and mass spectrometry of some trimethylsilyl derivatives of urinary glucuronides, *Anal. Biochem.*, 57, 173, 1974.

1242. **Thompson, R. M., Gerber, N., Seibert, R. A., and Desiderio, D. M.**, A rapid method for the mass spectrometric identification of glucuronides and other polar drug metabolites in permethylated rat bile, *Drug Metab. Dispos.*, 1, 489, 1973.

1243. **Ehrsson, H., Walle, T., and Wikström, S.**, Gas chromatography of ether glucuronides as methyltrifluoroactyl derivatives, *J. Chromatogr.*, 101, 206, 1974.

1244. **Tikkanen, M. J. and Adlercreutz, H.**, A method for the quantitative determination of estriol conjugates in human pregnancy, *J. Steroid Biochem.*, 3, 807, 1972.

1245. **Thompson, R. M.**, Gas chromatographic and mass spectrometric analysis of permethylated estrogen glucuronides, *J. Steroid Biochem.*, 7, 845, 1976.

1246. **Adlercreutz, H., Soltmann, B., and Tikkanen, M. J.**, Field desorption mass spectrometry in the analysis of a steroid conjugate, estriol-16α-glucuronide, *J. STeroid Biochem.*, 5, 163, 1974.

1247. **Speigelhalder, B., Rohle, G., Siekmann, L., and Breuer, H.**, Mass spectrometry of steroid glucuronides, *J. Steroid Biochem.*, 7, 749, 1976.

1248. **Millard, B. J.**, Newer developments in the mass spectrometry of drugs and metabolites, *Progr. Drug Metab.*, 1, 1, 1976.

1249. **Lyle, M. A., Ballante, S., Head, K., and Fenselau, C.**, Synthesis and characterisation of glucuronides of cannabinol, cannabidiol, Δ-(9)-tetrahydrocannabinol and Δ-(8)-tetrahydrocannabinol, *Biomed. Mass Spec.*, 4, 190, 1977.

1250. **Smith, R. G., Doyle, G., Daves, R. K., and Gerber, L. N.**, Hydantoin ring glucuronidation: characterisation of a new metabolite of 5,5-diphenylhydantoin in man and rat, *Biomed. Mass Spec.*, 4, 275, 1977.

1251. **Pook, K. H., Rominger, K. L., and Arndts, D.**, Mass spectral analysis of glucuronides from sympathomimetic hydroxyphenylalkylaminoethanols, *J. Pharm. Sci.*, 65, 1513, 1976.

1252. **Moss, M. S.**, The metabolism and urinary and salivary excretion of drugs in the horse and their relevance to detection of dope, in *Drug Metabolism from Microbe to Man*, Parke, D. V. and Smith, R. I., Eds., Taylor & Francis, London, 1977, 263.

1253. **MacDonnell, P. C., Ryder, E., Delvalle, J. A., and Greengard, O.**, Biochemical changes in cultured foetal rat liver explants, *Biochem. J.*, 150, 269, 1975.

1254. **Salmona, M., Saronio, C., Cantoni, L., Mussini, E., and Garattini, S.**, Gel-entrapped microsomes: a new tool for the study of drug metabolism, Abstr. 6th Int. Congr. Pharmacol., Helsinki, 1975, 63.

1255. **Fenselau, C., Head, K., McGlashan, D. W., Pallante, S., and Parikh, I.**, Synthesis of glucuronides by immobilized glucuronyl transferase, Abstr. 4th Pharmacol., Toxicol. Program Symp. 1975, Natl. Inst. Gen. Med. Sci., Wash., D.C., 7.

1256. **Brunner, G.**, Microsomal enzymes bound to artificial carriers, in *Artificial Liver Support*, Williams, R. and Murray-Lyon, I. M., Eds., Pitman Medical, London, 1975, 153.

1257. **Breckenridge, A., Bendling, M. R., and Brunner, G.**, Impact of drug monooxygenases in clinical pharmacology, in *Microsomes and Drug Oxidations*, Ullrich, V., Roots, I., Hildebrand, A., Conney, A. H., and Estabrook, R., Eds., Pergamon Press, Oxford, 1977, 385.

1258. **Bellet, H. and Raynaud**, An assay of bilirubin UDP-glucuronyltransferase on needle biopsies applied to Gilbert's syndrome, *Clin. Chim. Acta*, 53, 51, 1974.

1259. **Strickland, R. D., Gregory, D. M., and Lynch, J. L.**, A method for assaying hepatic UDP-glucuronyl transferase, *Biochem. Med.*, 11, 180, 1974.

1260. **Marniemi, J. and Hänninen, O.**, Radiochemical assay of UDP glucuronyltransferase, *FEBS Lett.*, 32, 273, 1973.

1261. **Lehnert, W., Linberg, J., and Kunzer, W.,** Empfindliche Mikromethode zur Bestimmung der UDP-Glucuonyltransferase-Activitat im Leberhomogenat mit [^{14}C]-Nitrophenol als Substrat, *Z. Klin. Chem. Klin. Biochem.,* 12, 23, 1974.

1262. **Temple, A. R., Done, A. K., and Clement, M. S.,** Studies of glucuronidation. III. The measurement of *p*-nitrophenyl glucuronide, *J. Lab. Clin. Med.,* 77, 1015, 1971.

1263. **Rao, G. S., Haueter, G., Rao, M. L., and Breuer, H.,** An improved assay for steroid glucuronyltransferase in rat liver microsomes, *Anal. Biochem.,* 74, 35, 1976.

1264. **Ziegler, J. M., Lisboa, B. P., Batt, A.-M., and Siest, G.,** Determination of UDP-glucuronyltransferase using UDP-[^{14}C]glucuronic acid, *Biochem. Pharmacol.,* 24, 1291, 1975.

1265. **Batt, A.-M., Ziegler, J. M., Loppinet, V., and Siest, G.,** Conjugation of hydroxyphenylhydantoin and hydroxyphenobarbital in rat liver microsomes. Induction by phenobarbital, *Biochem. Pharmacol.,* 26, 1354, 1977.

1266. **Hess, R. and Pearse, A. G. E.,** Histochemical demonstration of uridine diphosphate glucose dehydrogenase, *Experientia,* 17, 317, 1961.

1267. **Balogh, K., Jr. and Cohen, R. B.,** Histochemical localization of uridine diphosphoglucose dehydrogenase in cartilage, *Nature (London),* 192, 1199, 1961.

1268. **Stiller, D. and Gorski, J.,** Untersuchungen zur Histotopochemie der Uridindiphosphatglukose-Dehydrogenase, *Acta Histochem.,* 32, 356, 1969.

1269. **Stiller, D.,** Untersuchungen zur Topochemie des Entgiftungsstoffwechels durch Glucuronidierung. I. Nachweis der Uridindiphosphatglukose-Dehydrogenase in Leber und Niere verschiedener Species mittels einer verbesserten Gelinkubationsmethode, *Acta Histochem.,* 38, 376, 1970.

1270. **Fyffe, J. A.** The Induction of Certain Enzymes Involved with Glucuronidation, Ph.D. thesis, University of Dundee, Dundee, Scotland, 1975.

1271. **Meijer, A. E. F. H.,** Semipermeable membranes for improving the histochemical demonstration of enzyme activities in tissue sections, *Histochemie,* 30, 31, 1972.

1272. **Hanson, S. W. F., Mills, G. T., and Williams, R. T.,** A study of the determination of glucuronic acid by the naphthoresorcinol reaction, with the photoelectric absorptiometer, *Biochem. J.,* 38, 274, 1944.

1273. **Deichmann, W. B.,** The quantitative estimation of glucuronates in urine, *J. Lab. Clin. Med.,* 28, 770, 1943.

1274. **Fishman, W. H. and Green, S.,** Microanalysis of glucuronide glucuronic acid as applied to β-glucuronidase and glucuronic acid studies, *J. Biol. Chem.,* 215, 527, 1955.

1275. **Bitter, T. and Muir, H. M.,** A modified uronic acid carbazole reaction, *Anal. Biochem.,* 4, 330, 1962.

1276. **Yuki, H. and Fishman, W. H.,** A carbazole method for the differential analysis of glucuronate, glucosiduronate and hyaluronate, *Biochim. Biophys. Acta,* 69, 576, 1963.

1277. **Mazzuchin, A., Walton, R. J., and Thibert, R. J.,** Determination of total and conjugated glucuronic acid in serum and urine employing a modified naphthoresorcinol reagent, *Biochem. Med.,* 5, 135, 1971.

1278. **Whistler, R. L. and Feather, M. S.,** Carboxyl determinations. Uronic acid carboxyl and total carboxyl, *Methods Carbohydr. Chem.,* 1, 464, 1962.

1279. **Jones, J. K. N., and Pridham, J. B.,** A colorimetric estimation of sugars using benzidine, *Biochem. J.,* 58, 288, 1954.

1280. **Tomasič, J. and Keglević, D.,** Direct spectrophotometric assay of glucuronic acid in the presence of labile glucosiduronic acids, *Anal. Biochem.,* 45, 164, 1972.

1281. **Rodén, L., Baker, J. R., Cifonelli, J. A., and Matthews, M. B.,** Isolation and characterization of connective tissue polysaccharides, *Methods Enzymol.,* 28, 73, 1972.

1282. **Eisenberg, F., Jr.,** Determination of uronic acids by gas-liquid chromatography, *Methods Enzymol.,* 27, 168, 1972.

1283. **Eisenberg, F., Jr.,** Gas chromatographic assay of iduronic and glucuronic acids as aldonic acid butaneboronates, *Anal. Biochem.,* 60, 181, 1974.

1284. **Inoue, S.,** The determination of iduronic acid and glucuronic acid in heparin by a new technique. Reevaluation of the variable hexuronic acid content of mammalian heparins, *Biochim. Biophys. Acta,* 329, 264, 1973.

1284a. **Wolkoff, A. W., Scharschmidt, B. F., Plotz, P. H., and Berk, P. D.,** Purification of conjugated bilirubin: a new approach utilizing albumin-agarose gel affinity chromatography, *Proc. Soc. Exp. Biol. Med.,* 152, 20, 1977.

1285. **Wong, K. P.,** Measurement of nanogram quantities of UDP-glucuronic acid in tissues, *Anal. Biochem.,* 82, 559, 1977.

1286. **Shirai, Y., and Ohkubo, T.,** Measurement of glucuronide synthesis by tissue preparations, *J. Biochem. (Tokyo),* 41, 337, 1954.

1287. **Levvy, G. A. and Storey, I. D. E.,** The measurement of glucuronide synthesis by tissue preparations, *Biochem. J.,* 44, 295, 1947.

1288. **Rao, G. S., Haueter, G., Rao, M. L., and Breuer,** An improved assay for steroid glucuronyltransferase in rat liver microsomes, *Anal. Biochem.,* 74, 35, 1976.

1289. **Neale, M. G. and Parke, D. V.,** Effects of pregnancy on the metabolism of drugs in the rat and rabbit, *Biochem. Pharmacol.,* 22, 1451, 1973.

1290. **Levvy, G. A. and McAllan, A.,** Mammalian fucosidases. 3. β-D-Fucosidase activity and its relation to β-D-galactosidase, *Biochem. J.,* 87, 361, 1963.

1291. **Dodgson, K. S. and Spencer, B.,** Studies on sulphatases. 15. The aryl sulphatases of human serum and urine, *Biochem. J.,* 65, 668, 1957.

1292. **Haddock, B. A. and Garland, P. B.,** Effect of sulphate-limited growth on mitochondrial electron transfer and energy conservation between reduced nicotinamide-adenine dinucleotide and the cytochromes in *Torulopsis utilis, Biochem. J.,* 124, 155, 1971.

1292a. **Onishi, S., Itoh, S., Kawade, N., Isobe, K., and Sugiyama, S.,** An accurate and sensitive analysis by high-pressure liquid chromatography of conjugated and unconjugated bilirubin IX-α in various biological fluids, *Biochem. J.,* 185, 281, 1980.

1292b. **Johnson, D. B., Swanson, M. J., Barker, C. W., Fanska, C. B., and Murrill, E. E.,** Glucuronidation of lipophilic substrates: preparation of 3-benzo (α) pyrenyl-β-D-glucopyranosiduronic acid in multimilligram quantities by microsomal UDP-glucuronyl transferase, *Prep. Biochem.,* 9, 391, 1979.

1292c. **Moldeus, P.,** Paracetamol metabolism and toxicity in isolated hepatocytes from rat and mouse, *Biochem. Pharmacol.,* 27, 2859, 1978.

1292d. **Diamond, G. and Quebbemann, A. J.,** Rapid separation of *p*-nitrophenol and its glucuronide and sulphate conjugates by reversed-phase high performance liquid chromatography, *J. Chromatogr.,* 177, 368, 1979

1292e. **Lim, C. K.,** The separation of conjugated and unconjugated bilirubin in bile by high performance liquid chromatography, *J. Liquid Chromatogr.,* 2, 37, 1979.

1292f. **Batt, A. M., Mackenzie, P., Hanninen, O., and Vainio, H.,** Glucuronidation of 3-hydroxybenzo(a)pyrene in liver microsomes, *Med. Biol.,* 57, 281, 1979.

1292g. **Singh, J. and Wiebel, F. J.,** A highly sensitive and rapid fluorimetric assay for UDP-glucuronyltransferase using 3'-hydroxybenzo(a)pyrene as substrate, *Anal. Biochem.,* 98, 394, 1979.

1293. **Bellemann, P., Gebhardt, R., and Mecke, D.,** An improved method for the isolation of hepatocytes from liver slices. Selective removal of Trypan Blue-dyeable cells, *Anal. Biochem.,* 81, 408, 1977.

1294. **Gebhardt, R., Bellemann, P., and Mecke, D.,** Metabolic and enzymatic characteristics of adult rat liver parenchymal cells in non-proliferating primary monolayer cultures, *Exp. Cell Res.,* 112, 431, 1978.

1295. **Paul, J.,** *Cell and Tissue Culture,* 4th ed., Livingstone, Edinburgh, 1970.

Index

INDEX

A

Absorbtion, role of glucuronides in, 10—11
Acetamide, 12
2-Acetamidofluorene, 128
4-Acetamidophenol, 3
4-Acetamidophenyl glucuronide, 3—4
Acetaminophen, 12, 138, 176
Acetone, 36, 192
2-Acetylaminofluorene, 6
N-Acetyl-4-aminophenol, 70, 132
N-Acetylglucosamine, 10, 73, 172
N-Acetylglucosaminidation, 172
N-Acetylglucosaminyltransferase, 172
N-Acetylimidoguinone, 11
Actinomycin D, 135
Activation, 11, 35, 37, 137—138
    UDPglucuronyltransferase, see
        UDPglucuronyltransferase, activation
Activity quotient (AQ), 100
Adenosine monophosphate, see AMP
Adenosine triphosphate, see ATP
Adrenal gland, glucuronidation in, 157
Age, effect of, 99—118
Albumin, 40, 43
Aldosterone, 9
Aldrin, 143
Alimentary tract, see also specific parts, 151—156
2-Aminobenzoate, 130, 192—194
4-Aminobenzoate, 45, 106, 192—194
2-Aminobenzoic acid, 38, 53, 80, 117
4-Aminobenzoic acid, 20
4-Amino-3-biphenyl, 162
2-Aminolevulinic acid, 10
2-Amino-1-napthol, 12
2-Aminophenol, 19, 23—24, 26—27, 34—35, 38,
        45, 48, 49—50, 53, 59—60, 64—65, 79, 81,
        106, 108, 110, 114, 117, 120—121, 123,
        125, 130—132, 136, 138, 145, 151,
        156—157, 194—195
3-Aminophenol, 70
4-Aminophenol, 70
2-Aminophenyl glucuronide, see also 2-
        Aminophenol, 20
3-Aminophenyl glucuronide, see also 3-
        Aminophenol, 20
Aminopyrene, 178
4-Aminosalicylic acid, 20
2-Aminothiophenol, 38, 49, 69—70, 93, 130
AMP, 50—51
Amphetamine, 137
Amphian, glucuronidation in, 126
Ampicillin, 156
Δ3,5-Androstadiene-3,17-dione, 3-enol β-
        glucuronide, 13
5α-Androstanediol, 10
5β-Androstanediol, 10
5β-Androstane-3α,17β-diol 17-glucuronide, 10

5-Androstenediol glucuronide, 10
Androstenedione, 72
Androsterone, 9
5β-Androsterone, 9—10
Aniline hydroxylase, 178
8-Anilino-1-naphthalene sulfonate (ANS), 143
APS-kinase, 174
Arachidic acid, 44
Arachidonic acid, 44
N-Arylacethydroxamic acid, 11
ATP, 104
    activation of UDPglucuronyltransferase,
        50—51
ATP-sulfurylase, 174

B

Barbital, 145
Barbiturate, see also specific compounds
    induction, 138—140
        precocious, 109—111
Behenic acid, 44
Benzoglycine, 4
Benzoic acid, 4, 128
Benzo(a)pyrene, 70, 110—111, 129, 140,
        145—146, 150, 156, 162, 179, 189
8-Benzopyrenol, 157
2-Benzothiazole, 93
Bile, see also specific components of, 186
    cholestatic, 78
    β-glucuronidase in, see also β-glucuronidase, 95
    salt, 71—72
    secretion, 153—155
Bilirubin, see also Bile, 5, 23—25, 38, 45, 48, 50,
        54, 59—60, 65, 74, 77, 80, 106—110,
        113—114, 117, 120—121, 125—128, 132,
        136, 138, 143, 151—152, 155, 158, 161,
        165—166, 172, 189, 195—197
    conjugation, 6
    dietary influence on glucuronidation, 121
    diglucuronide, 21, 74—77
    glucosidation, 171—172
    mixed conjugate, 77—78
    monoglucuronide, 23, 74—77
    strains with low excretion of conjugates, 133
Biliverdin, 126
    substrate of UDPgluronyltransferase, 78
Biphenyl, 178
Bird, glucuronidation in, 126—127
Borneol, 17
Bornyl β-glucoside, 17
Bornyl glucuronide, 17
Breast milk factor, 122
Brij-35®, 46
Bromine, 70
Bromosulfophthalein, 132
Buprenorphine, 153

U

V

Volazocine, 71

W

Wistar rat, 130—132

X

Xanthurenic acid, 72, 74
Xenobiotic, see also specific compounds, 5, 137

activation, 137—138
competition by, 7
induction, 137—147
structures glucuronidated, 9
X-irradiation, 166
Xylosidation, 172—173

Z

Zinc ion, 47
Zoxazolamine, 95